# DYNAMICAL EVOLUTION OF STAR CLUSTERS –
CONFRONTATION OF THEORY AND OBSERVATIONS

INTERNATIONAL ASTRONOMICAL UNION

UNION ASTRONOMIQUE INTERNATIONALE

# DYNAMICAL EVOLUTION OF STAR CLUSTERS – CONFRONTATION OF THEORY AND OBSERVATIONS

PROCEEDINGS OF THE 174TH SYMPOSIUM OF THE
INTERNATIONAL ASTRONOMICAL UNION,
HELD IN TOKYO, JAPAN, AUGUST 22–25, 1995

EDITED BY

PIET HUT

*Institute for Advanced Study,
Princeton, New Jersey, U.S.A.*

and

JUNICHIRO MAKINO

*The University of Tokyo,
Tokyo, Japan*

Kluwer Academic Publishers

Dordrecht / Boston / London

A C.I.P. Catalogue record for this book is available from the Library of Congress.

ISBN 0-7923-4069-8

*Published on behalf of*
*the International Astronomical Union*
*by*
*Kluwer Academic Publishers, P.O. Box 17, 3300 AA Dordrecht, The Netherlands.*

*Kluwer Academic Publishers incorporates*
*the publishing programmes of*
*D. Reidel, Martinus Nijhoff, Dr W. Junk and MTP Press.*

*Sold and distributed in the U.S.A. and Canada*
*by Kluwer Academic Publishers,*
*101 Philip Drive, Norwell, MA 02061, U.S.A.*

*In all other countries, sold and distributed*
*by Kluwer Academic Publishers Group,*
*P.O. Box 322, 3300 AH Dordrecht, The Netherlands.*

*Printed on acid-free paper*

*All Rights Reserved*
*©1996 International Astronomical Union*

No part of the material protected by this copyright notice may be reproduced or utilized in any form or by any means, electronic or mechanical including photocopying, recording or by any information storage and retrieval system, without written permission from the publisher.

*Printed in the Netherlands*

# Table of Contents

| | |
|---|---|
| Preface | xi |
| The Organizing Committee | xiv |
| List of Participants | xv |

## August 22

### Introduction

D. Sugimoto
A Comparative Study of Globular Clusters — 1

S. G. Djorgovski
Correlations of Globular Cluster Properties as Constraints for Dynamical Evolution Models — 9

### New Observations with HST

P. Guhathakurta, B. Yanny, D. P. Schneider & J. N. Bahcall
Hubble Space Telescope Studies of the Dense Central Regions of Globular Clusters — 19

I. R. King
Dynamical Implications of Deep HST Imaging of Globular Clusters — 29

G. G. Fahlman, H. B. Richer, R. I. Ibata, N. C. Ivanans, G. Mandushev, J. E. Hesser, P. B. Stetson, M. A. Wood, D. A. Vandenberg, C. Pryor, W. E. Harris, H. E. Bond, M. Bolte & R. A. Bell
HST Observations of the Globular Cluster M4 — 39

M. M. Shara
Searches for Cataclysmic Variables in Globular Cluster Cores — 49

D. Richstone, K. Gebhardt, A. Dressler, S. Faber, C. Grillmair, J. Kormendy, Y. -I. Byun, T. Lauer, E. Ajhar & S. Tremaine
The Centers of Galaxies — 53

G. Meylan
Studies Through Radial Velocity Measurements of the Peculiar Motions of Stars in Galactic Globular Clusters — 61

G. Piotto, A. M. Cool & I. R. King
Stellar Luminosity and Mass Functions of Globular Clusters — 71

## August 23

### Observations of Fundamental Properties — Continued

R. Genzel
The Nuclear Star Cluster of the Milky Way: Star Formation,
Stellar Collisions and Central Dark Mass ... 81

### Stellar Dynamics, Models

K. Takahashi
Two-Dimensional Fokker-Planck Models ... 91

M. Giersz
Monte-Carlo Simulations ... 101

R. Spurzem
Fluid Techniques and Evolution of Anisotropy ... 111

P. Hut
The Role of Binaries in the Dynamical Evolution of the Core of a
Globular Cluster ... 121

D. C. Heggie
Statistics of Small-$N$ Simulations ... 131

### Large N-body Simulation

M. Taiji, J. Makino, T. Fukushige, T. Ebisuzaki & D. Sugimoto
GRAPE-4: A Teraflops Machine for $N$-Body Simulations ... 141

J. Makino
Gravothermal Oscillations ... 151

S. J. Aarseth
Star Cluster Simulations on HARP ... 161

## August 24

### X Ray and Compact Objects in Globular Clusters

J. E. Grindlay
High Resolution Studies of Compact Binaries in Globular Clusters
with HST and ROSAT ... 171

S. R. Kulkarni & S. B. Anderson
Pulsars in Globular Clusters ... 181

F. Verbunt
Comparison of X-Ray Sources in Old Open and in Globular Clusters ... 183

C. Pryor, J. M. Fletcher, J. E. Hesser, R. D. Mcclure, P. B. Stetson, H. B. Richer, G. G. Fahlman, R. A. Ibata, N. C. Ivanans, G. Mandushev, R. A. Bell, M. Bolte, H. E. Bond, W. E. Harris, D. A. Vandenberg & M. A. Wood
Searches for Binary Stars in Globular Clusters ... 193

R. A. M. J. Wijers
Diagnosing Structure and Evolution of Clusters with Neutron Star
Binaries 203

P. P. Eggleton
Combining Stellar Evolution and Stellar Dynamics 213

S. L. W. McMillan
Gravitational Scattering Experiments 223

L. G. Kiseleva
Lives of Hierarchical Triple Systems in Clusters and in the Field 233

M. B. Davies
Stellar Encounters in Dense Systems 243

F. A. Rasio
SPH Calculations of Collisions Between Main-Sequence Stars 253

## August 25

### X ray and Compact Objects in Globular Clusters — Continued

D. F. Chernoff & X. Huang
Frequency of Stellar Collisions in Three-Body Heating 263

R. A. Mardling
Tidal Capture in Star Clusters 273

### Galactic Connection and Environmental Effect

D. N. C. Lin & S. D. Murray
The Formation of Globular Clusters and of The Stars Within
Them 283

H. M. Lee
Dynamics of Galactic Nuclei Containing Massive Remnant Stars 293

D. C. Heggie & P. Hut
Dark Matter in Globular Clusters 303

V. Surdin
Properties of Globular Cluster System: Primordial or Evolutional? 313

### Panel Discussion

I. R. King, G. Meylan, F. Verbunt, P. Hut & D. Sugimoto
Panel Discussion 319

## Posters

### Observations of Star Clusters

B. Brandl, B. J. Sams, F. Bertoldi, A. Eckart, R. Genzel, S. Drapatz, R. Hofmann, M. Löwe & A. Quirrenbach
Adaptive Optics NIR Imaging of R136 in 30 Doradus: The Stellar Population of a Nearby Starburst ... 331

P. Guhathakurta, B. Yanny, J. N. Bahcall & D. P. Schneider
Preliminary Study of the Stellar Populations and Density Profile of NGC 6624 Using HST ... 333

M. Kontizas, D. Gouliermis & E. Kontizas
Central Densities of Star Clusters in the Magellanic Clouds ... 335

J. W. Menzies
A Search for Variables in the Central Regions of Southern Globular Clusters ... 337

A. K. Pandey, A. K. Durgapal, B. C. Bhatt, V. Mohan & H. S. Mahra
CCD Photometry of the Open Cluster BE 69 ... 339

A. Rosenberg, I. Saviane, G. Piotto, S. Zaggia & A. Aparicio
Pal 1: Another young globular cluster? ... 341

C. Sosin & I. R. King
Deep HST/FOC Observations of the Center of M15 ... 343

S. R. Zaggia
$M_V - \sigma$ Relation: A Universal Law? ... 345

### Stellar Evolution in Star Clusters

G. A. Drukier
On the Retention of Globular Cluster Neutron Stars ... 347

J. E. Grindlay & A. M. Cool
HST/FOS Discovery of Probable CVs in NGC 6397 ... 349

K. N. Kong & H. M. Lee
Binary-Single Encounters of 10 Solar Mass Black Holes Including the Effects of Gravitational Radiation ... 351

H. Negoro & K. Kawashima
ASCA observation of $\omega$ Centauri ... 353

S. F. **Portegies Zwart**
The Depletion of Giants in Globular Clusters ... 355

I. Saviane, G. Piotto, M. Capaccioli & F. Fagotto
Environmental Influence on Stellar Evolution: The Horizontal Branch of NGC 1851 ... 357

G. Szécsényi-Nagy
On the Role of dK and dM Stars in Piling up the Total Mass of
Star Clusters ... 359

H. Yamaoka
The Fate of Binary Systems After the Explosion of SNe 1993J and
1994I ... 361

## Stellar Dynamics, Models

C. Einsel & R. Spurzem
Fokker-Planck Models for Rotating Stellar Systems ... 363

T. Fukushige & D. C. Heggie
Pre-Collapse Evolution of Galactic Globular Clusters ... 365

Y. Funato, J. Makino, P. Hut & S. McMillan
Time-Symmetrized Kustaanheimo-Stiefel Regularization ... 367

D. C. Heggie
A Numerical Approximatation for Hierarchical Triples ... 369

D. C. Heggie, P. Hut & S. L. W. Mcmillan
Exchange Cross Sections For Hard Binaries ... 371

S. -I. Inutsuka
A Numerical Scheme for Collisions of Stars: Godunov-type Particle Hydrodynamics ... 373

P. -Y. Longaretti & C. Lagoute
Dynamical Evolution of Rotating Globular Clusters ... 375

P. -Y. Longaretti, R. Taillet & P. Salati
The Dark Side of Globular Clusters ... 377

S. L. W. McMillan & K. A. Engle
Are Gravothermal Oscillations Gravothermal? ... 379

N. Ramamani, D. C. Heggie & S. J. Aarseth
Galactic Disk Shocks on Globular Clusters ... 381

F. A. Rasio & D. C. Heggie
The Orbital Eccentricities of Binary Millisecond Pulsars in Globular Clusters ... 383

T. Tsuchiya, N. Gouda & T. Konishi
Itinerancy of Quasiequilibria in One-Dimensional Gravitating Systems ... 385

Y. Zhou & S. M. Miyama
The Energy Exchange between Different Masses in an Expanding Gravitating System ... 387

## Galactic Connection and Environmental Effects

V. M. Danilov
On the Dynamics of Young Open Star Clusters in the Joint Field of the Galaxy and of a Star Formation Region ... 389

E. Kim, M. G. Lee & D. Geisler
Spatial Structure of the Globular Cluster System around NGC 4472 ... 391

M. G. Lee, E. Kim & D. Geisler
Metallicity and Luminosity Functions of the Globular Clusters in NGC 4472 ... 393

M. Mori, Y. Yoshii, T. Tsujimoto & K. Nomoto
GRAPE-SPH Simulations of the Chemodynamical Evolution of Dwarf Galaxies ... 395

N. Nakasato, M. Mori, T. Tujimoto, G. Mathews & K. Nomoto
GRAPE-SPH Simulations of Globular Cluster Formation ... 397

C. Theis
Dynamics of Collapsing Shells ... 399

E. Vesperini
Evolution of the Galactic Globular Cluster System ... 401

E. Vesperini & D. F. Chernoff
Truncation of the Binary Distribution Function in Globular Cluster Formation ... 403

**AUTHOR INDEX** ... 405

# PREFACE

This symposium was first proposed in August 1993 when some members of my international collaboration on gravitational $N$-body problems gathered in Tokyo. The timing of the proposed Symposium was set as 1995 considering that a Tera-flops computer dedicated to $N$-body problems would become operational. At the same time the Symposium was intended to be situated in the third of a series of similar IAU Symposia.

The first one was IAU Symposium No.69 on Dynamics of Stellar Systems that was held in Besançon in 1974. Therein, the gravothermal collapse of globular cluster cores became one of the interesting topics for further elaboration. Before the second one, Symposium No. 113 on Dynamics of Star Clusters held in Princeton in 1984, the gravothermal collapse of the cores of globular clusters was worked out. Some globular clusters were observationally found to have collapsed cores and models of the gravothermal collapse were theoretically developed including similarity solutions of the collapse. So the problem of so-called post collapse evolution was selected as one of the main topics in the second Symposium. It became also one of the main subjects for theoretical investigations for the following decade, into the 1990's.

During this decade high resolution observations have made great progress with the help of CCD cameras and Hubble Space Telescope. Thus, we are now able to investigate densely populated regions of star clusters and galaxies as well as to classify every type of star. On the other hand, the progress in numerical modeling of the stellar systems should be noted. Now, it is the time to integrate such observational data into models where stellar evolution is combined with stellar dynamics including the evolution past core collapse. It will be one of the main aims of the present Symposium to open a new phase in the study of the evolution of stellar systems.

In this context it will be interesting to recall what was discussed in the preceding Symposium, No. 113, at the session of the Panel Discussion entitled *What Next? Priorities in Theory and Observations*. The panelists pointed out the following. Lyman Spitzer stressed the importance of investigating the effects of finite size of the stars, and the origin and history of actual globular clusters. Alar Toomre was interested in studying further how globular clusters came into being and how they end up. Tjeerd Van Albada pointed out the possible importance of the absence of high density globular clusters in the outer regions of the Galaxy and of the interaction between dynamical evolution and stellar evolution. Scott Tremaine discussed the evolution of the outer parts of clusters as a result of evaporation of the stars and of tidal interaction with the external gravitational field. Simon White recommended further study of the influence of the external field, and of the origin of globular clusters. Ivan King discussed the importance of studying the effect of mass mixtures in globular clusters and mass segregation due to their dynamical evolution. At the same time he pointed out, quite correctly, the importance of the expected progress in ground based CCD observations as well as Space Telescope observations.

In these four days of the present Symposium, we will see how such subjects have made progress. In my feeling, the progress in observations has been enormous while that in theoretical modeling seems somewhat left behind. One of the reasons has been that the problems are too compute-intensive for the existing supercomputers. For instance, modeling the post-collapse gravothermal oscillations in a direct $N$-body simulation remained inconclusive in simulations including several thousand bodies, because of the large statistical fluctuations present. Now, we can, however, have supercomputers with great amounts of memory to perform 3-dimensional hydrodynamic calculations of stellar collisions, and we have a dedicated computer which is several tens of times faster than general-purpose supercomputers. Therefore, I expect that the present Symposium will provide us with the future prospect of how observational and computational astronomy can be combined to clarify the evolution of realistic stellar systems.

The present IAU Symposium No. 174 was sponsored by IAU Commission 37 (Star Clusters and Associations), and co-sponsored by Commissions 28 (Galaxies), 30 (Radial Velocities), and 35 (Stellar Constitutions). The grant from IAU was distributed among relatively young astronomers from different countries. In addition, the Symposium was financially supported, in part, by the Japanese Ministry of Education, Science, Sports and Culture, and by the Nishina Memorial Foundation. The organizing committees made great effort in preparing the scientific program and in organizing the meeting itself. I would like to express my gratitude to all of them for making this Symposium possible. Last, but not least, I would like to thank the participants who delivered the important oral and poster papers.

Daiichiro Sugimoto, Chair, Scientific Organizing Committee

These proceedings are dedicated to the memory of Prof. Subramanyan Chandrasekhar, in recognition of his fundamental research on the dynamics of star clusters, and his many other profound contributions to astrophysics.

# THE ORGANIZING COMMITTEE

## SCIENTIFIC

George Djorgovski (USA)
Ken Freeman (Australia)
Hyung-Mok Lee (Korea)
Douglas C. Heggie (UK)
Piet Hut (USA)
Shogo Inagaki (Japan)
Steve McMillan (USA)
Georges Meylan (Germany)
Harvey B. Richer (Canada)
Daiichiro Sugimoto (Japan, Chairperson)
Frank Verbunt (Netherlands)

## LOCAL

Izumi Hachisu (University of Tokyo)
Junichiro Makino (University of Tokyo, chairperson)
Makoto Taiji (University of Tokyo)

The symposium was sponsored by IAU Commission 37 and co-sponsored by IAU Commissions 28, 30 and 35.

# Lists of Participants

| | |
|---|---|
| S. J. Aarseth | Institute of Astronomy, Madingley Road, Cambridge CB3 0HA, UK |
| B. Brandl | MPE, Giessenbach Str. 85748 Garching, Germany |
| R. Cannon | Anglo Australian Observatory, P.O. Box 296, Epping, NSW 2121, Australia |
| D. F. Chernoff | 602 CRSR Cornell University, Ithaca, NY 14853, USA |
| Y. Chikada | National Astronomical Observatory, 2-21-1 Osawa, Mitaka 181, Tokyo |
| A. M. Cool | Astronomy Dept., University of California, Berkeley, CA 94720, USA |
| V. M. Danilov | Astronomical observatory, Ural State University, Lenin ave 51, Ekaterinburg 620083, Russia |
| M, B. Davies | Institute of Astronomy, Madingley Road, Cambridge CB3 0HA, UK |
| G. Djorgovski | Astonomy, Caltech, Pasadena, CA 91125, USA |
| G. Drukier | Institute of Astronomy, Madingley Road, Cambridge, CB3 0HA, UK |
| P. P. Eggleton | Institute of Astronomy, Madingley Road, Cambridge, CB3 0HA UK |
| C. Einsel | Institut fur Astronomie, Olshavsenstrasse 40, D-24098 Kiel, Germany |
| G. Fahlman | University of British columbia, Vancouver, B.C., Canada V6T 1Z4 |
| T. Fukushige | Department of Earth Science and Astronomy, College of Arts and Sciences, University of Tokyo, 3-8-1 Komaba, Meguro ku, Tokyo 153, Japan |
| Y. Funato | Department of Earth Science and Astronomy, College of Arts and Sciences, University of Tokyo, 3-8-1 Komaba, Meguro-ku, Tokyo, 153, Japan |
| R. Genzel | MPE, Postfach 1603, D-85740 Garching, Germany |
| M. Giersz | Nicolaus Copernicus Astronomical Centre, Bartycka 18, 00-716 Warsaw, Poland |
| D. Gouliermis | National Observatory of Athens, Astronomical Institute, P.O.BOX 2048, 118 10 Athens, GREECE |
| J. E. Grindlay | Department of Astronomy, Harvard University, 60 Garden Street, Cambridge, MA 02138 USA |
| P. Guhathakurta | UCO/Lick Observatory, University of California, Santa Cruz, CA 95064, USA |
| I. Hachisu | Department of Earth Sciences and Astronomy, College of Arts and Sciences, University of Tokyo, 3-8-1 Komaba, Meguro-ku, Tokyo 153, Japan |

| | |
|---|---|
| D. C. Heggie | University of Edinburgh, Department of Mathematics and Statistics, King's Buildings, Edinburgh EH9 3JZ, U.K. |
| P. Hut | Institute for Advanced Study, Princeton, NJ 08540, USA |
| S. Inagaki | Department of Astronomy, Kyoto University |
| S. Inutsuka | Division of Theoretical Astrophysics, National Astronomical Observatory, Mitaka, Tokyo 181, JAPAN |
| Y. Kanya | Department of Physics, Kyoto University, Kitashirakawa, Sakyo-ku, Kyoto, 606-01, Japan |
| S. Kim | Pusan National University, Research Institute for Basic Sciences Pusan, 609-735 Korea |
| E. Kim | Department of Astronomy, Seoul National University, Seoul, Korea 151-742 |
| Ivan R. King | Astronomy Department, University of California, Berkeley, CA 94720-3411, U.S.A. |
| L. Kiseleva | Institute of Astronomy, Cambridge CB3 0HA, UK |
| S. Kulkarni | Division of Physics, Mathematics, and Astronomy, California Institute of Technology, 150-24, Pasadena, CA 91125, USA |
| M. G. Lee | Department of Astronomy, Seoul National University, Seoul, Korea 151-742 |
| H. M. Lee | Pusan University, Department of Earth Sciences, Pusan, 609-735 Korea |
| D. N. C. Lin | Lick Observatory, Univ. of California, Santa Cruz, CA 95064, USA |
| P.-Y. Longaretti | Observatoire de Grenoble, Laboratoire d'Astrophysique, BP 53X, Grenoble Cedex 9, FRANCE |
| G. MEYLAN | European Southern Observatory, Karl-Schwarzschild-Strasse 2, D-85748 Garching bei Muenchen, Germany |
| J. Makino | Department of Information Science and Graphics, College of Arts and Sciences, University of Tokyo, 3-8-1 Komaba, Meguro-ku, Tokyo 153, Japan |
| R. A. Mardling | Institute of Astronomy, Madingley Road, Cambridge CB3 0HA, UK |
| S. L. W. McMillan | Drexel University, Philadelphia, PA 19104 USA |
| J. W. Menzies | SAAO, P.O. Box 9, Observatory 7935, South Africa |
| M. Mori | Department of Astronomy, School of Science, University of Tokyo, 16-11-2 Yayoi, Bunkyo-ku, Tokyo 113, Japan |
| N. Nakasato | Department of Astronomy, School of Science, University of Tokyo, 16-11-2 Yayoi, Bunkyo-ku, Tokyo 113, Japan |
| H. Negoro | Dep. of Earth and Space Science, Faculty of Science, Osaka Univ., 1-1 Machikaneyama-cho, Toyonaka, Osaka 560, Japan |
| K. Nomoto | Dept. Astronomy, University of Tokyo |
| A. K. Pandey | U.P. State Observatory, Manora Peak, Naini Tal 263 129, INDIA |

| | |
|---|---|
| G. Piotto | Dipartimento di Astronomia, Vicolo dell'Osservatorio, 5 I-35122 Padova, Italy |
| C. Pryor | Dept. of Physics & Astronomy, Rutgers Univ. Piscataway, NJ 08855-0849, USA |
| F. Rasio | Department of Physics, M.I.T. 6-201, Cambridge, MA 02139, USA |
| D. Richstone | Dept of Astronomy, Univ of Michigan, Ann Arbor, MI, 48109 USA |
| D. Sugimoto | Department of Earth Science and Astronomy College of Arts and Sciences, University of Tokyo, 3-8-1 Komaba, Meguro-ku, Tokyo 153, Japan |
| G. Szecsenyi-Nagy | ELTE Csillagaszati Tanszek, Budapest Ludovika ter 2. H-1083 Department of Astronomy, Lorand Eotvos University HUNGARY |
| M. Shara | Space Telescope Science Institute, 3700 San Martin Drive Baltimore MD 21218, USA |
| M. Shimada | Department of Astronomy, Faculty of Science, Kyoto Univ. Kitashirakawa Sakyo-ku, Kyoto 606-01, Japan |
| C. Sosin | Department of Astronomy, 601 Campbell Hall, University of California, Berkeley, CA 94720, USA |
| R. Spurzem | Institut fur Astronomie, Olshavsenstrasse 40, D-24098 Kiel, Germany |
| V. G. Surdin | Sternberg Astronomical Institute, Moscow, Russia |
| T. Tsuchiya | Theoretical Astrophisics Division, National Astronomical Observatory, 2-21-2, Osawa, Mitaka, 181, Japan |
| M. Taiji | Department of Earth Science and Astronomy, College of Arts and Sciences, University of Tokyo 3-8-1, Komaba, Meguro-ku, Tokyo 153, Japan |
| K. Takahashi | 1-1 Kaneyama-Cho, Toyonaka-Shi, Osaka 560, Japan |
| C. Theis | Institut fur Astronomie, Olshavsenstrasse 40, D-24098 Kiel, Germany |
| H. UMEHARA | Division of Theoretical Astrophysics, 2-21-1, Osawa, Mitaka, Tokyo 181, Japan |
| T. Umemoto | Nobeyama Radio Observatory, Nobeyama, Minamimaki, Minamisaku, Nagano 384-13, Japan |
| F. Verbunt | Astronomical Institute, Postbox 80000, 3508 TA Utrecht, Neth. |
| E. Vesperini | Scuola Normale Superiore, Piazza dei Cavalieri 7, 56126 Pisa Italy |
| R. Wijers | Institute of Astronomy, Madingley Road, Cambridge CB3 0HA, UK |
| H. Yamaoka | Dept. of Physics, Kyushu Univ., 4-2-1 Ropponmatsu, Chuo-ku, Fukuoka 810, Japan |

| | |
|---|---|
| Y. Yokono | National Astronomical Observatory, 2-21-1 Ohsawa, Mitaka, Tokyo 181, Japan |
| S. Zaggia | Dipartimento di Astronomia, Univ. di Padova vicolo Osservatorio, 5 I-35100 Padova, Italy |
| Y. Zhou | Division of Theoretical Astrophysics, National Astronomical Observatory, Mitaka, Tokyo 181, Japan |

# A COMPARATIVE STUDY OF GLOBULAR CLUSTERS

DAIICHIRO SUGIMOTO

*Department of Earth Science and Astronomy,*
*College of Arts and Sciences, University of Tokyo*
*3-8-1 Komaba, Meguro-ku, Tokyo 153, Japan*

**Abstract.** This paper is intended for an introduction to the Symposium. In 47 Tuc many milli-second pulsars are found while none has been reported in $\omega$ Cen. One might think that in 47 Tuc they have been formed in the collapsed core through tidal capture of a main sequence star by a neutron star. If we use the standard model of gravothermal collapse of globular clusters to integrate the squared stellar density over the core and over its time history, we find, however, the accumulated probability of tidal capture is lower in 47 Tuc than $\omega$ Cen. Such contradiction suggests that it will be important to take account of mass segregation as well as stellar evolution in modelling dynamical evolution of star clusters.

## 1. Introduction

As an introduction to this Symposium I will compare two typical globular clusters, 47 Tuc and $\omega$ Cen with each other. The former, 47 Tuc, has a collapsed core with a density cusp. It has 11 milli-second pulsars (Manchester et al. 1991) and 21 blue stragglers (Paresce et al. 1991). In contrast, the latter, $\omega$ Cen, which is the largest globular cluster in the Galaxy has not been reported to contain any pulsars or blue stragglers, though there is a binary star NJL 5 as a cluster member (Margon and Cannon 1990). The existence of pulsars is also reported for the globular clusters having collapsed cores such as M15 and M5, for instance, as summarized by Lyne (1994).

A naive explanation of such observational facts was as follows. The pulsars may have originated from binary stars. The binary star may have been formed by tidal capture of a star encountering close to another star.

Therefore, more number of binary stars would have been formed in a denser core of the globular cluster.

I will show that it can not be so simple, however. If we look into the theory of a standard gravothermal collapse and integrate the binary formation rate over space and time, we will find that the number of accumulated binaries should be larger in $\omega$ Cen than in 47 Tuc because the former has a larger number of stars in its core. Something must be wrong in the model and the integration above.

## 2. Model of Gravothermal Collapse

Here we use the self-similar solution obtained by Lynden-Bell and Eggleton (1980). The radial density distribution of mass outside the core is expressed as

$$\rho(r,t) = \rho_c(\tau)\rho_*(\xi). \tag{1}$$

Here, $\xi$ is the spatial coordinate normalized with the core radius $r_c$ as

$$\xi = \frac{r}{r_c(\tau)}, \tag{2}$$

and $\tau$ is the time remaining to the complete collapse which is expressed using the present time $t$ and the time at the complete collapse $t_{\text{coll}}$ as

$$\tau = t_{\text{coll}} - t. \tag{3}$$

Similar expressions hold for other quantities such as the mass $M(r,t)$ contained within the sphere of radius $r$, and for the *rms* velocity of the stars $v(r,t)$ which is proportional to square root of the temperature in a gaseous model.

The similarity solution (Lynden-Bell and Eggleton 1980) is described for their spatial distributions in the halo as

$$\rho_*(\xi) \sim \xi^{-\alpha} = \xi^{-2.21}, \tag{4}$$

$$M_*(\xi) \sim \xi^{3-\alpha} = \xi^{-0.79}, \tag{5}$$

$$v_*(\xi) \sim \left(\frac{M_*}{\xi}\right)^{\frac{1}{2}} \sim \xi^{-\frac{\alpha-2}{2}} = \xi^{-0.11}, \tag{6}$$

where $\alpha = 2.21$ is used to obtain the extreme right-hand sides of the above relations. The temporal changes of the core radius and the other quantities are expressed as

$$r_c(\tau) \sim \tau^{\frac{2}{6-\alpha}} = \tau^{0.53}, \tag{7}$$

$$\rho_c(\tau) \sim \tau^{-\frac{2\alpha}{6-\alpha}} = \tau^{-1.17}, \tag{8}$$

$$M_c(\tau) \sim \tau^{\frac{6-2\alpha}{6-\alpha}} = \tau^{0.42}, \qquad (9)$$

$$v_c(\tau) \sim \tau^{-\frac{\alpha-2}{6-\alpha}} = \tau^{-0.055}. \qquad (10)$$

## 3. Formation Rate of Binaries

The local formation rate of binaries is given by Spitzer (1987, eq. 6-43) as

$$\frac{dn_b}{dt} = 10k \left(\frac{n}{10^4 \, \mathrm{pc}^{-3}}\right)^2 \left(\frac{M}{M_\odot}\right)^{1+\frac{\mu}{2}}$$

$$\times \left(\frac{R}{R_\odot}\right)^{1-\frac{\mu}{2}} \left(\frac{10 \, \mathrm{km \, s}^{-1}}{v}\right)^{1+\mu} \mathrm{pc}^{-3} \mathrm{Gyr}^{-1}. \qquad (11)$$

Here, $n$ is the number density of the stars which is proportional to $\rho$, and $M$ and $R$ are the mass and the radius of the individual star. The parameters $k$ and $\mu$ depend only on the stellar structure. We shall use the values for a polytrope of index 1.5, i.e., $k=2.1$ and $\mu=0.12$. From equations (4) and (11) we see that

$$\left(\frac{dn_b}{dt}\right) 4\pi r^3 \sim r^{-2\alpha+3+\frac{(1+\mu)(\alpha-2)}{2}} = r^{-1.30}, \qquad (12)$$

implying that the formation of binaries takes place mainly in the core rather than in the outer region of the globular cluster.

After some calculations using the relations above, the binary formation rate in the core is estimated as

$$\left(\frac{dN_b}{dt}\right)_{core} \simeq 30 \left(\frac{N_c}{10^5}\right) \left(\frac{n_c}{10^3 \, \mathrm{pc}^{-3}}\right) \left(\frac{10 \, \mathrm{km \, s}^{-1}}{v_c}\right)^{1.12} \mathrm{Gyr}^{-1}$$

$$\sim \tau^{a-1} = \tau^{-0.69}, \qquad (13)$$

where the power index to $\tau$ is expressed as

$$a - 1 = \frac{6 - 4\alpha + (1+\mu)(\alpha-2)}{6-\alpha}. \qquad (14)$$

The negative value of $a-1 = -0.69$ implies that more binaries are formed in the later stages when evaluated *per unit time*. However, the positive value of $a = 0.31$ implies that the later stages are shorter and that the binaries are *accumulated* rather during the earlier stages while $\tau$ is still large.

Integrating equation (13) backward from the present $\tau_0$ to the birth of the globular cluster $\tau_1$, i.e., over the age of the globular cluster,

$$t_{age} = \tau_1 - \tau_0, \qquad (15)$$

we obtain the number of binaries formed so far,

$$N_{\rm b} = \int_{\tau_0}^{\tau_1} \left(\frac{dN_{\rm b}}{dt}\right) d\tau$$

$$= \left(\frac{dN_{\rm b}}{dt}\right)_0 \tau_0 \int_0^{\ln(\tau_1/\tau_0)} \left(\frac{\tau}{\tau_0}\right)^a d\ln(\tau/\tau_0) = \left(\frac{dN_{\rm b}}{dt}\right)_0 \tau_{\rm eff}, \quad (16)$$

where the effective time $\tau_{\rm eff}$ is expressed as

$$\tau_{\rm eff} = \frac{\tau_0}{a}\left[\left(1 + \frac{t_{\rm age}}{\tau_0}\right)^a - 1\right]$$

$$= \begin{cases} \left(\dfrac{t_{\rm age}}{a}\right)\left(\dfrac{\tau_0}{t_{\rm age}}\right)^{1-a}, & \text{for } t_{\rm age} \gg \tau_0, \\ t_{\rm age}, & \text{for } t_{\rm age} \ll \tau_0. \end{cases} \quad (17)$$

## 4. Comparison between 47 Tuc and $\omega$ Cen

For illustrative purpose we apply the results in the preceding section to Model No. 14 of 47 Tuc by Meylan (1988) and to Model No. 13 of $\omega$ Cen by Meylan (1987). Though there are other models (*e.g.*, Chernoff and Djorgovsky 1989), and though the estimates given in the preceding section are rather rough, the following discussions will not be altered so far as the qualitative conclusions are concerned. The relevant quantities of Meylan's models are summarized in Table 1. Here, the notations have the following meanings; $M_{\rm tot}$ the total mass of the globular cluster, $\rho_{\rm h}$ the density at the half-mass radius, $v_{\rm s}$ the velocity scale corresponding to the velocity dispersion, $t_{\rm r,h}$ the two-body relaxation time at the half-mass, and $t_{\rm r,c}$ the relaxation time in the core of the globular cluster.

Quantities derived using the relations in the preceding sections are summarized in Table 2. Here, the time left to the complete collapse $\tau_0$ is calculated using equation (4-17) of Spitzer (1987), *i.e.*, using

$$\tau_0 = 190\, t_{\rm r,c}. \quad (18)$$

To convert the mass density to the number density of the stars we used the mass at the main sequence turn-off which is 0.79 $M_\odot$ both for 47 Tuc and $\omega$ Cen. We see that $\tau_{\rm eff}$ is much shorter than $t_{\rm age}$ for 47 Tuc while it is almost equal to $t_{\rm age}$ for $\omega$ Cen. This implies that the binaries are being formed much faster than the past in 47 Tuc while almost at the same formation rate in $\omega$ Cen. We see in Table 2 that the number of binaries formed so far, *i.e.*, $N_{\rm b}$, is calculated to be too large.

Table 1. Model of globular clusters by Meylan (1988, 1987)

| cluster | 47 Tuc | ω Cen |
|---|---|---|
| Model No. | 14 | 13 |
| $t_{\rm age}$ (Gyr) | 16 | 15 |
| $M_{\rm tot}$ ($10^6 M_\odot$) | 0.67 | 3.33 |
| $r_c$ (pc) | 0.56 | 3.9 |
| $\rho_c$ ($10^3\ M_\odot\ {\rm pc}^{-3}$) | 77 | 3.2 |
| $\rho_h$ ($10^3\ M_\odot\ {\rm pc}^{-3}$) | 0.58 | 0.30 |
| $v_s$ (km s$-1$) | 12.1 | 17.1 |
| $t_{r,h}$ (Gyr) | 2.4 | 16 |
| $t_{r,c}$ (Gyr) | 0.0016 | 0.78 |

To compare $N_b$ with observations of pulsars we have to take account of the followings. The progenitor of the neutron star is a massive star. Taking account of the Salpeter's initial mass function (Salpeter 1955), we estimate the number of binary stars containing neutron star as one of its component stars as

$$N_{\rm b,n} = N_{\rm b} \left(\frac{16}{0.79}\right)^{-1.4}, \qquad (19)$$

where the typical mass of the progenitor is taken as 16 $M_\odot$. This yields $N_{\rm b,n} = 290$ (47 Tuc) and 1030 ($\omega$ Cen). The neutron star will be spun-up to become a milli-second pulsar. If we assume the lifetime of the millisecond pulsar to be $t_{\rm pul}$, then the number of the milli-second pulsars is estimated as

$$N_{\rm pul} = N_{\rm b,n} \left(\frac{t_{\rm pul}}{\tau_{\rm eff}}\right). \qquad (20)$$

Table 2. Parameters for and accumulated numbers of binary formations

| cluster | 47 Tuc | ω Cen |
|---|---|---|
| $N_{\rm tot}$ ($10^3$) | 850 | 4200 |
| $N_c$ ($10^3$) | 37 | 520 |
| $N_c n_c$ ($10^9$ pc$^{-3}$) | 3.6 | 2.1 |
| $\tau_0$ (Gyr) | 0.3 | 150 |
| $t_{\rm age}$ (Gyr) | 16 | 15 |
| $\tau_{\rm eff}$ (Gyr) | 2.4 | 14.5 |
| $N_b$ | 19000 | 68000 |

This yields $N_{\text{pul}} = 120$ (47 Tuc) and 71 ($\omega$ Cen) for $t_{\text{pul}} \simeq 1$ Gyr. These numbers are still too large.

During the gravothermal collapse the mass segregation must have proceeded in 47 Tuc since its half-mass relaxation time is as short as 2.4 Gyr which should be compared with the age of the cluster, 16 Gyr. Because the mass of the neutron star progenitor is heavier than the average mass of the stars, the time scale of the mass segregation becomes of the order of 0.1 Gyr. This implies that the progenitor star had become a neutron star before it sank down to the core. However, the neutron stars were heavier than the average mass stars so that they have sunk down into the core thereafter.

The occurrence of such mass segregation in 47 Tuc can easily be imagined from the following facts. According to Meylan (1988) a mass function, which is much flatter than Salpeter's (1955) one, better fits both to observation (over smaller mass range) and to theoretical model (over larger mass range) for 47 Tuc. Moreover, the half-mass relaxation time as listed in Table 1 indicates that the time to complete core collapse is as long as $t_{\text{coll}} = 15\, t_{r,h} = 36$ Gyr (see, e.g., Spitzer 1987, p. 93), i.e., appreciably longer than $t_{\text{age}}$. In spite of such estimate 47 Tuc has an almost collapsed core (Meylan 1988). This implies that the core collapse should have taken place more rapidly than such estimate. As a result of mass segregation, models containing multi-mass components are shown to collapse more rapidly than the single-mass component models, if they are compared with the models that consist of stars of mass corresponding to the average mass of the multi-mass component models (Inagaki and Wiyanto 1984; Funato et al. 1993).

We shall assume that all the neutron stars sank down into the core in 47 Tuc while no mass segregation has taken place at all in $\omega$ Cen. Then, the number of milli-second pulsars should be increased by a factor of $N_{\text{tot}}/N_c$ for 47 Tuc to yield $N_{\text{pul}} = 2800$. When this is compared with $N_{\text{pul}} = 71$ for $\omega$ Cen, the ratio of them may not be inconsistent with observation. However, their absolute values are too large. This may imply that some of the neutron stars should have escaped from the cluster at the time of their formation by sling shot effect or escaped from the core in later stages by exchange collision. In this relation it is interesting to see that the depth of the core gravitational potential, which is proportional to $N_c/r_c$, is about a half for 47 Tuc of the value for $\omega$ Cen because of the smaller total and core masses of the former. This implies that the stars escaped more easily from the core of 47 Tuc than $\omega$ Cen. This will reduce the ratio of the expected number of pulsars to some extent. Though it reduces also the absolute values of the numbers of pulsars at the same time, there will be enough margin for them.

## 5. Conclusion

We are surprised how many uncertain points and wide uncertainties are involved in the problem. Even with the rough estimate and discussion that I have given in this talk, we see that the following points await for further studies to understand the evolution of actual globular clusters.

Theoretical model should include the mass-spectrum of the constituent stars to follow the mass segregation and evolution of each star. Such a model requires much larger $N$-body simulations and, correspondingly, much higher speed of computers, since each mass-bin and higher-mass bins, in particular, should have appreciable number of stars. It is desirable that young globular clusters as exist in the Magellanic Clouds are studied more extensively to clarify the dependence of the local mass functions on the radial distance from the center of the cluster (Kontizas 1994). In order to clarify the escape of the stars from the cluster it will be necessary to think about a scaling law to be applied to $N$-body simulations. From the side of the origin of pulsars we need to investigate more about their spin-up process and their lifetime. Comparison between pulsars and blue stragglers will be useful not only for the difference in their histories of origin but also for their roles as independent indicators of dynamical evolution of the cluster. Now, I think, it is matured to tackle such problems. We now have a plenty of information from observations on the one hand, and a tera-flops computer dedicated to $N$-body problem (Taiji 1995) on the other hand.

I would like to thank Professor Frank Verbunt with whom I enjoyed in 1992 such discussions as given in this talk. This work was partially supported by the Grant-in-Aid for Specially Promoted Research (04102002) of the Ministry of Education, Science, Sports and Culture.

## References

Chernoff, D.F. and Djorgovsky, G. (1989) *Astrophys. J.* **339**, 904.
Funato, Y., Makino, J., and Ebisuzaki, T. (1993) *Publ. Astron. Soc. Japan* **45**, 289.
Inagaki, S. and Wiyanto, P. (1984) *Publ.Astron. Soc. Japan*
Kontizas, M. 1994, private communications.
Lynden-Bell, D. and Eggleton, P.P. (1980) *Mon. Not. R. astr. Soc.* **191**, 483.
Lyne, A.G. (1994) *AIP Conf. Proc., The Evolution of X-Ray Binaries*, eds. S.S. Holt and C.S. Day, AIP Press, **308**, 331.
Manchester, R.N., Lyne, A.G., Robinson, C., D'Amico, N.D., Bailes, M., and Lin, J. (1991) *Nature* **352**, 219.
Margon, B. and Cannon, R. (1990) *Observatory* **109**, 82.
Meylan, G. (1987) *Astron. Astrophys.* **184**, 144.
Meylan, G. (1988) *Astron. Astrophys.* **191**, 215.
Paresce, F. and the other 20 authors (1991)*Nature* **352**, 297.
Salpeter, E.E. (1955) *Astrophys. J.* **121**, 161.
Spitzer, L., Jr. (1987)*Dynamical Evolution of Globular Clusters*, Princeton Univ. Press.
Taiji, M. (1955) in this volume.

# CORRELATIONS OF GLOBULAR CLUSTER PROPERTIES AS CONSTRAINTS FOR DYNAMICAL EVOLUTION MODELS

S.G. DJORGOVSKI
*Palomar Observatory, Caltech*
*Pasadena, CA 91125, USA*

**Abstract.**
Correlations between globular cluster (GC) properties are reviewed, including some new work on cluster tidal radii and densities. Most of the observed correlations can be interpreted within the framework of our current understanding of their dynamical evolution. These correlations provide empirical constraints for models of GC formation and evolution.

## 1. Introduction

Distributions of, and correlations between various global properties of globular clusters (GCs) can provide important observational constraints for models of their evolution, or even formation. Their studies may be the only way in which we can address the *global* picture of GC evolution.

The subject has been covered extesively in several recent papers: van den Bergh (1994ab, 1995), Djorgovski & Meylan (1994; hereafter DM), Djorgovski (1995), and Bellazzini *et al.*(1995); see also Surdin, this volume. Instead of repeating much of what was covered there, we will only give a brief summary of the results from these studies in Sect. 2, and discuss new correlations of tidal-radius related quantities in Sect. 3. The reader may wish to start with the paper by DM, and references therein, and consider this review as an update to it. A comparison of GC properties with those of early-type galaxies was presented by Djorgovski (1993a).

Most GC properties in general, and relaxation times in particular, span a large dynamical range, e.g., a factor of $\sim 10^5$ in the central relaxation time, $t_{rc}$. That practically guarantees that a large range of dynamical evolution states will be present in the sample, which makes it possible to use

GC properties and their correlation as direct probes of their dynamical evolution. Likewise, GCs cover a range of a factor of $\sim 10^2$ in Galactocentric radius, $R_{GC}$; given the flat rotation curve for the Galaxy, that translates in a comparable range in disk or bulge crossing frequencies. That makes it possible to study the differential effects of tidal shocks from disk and bulge passages over the sample as a whole.

The observed properties of GCs today are a complicated product of the initial conditions (reflecting the formative processes of GCs), and some 15 Gyr of dynamical evolution, which can be driven both by the internal instabilities, such as the core collapse and mass segregation, and the external effects of the Galactic tidal field. Internal dynamical evolution can be modulated (typically accelerated) by the tidal shocks. One thus expects a complex picture. Physical processes which affect more than one observable property will generate correlations; for example, core collapse will simultaneously change the core radii, $r_c$, central densities, $\rho_0$, or surface brightness, $\mu_0$, and concentrations, $c$. A complex interplay of physical effects can then result in multivariate correlations.

Nevertheless, it is possible to interpret and disentangle some of these effects. Core properties are presumably dominated by the effects of core collapse, since $\langle t_{rc} \rangle \ll$ cluster ages. Cores thus have no memory of the initial conditions. On the other hand, theory suggests that most half-light properties change little during the core collapse. Tidal effects may be found in correlations with $R_{GC}$, or the distance to the Galactic plane, $Z_{GP}$. Cluster luminosities (absolute magnitudes, $M_V$), orbits (i.e., $R_{GC}$), and velocity dispersions, $\sigma$, may be changed very little by the dynamical evolution for most clusters, and thus may reflect the formation processes (Murray & Lin 1992). Certainly the metallicities, [Fe/H], are primordial.

## 2. A Brief Summary of the Previous Work

DM have analysed a subset of the data compiled by Djorgovski (1993b). Some of their results are as follows:

The dynamical range the core parameters, which are presumably more affected by dynamical evolution, is much larger than for the corresponding half-light parameters. The more rapid evolution of core regions breaks the homology of cluster structures, and introduces the shape parameter in the GC sequence, the King $c$. This process can be contemplated in looking at Fig. 1. Presumably GCs evolve by sliding down and to the left on these correlations. Note that for a typical GC, the time between disk (or bulge) passages is $\sim 10^8$ yr, which is roughly the value of the half-mass relaxation time, $t_{rh}$, where the deviation $t_{rc}/t_{rh} \to 0$ really sets in. From the distribution of GC relaxation times, one can estimate the present rates of GC core

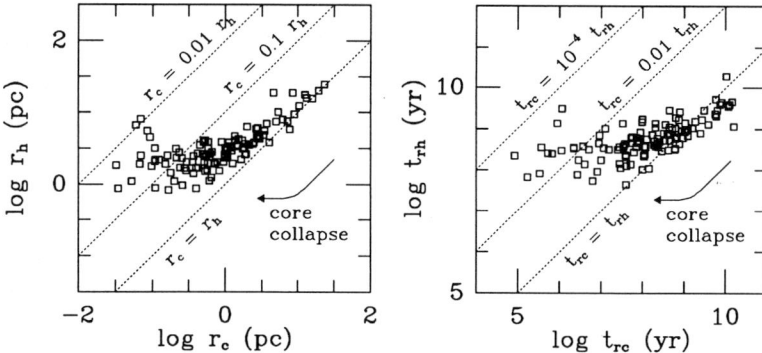

*Figure 1.* A comparison of core and half-light radii (left) and relaxation times (right) for Galactic GCs. At the slow evolution end (large radii and relaxation times), the core and the half-hight parameters of clusters are similar. They deviate from this ever more strongly as the relaxation times decrease, as clusters approach the core collapse.

collapses and GC evaporation in the galaxy, both being of the order of a few GCs per Gyr (Hut & Djorgovski 1992).

The best observed correlations are between the core parameters: core radius, $r_c$, central surface brightness, $\mu_0$, and the concentration, $c$, as well as any derived quantities, such as the central relaxation time, $t_{rc}$. Two of them are shown in Fig. 2. The correlations are exactly as what may be expected from a population of GCs approaching the core collapse. The observed scaling relations are:

$$I_0 \sim r_c^{-1.8 \pm 0.2} \qquad \rho_0 \sim r_c^{-2.6 \pm 0.15}$$

where $I_0$ is the central surface brightness in linear units, rather than magnitudes, and $\rho_0$ is the central luminosity density, both in the V band. Note that for a single-mass-species cluster, the exponent in the second relation should be $-2.23$ from the standard core collapse theory (Lynden-Bell & Eggleton 1980, Cohn 1980), marginally different from what is observed; perhaps this reflects the real mass spectrum of the stars. These scaling relations imply $L_{core} \sim r_c^{0.3 \pm 0.2}$, possibly reflecting the accelerated evaporation or ejection of stars from a collapsing core.

The core parameters and concentrations also correlate with the position in the Galaxy, with clusters closer to the Galactic center or plane being more concentrated and having smaller and denser cores (cf. also Chernoff & Djorgovski 1989, and Djorgovski 1993a). These trends are *much* more pronounced for the fainter (less massive) clusters. This is in an agreement with a picture where tidal shocks form disk or bulge passages accelerate dynamical evolution of clusters (cf. Chernoff & Weinberg 1990, Aguilar

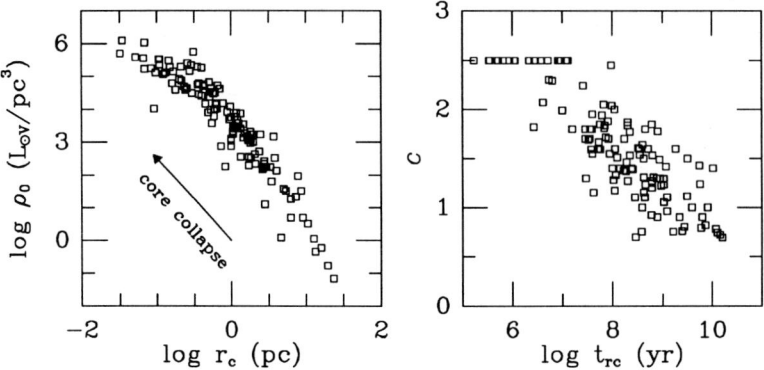

*Figure 2.* Left: The correlation between core radii ($r_c$) and central luminosity densities ($\rho_0$). Right: Dependence of cluster concentrations (King $c$) on the central relaxation time ($t_{rc}$). PCC clusters have been set to $c = 2.5$, and their $t_{rc}$'s are greatly uncertain. Both trends are expected in a simple picture of globular clusters as a collapsing sequence of King models.

1993, and references therein). The mean trends for the binned data are:

$$r_c \sim R_{GC}^{1.1 \pm 0.5} \qquad \rho_0 \sim R_{GC}^{-2.8 \pm 1.6}$$

Since for a flat Galaxy rotation curve the frequency of disk passages scales as $R_{GC}^{-1}$, these scaling relations imply shrinking of the GC cores roughly in a direct proportion to the frequency of tidal shocks.

It is also worth noting that for almost all clusters, half-mass relaxation times, $t_{rh}$ are comparable to, or greater than the time intervals between the disk passages, $t_{DC}$, as estimated from $R_{GC}$ by assuming a flat Galaxy rotation curve.

There are no good correlations of other parameters with luminosity, although more luminous clusters tend to be more concentrated (cf. also Djorgovski 1991, and van den Bergh 1994c). When data are binned in luminosity, several trends emerge: more luminous clusters tend to have smaller and denser cores. The corresponding scaling relations are:

$$r_c \sim L^{-0.5 \pm 0.25} \qquad \rho_0 \sim L^{2 \pm 1}$$

where $L$ is the cluster total luminosity in the $V$ band. Possibly this may reflect the initial conditions (Bellazzini *et al.* 1995).

Cluster metallicities do not correlate with any other parameter, including luminosity and velocity dispersion (a behavior strikingly different from that of elliptical and dwarf galaxies; see Djorgovski 1993a). The only detectable trend is with the position in the Galaxy, probably reflecting Zinn's

(1980) disk-halo dichotomy. Along with their great chemical homogeneity and the narrowness of the main sequences, this suggests that GCs were not self-enriched systems, but that they formed from a pre-enriched material within larger structures (former dwarf galaxies or protogalactic fragments).

Central velocity dispersions, $\sigma$, show excellent correlations with luminosity and surface brightness. Their origin is not well understood, but they may well reflect initial conditions of cluster formation. The corresponding scaling relations are:

$$\sigma \sim L^{0.60\pm0.15} \qquad \sigma \sim I_0^{0.50\pm0.10} \qquad \sigma \sim I_h^{0.45\pm0.05}$$

where $I_h$ is the mean central surface brightness within the $r_h$, in linear units. Core radii and concentrations play a role of a "second parameter" in these correlations. Djorgovski (1995) has extended this analysis to obtain bivariate correlations analogous to the "Fundamental Plane" (FP) of elliptical galaxies. For the core parameters, the bivariate scaling relation is:

$$r_c \sim \sigma^{2.0\pm0.1} I_0^{-1.1\pm0.1}$$

which is exactly what may be expected from the virial theorem, if all GC cores have a similar structure (which they do), and uniform $(M/L)$ ratios. For the half-light parameters, the bivariate scaling relation is:

$$r_h \sim \sigma^{1.45\pm0.2} I_h^{-0.85\pm0.1}$$

which is very close to the corresponding FP relation for elliptical galaxies. Since the $(M/L)$ ratios are unlikely to vary a lot among GCs, the culprit is clearly their variety of density profiles, with the concentration correlated with other parameters. This is an important lesson for the physical interpretation of the FP of ellipticals itself.

A multivariate statistical analysis of the entire data set shows that the global manifold of cluster properties has a high statistical dimensionality, $D > 4$ (cf. also Djorgovski 1981). However, a subset of structural, photometric, and dynamical parameters forms a statistically 3-dimensional family, as expected from objects following King (1966) models, which DM propose to call the King Manifold.

Clusters with post-core-collapse (PCC) morphology (Djorgovski & King 1986) participate in the same correlations as the clusters with King model (KM) morphology. Operationally, their core radii were set to the observed HWHM in the surface brightness profiles, which is just an upper limit. It may be fortuitous that seeing effects move the data points roughly along the observed correlations.

This analysis did not reveal presence of any distinct subgroups within the Galactic GC system, beyond the disk-halo dichotomy, and the possible

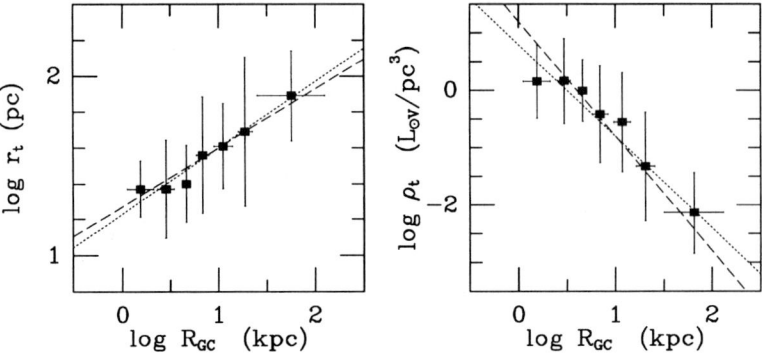

*Figure 3.* Median estimated tidal radii, $r_t$, and mean luminosity densities within the $r_t$, $\rho_t$, in bins of 21 clusters each (fewer in the last bin). The "error" bars indicate quartile-estimated sigma for each bin, in both coordinates. The best fit scaling laws, $r_t \sim R_{GC}{}^{0.37}$ and $\rho_t \sim R_{GC}{}^{-1.6}$ are indicated with dotted lines. Alternative, nearly as good fits, $r_t \sim R_{GC}{}^{1/3}$ and $\rho_t \sim R_{GC}{}^{-2}$ are indicated with dashed lines.

groups identified by van den Bergh (1993). This is curious, given that many GCs may have been acquired or produced in accretion events in Galaxy's history. Apparently, dynamical evolution processes, which should be universal, drive most of the observed correlations.

It is now becoming possible to conduct similar global statistical studies for the GC systems in the Magellanic Clouds (cf. Meylan & Djorgovski 1987), in M31, where HST resolution is necessary (cf. Fusi Pecci et al.1994), or even beyond. Comparative studies of GC systems in other, nearby galaxies can add considerably to our understanding gained so far from the Galactic GC system alone.

## 3. Correlations Involving Tidal Radii and Densities

There are very few measured tidal radii for globular clusters: this is an extremely difficult task. A recent attempt to measure some was made by by Grillmair et al.(1995). Lacking real measurements, very rough estimates may be obtained as $\log r_t = \log r_c + c$, using the data from Djorgovski (1993b).

Mean cluster densities $\rho_t$ within $r_t$ can then be obtained from their total luminosities. These must be higher than the local dark halo density at the perigalacticon:

$$\rho_t^{\text{cluster}} \approx \alpha \, \rho^{\text{halo}}(R_{peri})$$

where $\alpha \approx 5.5$ (see below). Innanen et al.(1983) find that:

$$\frac{r_t}{R_{peri}} = \frac{2}{3} \left[ \frac{m_{cluster}}{(3+\epsilon)M(R_{peri})} \right]^{1/3}$$

where $\epsilon$ is the orbital eccentricity, and $M(R_{peri})$ is the enclosed Galactic mass within $R_{peri}$. For a Galaxy with a flat rotation curve $V = const.$, $M(R)/R = V^2/G$. Assuming $\epsilon \approx 0$, it is easy to derive:

$$R_{peri} = \alpha V (4\pi G)^{-1/2} \rho_t^{-1/2}$$

where $\alpha = (243/8)^{1/2}$. For simple estimates,

$$R_{peri} \approx 0.95 \, \alpha \, \rho_t^{-1/2} \approx 5.2 \, \rho_t^{-1/2}$$

where $R_{peri}$ is in kpc, and $\rho_t$ is in $M_\odot/pc^3$.

Trends of $r_t$ and $\rho_t$ with the Galactocentric radius are shown in Fig. 3 for binned data. Tidal radii and the mean densities within $r_t$ decline at smaller Galactocentric radii, as expected. The mean trends are:

$$r_t \sim R_{GC}{}^{0.37\pm0.05} \quad (\approx R_{GC}{}^{1/3})$$

$$\rho_t \sim R_{GC}{}^{-1.6\pm0.2} \quad (\approx R_{GC}{}^{-2}?)$$

The second relation nearly mimics the average density law of the galaxy, implied by the flat rotation curve, $\rho \sim R_{GC}{}^{-2}$, which is gratifying to see. Also, at a given $R_{GC}$, more luminous clusters tend to have larger $r_t$'s.

The ratios of the present Galactocentric distances $R_{GC}$ and the derived perigalactic radii $R_{peri}$ are a statistical measure of the shapes of globular cluster orbits: for nearly circular orbits, $R_{peri} \approx R_{GC}$; for eccentric/plunging orbits, $R_{peri} < R_{GC}$. Such an estimate may be very uncertain, especially for the PCC clusters, whose derived tidal radii, and thus the perigalactic radii are more uncertain, or the clusters near the Galactic center, say at $R_{GC} < 2$ kpc, where the errors in distances and the $R_{GC}/R_{peri}$ ratios become too high. A "clean" sample of clusters with $c < 2.1$ and $R_{GC} < 2$ kpc may be safer for this analysis.

¿From the distribution shown in Fig. 4, it appears that most clusters are on nearly circular orbits (or at least not very eccentric ones).

We can now explore correlations for clusters with statistically different orbits. Using the "clean" sample of clusters with $c < 2.1$ and $R_{GC} < 2$ kpc, and dividing the sample as: (1) $R_{GC}/R_{peri} \leq 1.5$, clusters with nearly circular orbits, and (2) $R_{GC}/R_{peri} > 1.5$, clusters with plunging/eccentric orbits, we find the following trends:

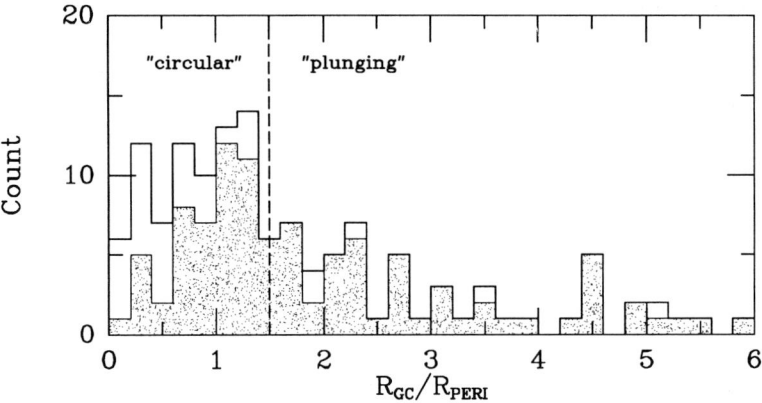

*Figure 4.* Distribution of the estimated $R_{GC}/R_{peri}$ ratios for Galactic GCs. Clusters on more elongated orbits should have on the average larger $R_{GC}/R_{peri}$. The open histogram is for all GCs in the samlpe, and the shaded histogram for the "clean" sample with $c < 2.1$ and $R_{GC} < 2$ kpc.

At a given $R_{GC}$, "plunging" clusters tend to have smaller radii and higher densities, both core and half-light, and tend to be slightly more concentrated. Also, most correlations with $R_{GC}$ are better for the "circular" sample; i.e., the "plunging" sample shows more scatter. A possible interpretatin of these trends is that the effects of tidal shocks appear to be slightly stronger for the "plunging" clusters, both in terms of their tidal truncation, and the acceleration of their dynamical evolution towards the core collapse.

Correlations with velocity dispersion tend to be slightly better for the "circular" sample (there is even a correlation with the cluster concentration). There are no obvious differences in the (non)correlations involving cluster metallicities.

A similar analysis was presented by van den Bergh (1995), who claims that there are correlations between the cluster perigalactic radii and metallicities or ages. Van den Bergh (1994a) also suggested that GCs on retrograde orbits tend to have half-light radii lower than average at the same $R_{GC}$, whereas GCs with circular orbits tend to have larger half-light radii.

In this analysis we made gross approximations, such as the circular orbits. The reality may be much more complicated, as suggested, e.g., by the recent study by Dauphole et al.(1995). More realistic estimates of both GC tidal radii and perigalactica may reveal more interesting correlations.

## 4. Concluding Remarks

The situation seems encouraging: the observed correlations largely bear out the expectations from our theories, at least qualitatively, and in many cases quantitatively as well. What remains to be done is to compare the predictions with the observations in some detail. Modeling the effects of tidal shocks and GC evolution in a realistic tidal field of our Galaxy may be quite interesting. There are also correlations, most notably those involving velocity dispersions, which were not clearly expected from theory, and are not yet really understood.

Observationally, the greatest need remains for better distances to most clusters. Other parameters can also be substantially improved in many cases. Actual measurements of GC tidal radii and envelopes (as opposed, e.g., to crude estimates made in this work) are also a high priority; the methodology used by Grillmair et al.(1995) is very promising for this task. Direct observational constraints on GC orbits, e.g., from proper motions, will also be very useful for modeling of tidal effects.

The structure of collapsed cores remains murky, and core radii for highly evolved (concentrated) clusters are not well defined, even with the HST data. Cores for PCC clusters may never be well defined. It may be better then to institute a new radial scale for the "core" regions of GCs, which would be operationally well defined. One possibility is to use fixed fraction of the total light radii, e.g., radii enclosing 10% of the total light (in projection). Similarly, tidal radii, which may be equally difficult to define and measure, could be replaced with the radii enclosing 90% of the total light. Their ratios could define new concentration index, $c_* = \log(r_{90}/r_{10})$. This can be done, e.g., with the profiles compiled by Trager et al.(1995).

This review did not address the correlations between stellar populations and dynamical structure of GCs (Djorgovski et al.1991, Fusi Pecci et al.1993; see Djorgovski & Piotto 1993 for a review and references). This also includes the possible dependence of the formation rates of millisecond pulsars and LMXBs on the cluster parameters, the origins of blue stragglers, etc. This is a fascinating area of GC research, in which much progress yet remains to be made, both observationally (especially with the HST; see several excellent papers in this volume) and theoretically, through detailed simulations of stellar collisions and tidal interactions in dense cluster cores.

This work was supported in part by the NSF PYI award AST-9157412, by grants from NASA, and by the Bressler Foundation. The author would like to thank to the conference organizers, and especially Drs. Sugimoto and Makino, for their great efforts and hospitality.

# References

Aguilar, L. 1993, in Galaxy Formation: The Milky Way Perspective, ed. S. Majewski, ASPCS, 49, 155
Bellazzini, M., Vesperini, E., Fusi Pecci, F. & Ferraro, F. 1995, MNRAS, in press
Chernoff, D. & Djorgovski, S. 1989, ApJ, 339, 904
Chernoff, D. & Weinberg, M. 1990, ApJ, 351, 121
Cohn, H. 1980, ApJ, 242, 765
Dauphole, B., Geffert, M., Collin, J., Ducourant, C., Odenkirchen, M. & Tucholke, H.-J. 1995, A&Ap, in press
Djorgovski, S. & King, I.R. 1986, ApJ, 305, L61
Djorgovski, S. 1991, in Formation and Evolution of Star Clusters, ed. K. Janes, ASPCS, 13, 112
Djorgovski, S., Piotto, G., Phinney, E.S. & Chernoff, D.F. 1991, ApJ, 372, L41
Djorgovski, S. 1993a, in The Globular Cluster – Galaxy Connection, eds. G. Smith & J. Brodie, ASPCS, 48, 496
Djorgovski, S. 1993b, in Structure and Dynamics of Globular Clusters, eds. S. Djorgovski & G. Meylan, ASPCS, 50, 373
Djorgovski, S. & Piotto, G. 1993, in Structure and Dynamics of Globular Clusters, eds. S. Djorgovski & G. Meylan, ASPCS, 50, 203
Djorgovski, S. & Meylan, G. 1994, AJ, 108, 1292
Djorgovski, S. 1994, ApJ, 438, L29
Fusi Pecci, F., Ferraro, F., Bellazzini, M., Djorgovski, S., Piotto, G. & Buonanno, R. 1993, AJ, 105, 1145
Fusi Pecci, F., et al.1994, A&Ap, 284, 349
Grillmair, C., Freeman, K., Irwin, M. & Quinn, P. 1995, AJ, 109, 2553
Hut, P. & Djorgovski, S. 1992, Nature, 359, 806
Innanen, K., Harris, W. & Webbink, R. 1983, AJ, 88, 338
King, I.R. 1966, AJ, 71, 64
Lynden-Bell, D. & Eggleton, P. 1980, MNRAS, 191, 483
Meylan, G. & Djorgovski, S. 1987, ApJ, 322, L91
Murray, S. & Lin D. 1992, ApJ, 400, 265
Trager, S.C., King, I.R. & Djorgovski, S. 1995, AJ, 109, 218
van den Bergh, S. 1993, ApJ, 411, 178
van den Bergh, S. 1994a, ApJ, 432, L105
van den Bergh, S. 1994b, AJ, 108, 2145
van den Bergh, S. 1994c, ApJ, 435, 203
van den Bergh, S. 1995, AJ, 110, 1171
Zinn, R. 1980, ApJS, 42, 19

# HUBBLE SPACE TELESCOPE STUDIES OF THE DENSE CENTRAL REGIONS OF GLOBULAR CLUSTERS

PURAGRA GUHATHAKURTA
*Univ. of California, Lick Obs., Santa Cruz, CA 95064, USA*

BRIAN YANNY
*Fermi National Accelerator Lab., Batavia, IL 60510, USA*

DONALD P. SCHNEIDER
*Dept. of Astronomy & Astrophysics, Pennsylvania State Univ., University Park, PA 16802, USA*

AND

JOHN N. BAHCALL
*Inst. for Advanced Study, Princeton, NJ 08540, USA*

**Abstract.** We present results from an ongoing program to probe the dense central parts of Galactic globular clusters using multicolor Hubble Space Telescope images (WF/PC-I and WFPC2). Our sample includes the dense clusters M15, 47 Tuc, M30, NGC 6624, M3 and M13. The two main goals of our program are to measure the shape of stellar density profile in clusters (the slope of the density cusp in post core collapse clusters, in particular) and to understand the nature of evolved stellar populations in very dense regions and their variation as a function of radius. The latter includes studies of blue straggler stars and of the central depletion of bright red giants. Our recent WFPC2 study of M15 is described in detail.

## 1. Introduction

Many of the highlights of this symposium were provided by recent observations with the Hubble Space Telescope (HST). The high angular resolution and faint stellar detection threshold of HST make it an ideal instrument for the study of dense star clusters. It is instructive to quantify the benefits of high angular resolution in the context of the central region of a dense globular cluster (such as 47 Tuc or M15). The $1''$ seeing (FWHM) that is typical

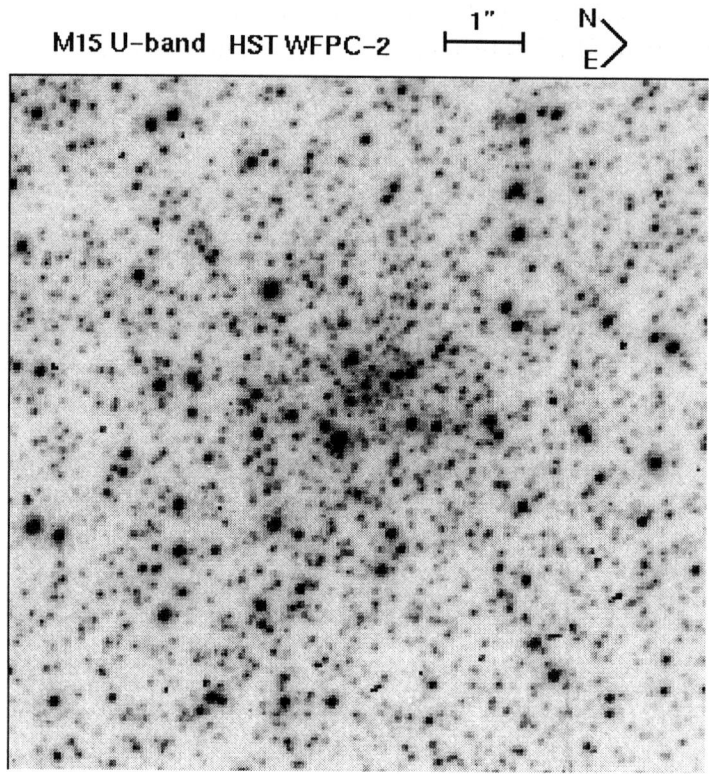

*Figure 1.* Short $U$ band HST WFPC2 exposure of the central $9'' \times 9''$ of M15. The surface density of stars rises towards the center all the way in to subarcsecond scales. The $0\rlap.{''}1$ resolution of images like the one displayed makes it possible to resolve individual stars down to the main sequence turnoff even near the dense cluster center.

of groundbased observatories causes the blurred images of the brightest stars, those on the upper red giant branch (RGB), to crowd the central region, limiting studies to stars brighter than the horizontal branch (Aurière and Cordoni 1981). The "dynamic range" of such observations—roughly defined to be the range of stellar brightnesses over which accurate photometry is possible—is limited to about 2.5 mag or a factor of 10. Under very good seeing conditions (FWHM $\sim 0\rlap.{''}5$), it is possible to study subgiants that are up to 5 mag fainter than the tip of the RGB, corresponding to a dynamic range of 100 (Bolte et al. 1993; Stetson 1994). The sharp $0\rlap.{''}1$ core of the pre-repair HST point spread function represented a great leap forward in our ability to perform complete star identifications down to faint limiting magnitudes, but its broad and complicated wings ($r_{\rm PSF} \sim 2\rlap.{''}5$)

made accurate photometry difficult, if not impossible. The dynamic range of short pre-repair HST exposures of crowded clusters is about 200, barely reaching down to the main sequence turnoff (Lauer et al. 1991; Paresce et al. 1991; Guhathakurta et al. 1992; Sosin and King 1995). This subject has received a tremendous boost thanks to post-repair HST imaging (WFPC2 and FOC+COSTAR) with its dynamic range of over 1000, enabling crowded field photometry of stars up to 2 mag below the turnoff even with exposure times $\lesssim 1$ min (cf. Yanny et al. 1994b; Fig. 1).

Over the last 5 years, we have analyzed HST snapshots of the central regions of several nearby dense Galactic globular clusters. This includes pre-repair WF/PC-I $V$ and $I$ images of 47 Tuc (Guhathakurta et al. 1992), M15 (Yanny et al. 1994a), M3 (Guhathakurta et al. 1994), and M13 (Cohen et al. 1996), as well as post-repair WFPC2 $U$, $B$, and $V$ images of M30 (Yanny et al. 1994b; Guhathakurta et al. 1996a), M15 (Guhathakurta et al. 1996b), and NGC 6624 (Yanny et al. 1996). This paper is based primarily on these datasets.

There are two main themes in our HST studies of globular clusters. The first is the characterization of the stellar surface density profile on sub-arcsecond scales (slopes, core radii). This paper focusses on new WFPC2 results on M15 (Sec. 2). The broader goal is to place empirical constraints on the physical nature of gravothermal collapse and on the shapes of stellar concentrations around massive black holes. The second theme involves studying the mix of stellar populations as a function of radius (and hence of ambient stellar density) in dense cluster cores, in order to investigate the effects of a dense environment on the evolution of individual stars and the effects of mass segregation. This includes ongoing work on centrally concentrated blue straggler stars (BSSs) in 47 Tuc, M30, and M15 and the depletion of extreme RGB stars near the center of M15 and other post core collapse clusters (Sec. 3).

## 2. Density Distribution

Visible stars in globular clusters are a convenient tracer of the underlying mass distribution. The shape of the stellar luminosity function is such that bright RGB stars contribute a significant fraction of the total cluster light while constituting a very small fraction of the total mass. For example, RGB stars brighter than the horizontal branch represent less than 10% of all post main sequence stars by number, but are responsible for about two-thirds of the light. Evolved stars in globular clusters have roughly the same mass ($M_{TO}$) despite the fact that their luminosities span a range of nearly 7 mag (factor of 500). For this reason, the number-weighted stellar density (counts) provides a more reliable measure of the mass density than

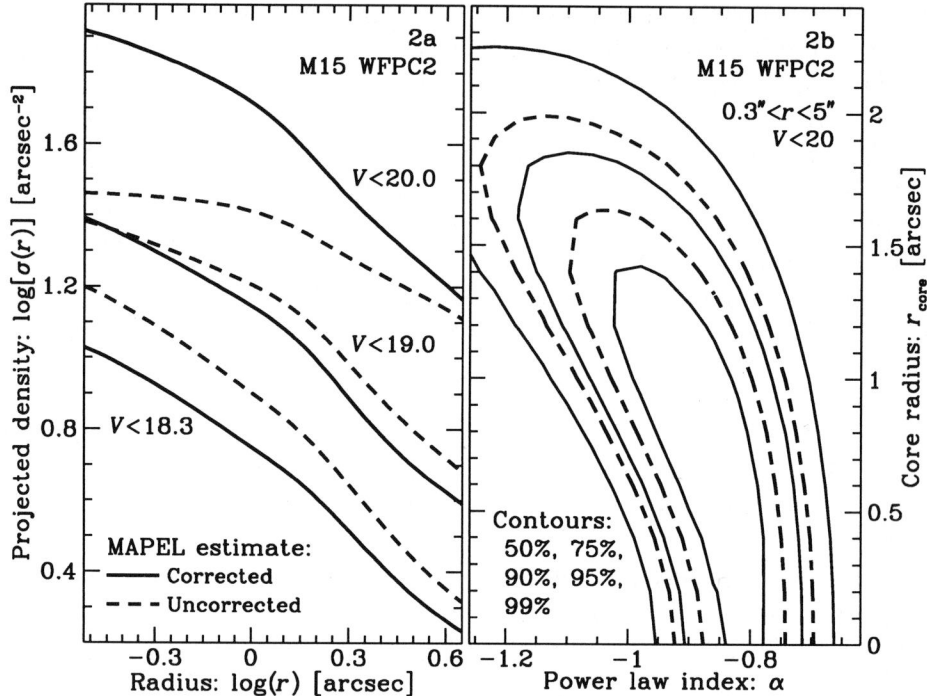

Figure 2. Left (a) Non-parametric (MAPEL) estimate of the stellar surface density profile in the central part of M15 based on HST WFPC2 data. The raw counts (dotted lines) are converted to the corrected profile (solid line) using correction factors derived from a simulation. The different stellar samples yield corrected profiles that are similar in shape. Right (b) The joint probability distribution of the core radius and asymptotic power law index $\alpha$ of M15. The tilt of the likelihood contours indicates significant covariance between the two parameters. The data are consistent with a simple power law of index $\alpha \sim -0.8$ ($r_{core} = 0$).

the surface brightness. Magnitude-limited star counts must be corrected for the effects of incompleteness and photometric error which depend strongly on the degree of crowding, which in turn varies with distance from the cluster center. These corrections are determined using detailed simulations of the data for each cluster; each simulation mimics the PSF shape, stellar luminosity function, and surface density distribution of stars in the cluster.

## 2.1. M15'S CENTRAL CUSP

M15 has a strongly peaked surface brightness profile and the highest central density of known globular clusters. These have long made it a promising post core collapse candidate (King 1975) or the site of a central massive black hole (Bahcall et al. 1975). Lauer et al. (1991) suggested that the diffuse residual light in a pre-repair HST image of M15 has a 2″ core, but realistic simulations of their data indicated that the diffuse light is not a reliable measure of the stellar density because photometric errors lead to inaccurate subtraction of the light of bright giants and because of the complicated nature of the aberrated PSF (Yanny et al. 1994a).

Short WFPC2 $V$ (F555W) exposures of M15 have been used to derive counts of stars above the turnoff. Within $r \lesssim 3''$, crowding is severe despite HST's 0.″1 resolution (Fig. 1). The brightest stars ($V < 16$) are easiest to count but their small number leads to large Poisson error in the density estimate. Photometric scatter and bias cause overcount in subgiant samples ($V < 18$) which lie in a particularly steep part of the stellar luminosity function. Fainter samples ($V < 20$) suffer primarily from incompleteness near the cluster center. We have experimented with various limiting magnitudes and independent stellar samples, using the simulation to correct for the above effects in each case, and they all yield consistent results for density profile of M15 indicating that the corrections are reliable (Fig. 2a).

The corrected surface density profile of post main sequence stars in M15 ($V \lesssim 19$) is well approximated by a power law, $\sigma(r) \propto r^\alpha$, with an index $\alpha = -0.82 \pm 0.12$ over the radial range $0.''3 \lesssim r \lesssim 6''$ (Fig. 2a). Interior to $0.''3$, uncertainties in the cluster centroid position, star count correction factor, and star counts (Poisson error) prevent reliable determination of the density. M15's density profile steepens to a slope of $\alpha \sim -1.3$ for $6'' < r < 30''$ and to $\alpha \sim -2$ beyond $r = 30''$. A non-parametric estimate of the spatial density profile of M15 (using the MAPEL method of Merritt and Tremblay 1994) suggests that the density of evolved $V < 19$ stars alone exceeds $10^5 \, \mathrm{pc}^{-3}$ in the inner 0.02 pc.

A power law is an adequate description of the density profile in M15's central 6″. The best fit $\alpha \sim -0.8$ index is consistent with that expected from models of core collapse (cf. Grabhorn et al. 1992) or with the predicted shape of the stellar concentration around a massive black hole (Bahcall and Wolf 1976, 1977). While the density profile shows no hint of flattening at small radii, a small but finite core cannot be ruled out by the data. The 99% and 90% upper limits to the core radius of M15 are $2.''2$ and $1.''8$ (0.1 pc or $0.03 \, r_{\mathrm{half-mass}}$), respectively (Fig. 2b). This is comparable to the core sizes expected from binary heating and core oscillations (Grabhorn et al. 1992).

## 2.2. RADIAL DENSITY PROFILES OF OTHER CLUSTERS

Short HST WFPC2 exposures have been used to study the central regions of the clusters M30 and NGC 6624 (Yanny et al. 1994b, 1996) in a manner similar to the M15 study described above. The density profiles of these two clusters are also well approximated by a single power law in their central

5" (∼ 0.2 pc). The cusp in M30 is somewhat shallower than the one in M15 ($\alpha = -0.4\pm0.15$) while that in NGC 6624 is comparably steep ($\alpha \sim -0.85$). The central stellar surface densities of both clusters are somewhat lower than that of M15. The groundbased study of Djorgovski and King (1986) placed all three clusters in the 'post core collapse' category, characterized by surface brightness profiles rising towards the center (to the ∼ 1" resolution limit). As in the case of M15, WFPC2 star counts in M30 and NGC 6624 yield reliable estimates of the density profile down to $r \sim 0\rlap.{''}3$, and these show no hint of flattening into a core. The upper limit on the core radius of the two clusters is comparable to that of M15, roughly 2" or 0.1 pc.

Not every cluster in our sample is of the post core collapse variety. Pre-repair WF/PC-I images of the 'King model' clusters M13, M3, and 47 Tuc have been/are being analyzed. All three have relaxed distributions with flat cores of radius roughly 30" (∼ 1 pc) and very high core stellar densities. Note, the radial distribution of evolved stars in 47 Tuc (measured relative to the number weighted cluster centroid or 'center of mass') is well fit by a King profile with $r_{\rm core} = 23'' \pm 2''$ (0.5 pc) in disagreement with the 8" core radius found by Calzetti et al. (1993) for ultraviolet selected stars. The lack of any detectable cusp places a 95% upper limit on the mass of a compact central object in 47 Tuc of about 2500 $M_\odot$ (Guhathakurta et al. 1992).

## 3. Stellar Populations

The superb angular resolution of HST makes it possible to carry out accurate photometry of faint stars in crowded regions ($1\sigma$ color errors $\lesssim$ 0.1 mag). This is essential in order to reliably define the various stellar types (e.g. red giant, horizontal branch, subgiant, and turnoff stars). The color–magnitude diagrams of the central regions of 47 Tuc and M15 presented in Fig. 3 demonstrate the power of HST. These data are ideal for studying blue straggler stars, in particular, since this population is subject to contamination due to the blending of turnoff stars and inaccurate photometry of subgiants/blue horizontal branch stars in low resolution images.

### 3.1. BLUE STRAGGLERS

The 47 Tuc data were obtained with the pre-repair HST's Planetary Camera and were reduced and analyzed in collaboration with R. Gilliland and P. Edmonds (STScI). The $U$ data consist of a series of 99 1000 sec exposures which were combined into a very deep, doubly oversampled image (1 pixel = $0\rlap.{''}023$). These data yield a clean and complete sample of BSS out to $r = 2\,r_{\rm core}$ in 47 Tuc (see Fig. 3a), extending the samples found by Paresce et al. (1991) and Guhathakurta et al. (1992) in earlier pre-repair HST studies.

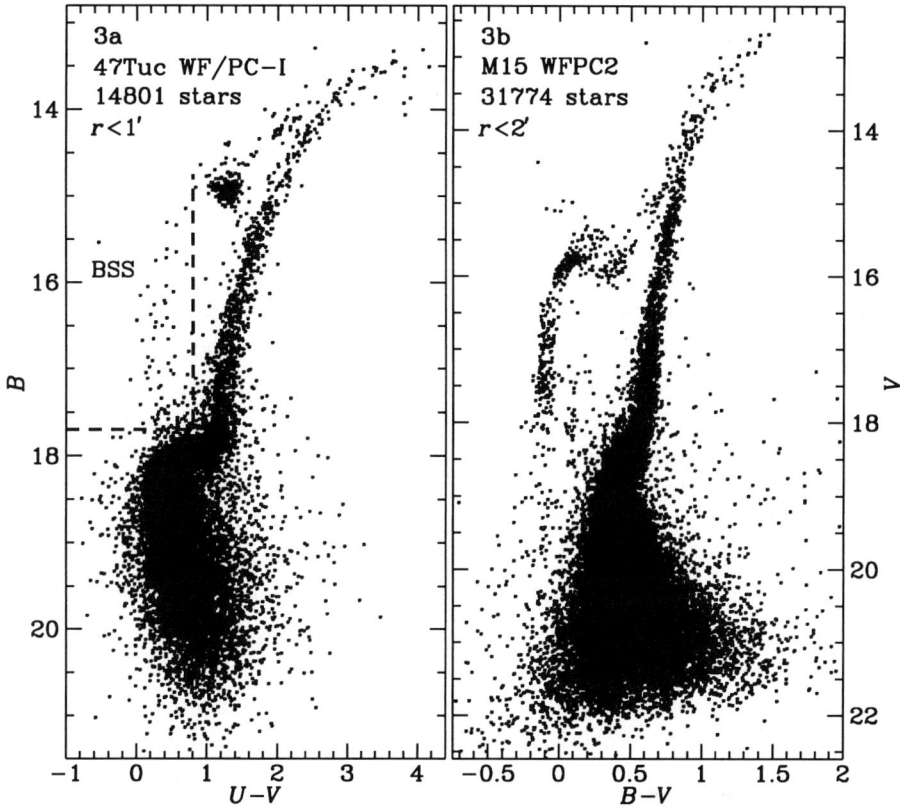

*Figure 3.* **Left** *(a)* A $B$ vs. $U - V$ color–magnitude diagram of the core of 47 Tuc derived from very deep pre-repair HST WF/PC-I data (in collaboration with Gilliland and Edmonds). Note the prominent sequence of blue straggler stars (BSS) extending above the main sequence turnoff. **Right** *(b)* A $V$ vs. $B - V$ color–magnitude diagram of the central part of M15 from HST WFPC2 stellar photometry. A small number of blue straggler candidates are visible slightly brighter and bluer than the main sequence turnoff (and to the red of the drooping tail of the blue horizontal branch). Over $3 \times 10^5$ stars are detected in the 5 arcmin$^2$ area of the WFPC2 mosaic.

Our HST WFPC2 data on M15 reveal a small number of BSS candidates, located 1–2 mag brighter and 0.1–0.4 mag bluer than the main sequence turnoff: $V_{\rm TO} \approx 19$, $(B - V)_{\rm TO} \approx 0.4$ (Fig. 3b). The region of M15's color–magnitude diagram bounded by the drooping blue horizontal branch, subgiants, and turnoff is relatively free of objects. The numerous "yellow stragglers" seen in Stetson's (1994) groundbased data were probably artifacts of blending (as the author suspected!).

We measure the specific frequency of BSS $F_{\rm BSS}$ (as defined by Bolte et al. 1993) in the central regions of several globular clusters: M30, NGC 6624,

M15, 47 Tuc, M13, and M3. Sosin and King (1995) point out that it is interesting that the scatter in BSS frequency from cluster to cluster is only about a factor of three, with a mean value of $F_{\rm BSS} \sim 0.12$, despite their $\gtrsim 10^4$ range in central stellar density. Within each of the above clusters though, the BSS are centrally concentrated relative to other evolved stars, with $F_{\rm BSS}$ increasing by about a factor of 2 between $r \sim 2$ pc and the center. This can be explained in terms of mass segregation: BSS in dense cluster cores are thought to be products of stellar collisions with $M_{\rm BSS} \sim 2\,M_{\rm turnoff}$. There are suggestions of a bimodal BSS population in M3 where $F_{\rm BSS}$ seems to go through a minimum at $r \sim 2'$ or $5\,r_{\rm core}$ (Ferraro et al. 1993). However, our HST WF/PC-I study of the central 1' of M3 (Guhathakurta et al. 1994) shows that the BSS sample derived from groundbased images of this crowded region suffers from the effects of blending and incompleteness, even though the images were obtained in $0''\!.5$ seeing (Bolte et al. 1993). Clearly, a uniformly high quality dataset (in terms of angular resolution and depth) is needed, covering the full radial range of the cluster, before definite conclusions can be drawn about the nature of the BSS in M3.

The observed number of BSS relative to other evolved stars serves as a diagnostic for theories of BSS formation. Observational studies of these stars in dense globular cluster cores are beginning to explore areas beyond that probed by BSS counts alone. These include accurate monitoring to study photometric variability caused by SX Phe-type pulsations and eclipsing binaries (cf. Edmonds et al. 1996) and comparison of the BSS luminosity function and color distribution to models of BSS formation and evolution (Bailyn and Pinsonneault 1995).

## 3.2. CENTRAL RED GIANT DEFICIENCY IN M15

Several post core collapse globular clusters display color gradients near their centers indicating changes in the mix of stellar populations in regions of very high density (Piotto et al. 1988; Cederbloom et al. 1992). These gradients are usually in the sense of the overall cluster light becoming bluer inwards and are attributed to a deficiency of the brightest RGB stars. Stetson (1994) convincingly demonstrated this RGB deficiency for the cluster M15. While the brightest giants are relatively easy to detect, even in these very dense regions, it is important to define a complete sample of fainter stars (e.g. subgiants) that can be used as the normalizing population relative to which the RGB deficiency is measured.

We use WFPC2 stellar photometry in the central $100''$ of M15 (see Fig. 3b) to study the radial variation of the relative numbers of bright RGB stars (defined to be those brighter than the horizontal branch or $V < 16$), horizontal branch stars, and subgiants ($16 < V < 17.5$). The

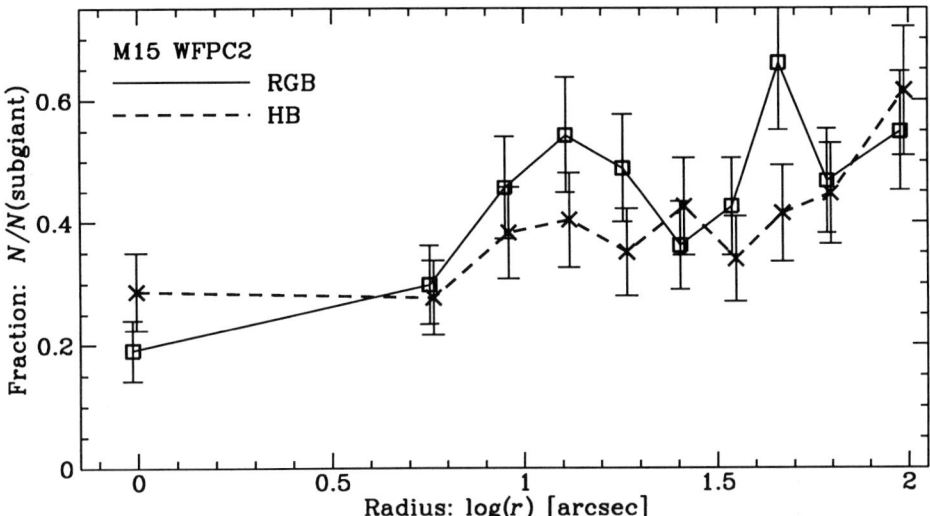

*Figure 4.* Radial gradients in the mix of evolved stars near M15's center derived from HST WFPC2 photometry. The fraction of bright red giant branch stars relative to subgiants (solid line) is about a factor of 2 lower in the inner $10''$ compared to the region further out ($r = 10''$–$100''$). The fraction of horizontal branch stars (dashed line) is roughly constant over the entire $r < 100''$ region; the 20% decrease in the inner $10''$ and the 50% increase beyond $r = 1'$ are not statistically significant ($\lesssim 2\sigma$ effects). There are comparable numbers of bright RGB stars and HB stars in the $r = 10''$–$100''$ region. Error bars show Poisson errors.

results are shown in Fig. 4. The fraction of bright RGB stars relative to subgiants appears to decrease with decreasing radius inside $r < 10''$ or 0.6 pc ($\gtrsim 3\sigma$ significance). The RGB fraction in the inner $10''$ is about half the value measured further out. The horizontal branch stars on the other hand exhibit no strong radial variation relative to subgiants. The apparent 20% decrease in the horizontal branch fraction in the inner $10''$ is only a $1\sigma$ effect and the possible increase beyond $r = 1'$ a $2\sigma$ effect; it is worth noting that both red and blue horizontal branch stars display these trends. The central RGB depletion and the constancy of the horizontal branch fraction is independent of the exact definition of the subgiant sample— i.e. the results are identical relative to faint or bright subgiant samples. We avoid using turnoff or main sequence stars ($V \gtrsim 18$) as the normalizing population since counts of these stars are affected by crowding near M15's center (Guhathakurta *et al.* 1996b).

There are roughly equal numbers of bright RGB and horizontal branch stars in M15's $r > 10''$ region (Fig. 4), suggesting that a star spends comparable amounts of time in these two evolutionary phases. The fact that bright RGB stars are depleted by a factor of two in the dense central 0.6 pc whereas stars on the horizontal branch (evolutionary descendants of RGB stars) are not indicates that the bright RGB phase is 'short circuited' in this region. If the RGB depletion were caused by destruction of giants, we would expect to see a similar central dip in the horizontal branch fraction, and this is not observed.

P.G. would like to thank his collaborators Peter Edmonds and Ron Gilliland for allowing him to present data from an unpublished study of 47 Tuc, the conference organizers for an enjoyable and fruitful week in Tokyo, and the editors for their patience in the months since.

## References

Aurière, M. and Cordoni, J.-P. (1981) *A. Ap. S.* **46**, 347.
Bahcall, J.N., Bahcall, N.A. and Weistrop, D. (1975) *Astrophys. Lett.* **16**, 159.
Bahcall, J.N. and Wolf, R.A. (1976) *Ap. J.* **209**, 214.
Bahcall, J.N. and Wolf, R.A. (1977) *Ap. J.* **216**, 883.
Bailyn, C.D. and Pinsonneault, M.H. (1995) *Ap. J.* **439**, 705.
Bolte, M., Hesser, J.E. and Stetson, P.B. (1993) *Ap. J.* **408**, L89.
Calzetti, D., de Marchi, G., Paresce, F. and Shara, M. (1993) *Ap. J.* **402**, L1.
Cederbllom, S.E., Moss, M.J., Cohn, H.N., Lugger, P.M., Bailyn, C.D., Grindlay, J.E. and McClure, R.D. (1992) *A. J.* **103**, 480.
Cohen, R.L., Guhathakurta, P., Yanny, B., Bahcall, J.N. and Schneider, D.P. (1996) *A. J.* in prep.
Djorgovski, S. and King, I.R. (1986) *Ap. J.* **305**, L61.
Edmonds, P.D., Gilliland, R.L., Guhathakurta, P., Petro, L.D., Saha, A. and Shara, M.M. (1996) *A. J.* in prep.
Ferraro, F.R., Fusi Pecci, F., Cacciari, C., Corsi, C., Buonanno, R., Fahlman, G.G. and Richer, H.B. (1993) *A. J.* **106**, 2324.
Grabhorn, R.P., Cohn, H.N., Lugger, P.M. and Murphy, B.W. (1992) *Ap. J.* **392**, 86.
Guhathakurta, P., Webster, Z., Yanny, B., Bahcall, J.N. and Schneider, D.P. (1996a) *A. J.* in prep.
Guhathakurta, P., Yanny, B., Bahcall, J.N. and Schneider, D.P. (1994) *A. J.* **108**, 1786.
Guhathakurta, P., Yanny, B., Schneider, D.P. and Bahcall, J.N. (1992) *A. J.* **104**, 1790.
Guhathakurta, P., Yanny, B., Schneider, D.P. and Bahcall, J.N. (1996b) *A. J.* **111**, 267.
King, I.R. (1975) in *Dynamics of Stellar Systems, IAU Symposium No. 69*, ed. A. Hayli, Reidel, Dordrecht, p. 99.
Lauer, T.R. et al. (1991) *Ap. J.* **369**, L45.
Merritt, D. and Tremblay, B. (1994) *A. J.* **108**, 514.
Paresce, F. et al. (1991) *Nature* **352**, 297.
Piotto, G., King, I.R., and Djorgovski, S. (1988) *A. J.* **96**, 1918.
Sosin, C. and King, I.R. (1995) *A. J.* **109**, 639.
Stetson, P.B. (1994) *P. A. S. P.* **106**, 250.
Yanny, B., Guhathakurta, P., Bahcall, J.N. and Schneider, D.P. (1994a) *A. J.* **107**, 1745.
Yanny, B., Guhathakurta, P., Schneider, D.P. and Bahcall, J.N. (1994b) *Ap. J.* **435**, L39.
Yanny, B., Guhathakurta, P., Schneider, D.P. and Bahcall, J.N. (1996) *A. J.* in prep.

# DYNAMICAL IMPLICATIONS OF DEEP HST IMAGING OF GLOBULAR CLUSTERS[1]

IVAN R. KING
*Astronomy Department, University of California*
*Berkeley, CA 94720-3411, U.S.A.*

**Abstract.** HST observations contribute in many ways to a better understanding of the dynamical nature of globular clusters. Unprecedentedly faint photometry gives new determinations of the numbers of low-mass stars. Cluster-to-cluster differences at the faint ends of the mass functions suggest differences in dynamical evolution. Mass segregation is clearly observed, from the envelope inward to the dense cluster center. The distribution of stars in the hitherto unresolved cores gives new data with which to test theories of core collapse, and these core profiles are also sensitive to the number of unseen remnant stars and binaries at the cluster center.

(*Note:* The results discussed in this paper come from a collaboration between the author, Adrienne Cool, Giampaolo Piotto, Craig Sosin, and Jay Anderson. Parts of this work are described elsewhere in this volume by Piotto *et al.*, by Cool *et al.*, and by Sosin & King.)

## 1. Introduction

The Hubble Space Telescope offers two great new advantages in the study of globular clusters: it allows observing faint stars in the center, and it goes much fainter than we have gone from the ground.

The best ground-based resolving power is characterized by a full width at half maximum of a little better than 0.4 arcsec. By contrast, the central peak of the diffraction-limited HST image has a FWHM of 0.045 arcsec, almost a factor of ten times better. To be sure, some of the best HST

---

[1] Based on observations with the NASA/ESA *Hubble Space Telescope*, obtained at the Space Telescope Science Institute, which is operated by AURA, Inc., under NASA contract NAS5-26555.

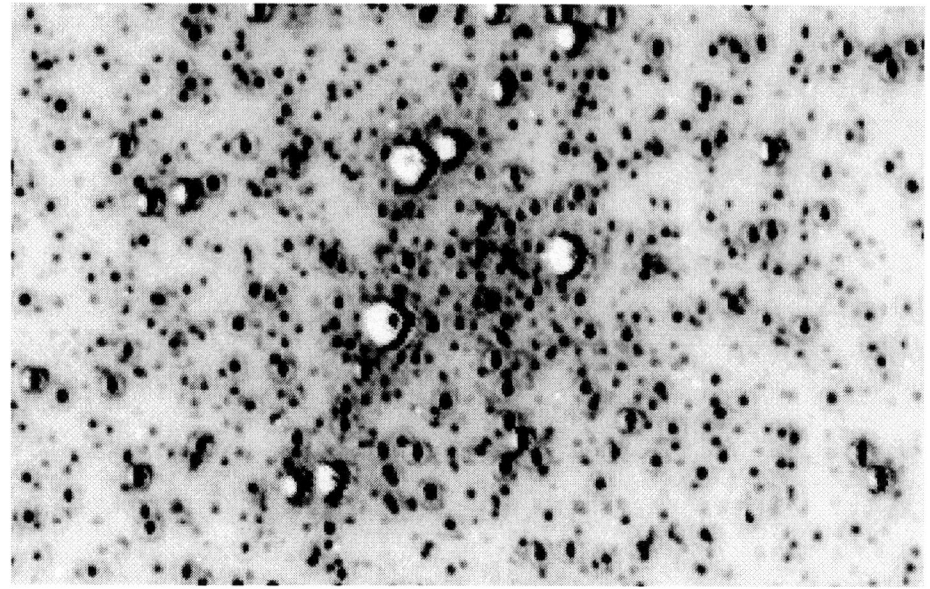

*Figure 1.* A 5&#8243;.6×3&#8243;.5 part of an FOC image of the center of M15. The pixel size is 0.014 arcsec. The white areas are FOC saturation in the bright stars.

imaging is done with the Wide Field Camera, whose pixel is 0.1 arcsec, but this is still considerably better than we can do from the ground. Ground-based observers are now experimenting with aperture-synthesis techniques that will allow them to reach the diffraction limit, but this is still in the future, and will cover a much smaller area than the WFC field. By far the best tool we have today for imaging globular clusters is HST.

As an example, Figure 1 shows the center of M15, as imaged with HST's Faint Object Camera. Here we absolutely have to have the resolution of the FOC; the Planetary Camera, with 3 times as large a pixel, just doesn't have a high enough resolution.

As for faintness, my favorite comparison is between our HST result and a deep study done with the CTIO 4-meter (Fig. 2). The color scales are different, but the vertical scale of $V$ magnitudes is the same in both diagrams. We can do a lot with those extra 3 or 4 magnitudes of depth.

Since this is a dynamics meeting, I will center my emphasis on dynamics; but the data with which I will be dealing are photometric, so I will begin with a few words about photometric techniques, followed by a brief discussion of some of our results.

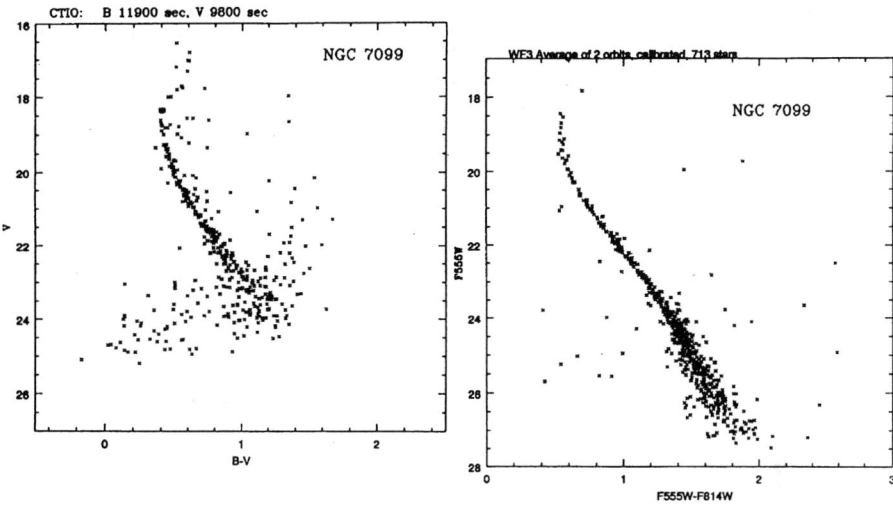

*Figure 2.* Comparison of ground-based (left) and HST CMDs of the same cluster. The color systems are different, but the two graphs have the same vertical $V$ scale, which shows how much fainter HST goes.

## 2. Photometric Results

Photometry with HST has turned out to be something of an art. We began by using DAOPHOT in what has become the standard way, but the results weren't very good. We then did a lot of experimenting and arrived at a method that was much more accurate. (For further details see Cool & King 1995, where it is explained that we later found that an unconventional use of DAOPHOT can give measurements at least as accurate as ours.) I can't resist mentioning that this is another of the virtues of globular clusters: the narrowness of your main sequence is a measure of how well you have measured.

As an example, in the paper by Cool, Piotto, & King, elsewhere in this volume, their Fig. 1 shows our color–magnitude diagram for NGC 6397, the globular cluster with the smallest apparent distance modulus. We get a remarkably narrow main sequence. Interestingly, the main sequence has striking bends, and it is a challenge for theories of stellar structure to match these. We have been in touch with some of the groups who calculate stellar models, and one of them has produced a quite good fit to our observations.

For a more extensive discussion of our results on NGC 6397, see Cool, Piotto, & King (1996).

A really fascinating problem is the bottom of the main sequence. In the region near the faint limit of our observations is the mass where the lower limit of hydrogen burning should be. Although this point is well defined in mass, it is very ill defined in luminosity, because what happens here is that the luminosity plunges very steeply over a quite small range in mass. But luminosity is already very sensitive to mass even above the limit, so the luminosity function is already dropping rapidly and is thus hard to follow.

Getting the faint end of the luminosity function is even harder because of the presence of numerous field stars. At its low Galactic latitude ($b = -12°$), this cluster has a rich foreground, but with its moderate extinction ($A_V = 0.56$) it also has a rich background. Nature seems truly malicious here, as our best chance of locating the hydrogen-burning limit gets swamped in field stars. There is a way of solving the problem, however. The proper motion of the cluster is known, and we have calculated that by early 1997, a time for which HST observing proposals are already under review, the displacement of cluster stars from their 1994 positions will be great enough to allow a star-by-star separation of cluster from field. (In the course of demonstrating the accuracy of astrometry on WFPC2 images, we actually detected the mean motion of cluster stars between August and October 1994—another demonstration of the unprecedented accuracy that HST offers.)

The hydrogen-burning limit of course has very little to do with dynamics, but the number of low-mass stars does matter, and I will return to that.

Cool *et al.* also clearly show a white-dwarf sequence in NGC 6397. The number of white dwarfs also matters to the dynamics, although our observations do not say much about that, because we see only the brightest white dwarfs, which are a small fraction of the total. Further study of white dwarfs will be of great interest to stellar-evolution theory, but for dynamical modeling we should continue to predict the WD number from assumptions about the mass function of the part of the main sequence that has already evolved away.

Another important result of the photometry is the luminosity function; for the dynamics of a cluster we certainly need to know the number of stars of each mass. Luminosity functions are discussed more extensively elsewhere in this volume by Piotto, Cool, & King; so here I will merely mention that the derivation of LFs is far from trivial and that there are some unreliable results in the literature, particularly for the faint stars.

And even when one has a good luminosity function, there are further problems. For dynamical models one needs masses rather than luminosities, and the conversion is far from trivial. There is no observed mass–luminosity relation for the metallicity of this cluster ($[Fe/H] = -1.9$), and the theoretical M–L relations for low-mass metal-poor stars are far from reliable. As already mentioned, photometric results such as ours will stimulate and

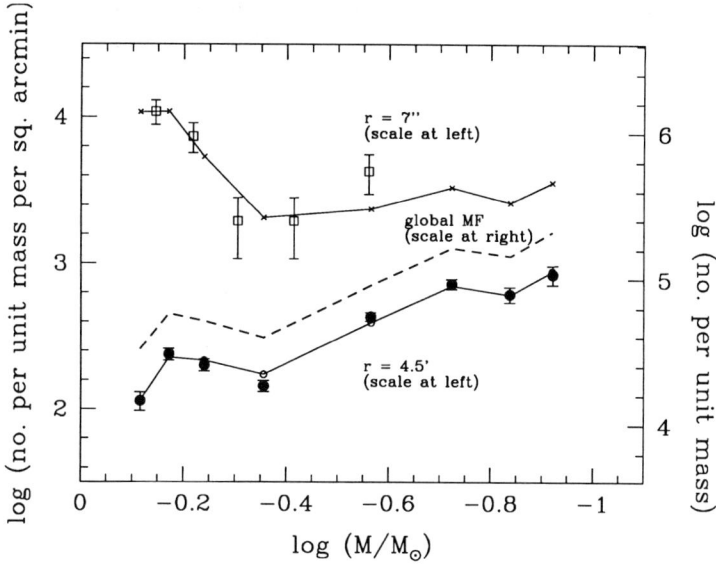

*Figure 3.* Mass functions in NGC 6397 at radii 7″ and 4.'6, as observed, in stars per arcmin$^2$. The numbers in the 7″ field are higher because of the higher density at the cluster center, but the mass function is quite different. The broken lines are from a dynamical model fitted to the cluster; the numbers have been fitted to the observations at 4.'5 but not at 7″. Also shown is the global mass function of the model; the right-hand ordinate scale applies to it.

guide stellar-evolution theory, and the consequent improvements should result in better M–L relations.

Even a perfect mass function would still leave one more problem: it is the local mass function of one region in the cluster, and what we need for dynamics is the global mass function of the cluster. For this a dynamical model is needed, and I will return to this question.

## 3. Mass segregation

Differences in the radial distributions of stars of different mass have been seen a number of times from the ground, significantly but weakly (Sandage 1954, Oort & van Herk 1959, Richer & Fahlman 1989, Drukier et al. 1993, and many other places). In HST images, however, we can see the faint stars all the way in to the cluster center, and the effects are tremendous. They have already been seen and commented on (Shara 1995; Paresce, De Marchi & Jedrzejewski 1995). What we have done is to quantify the effect by fitting a cluster model (King, Sosin, & Cool 1995).

Our first impression of the images of NGC 6397 was astonishment; there were almost no faint stars at the center. But when we fitted a dynamical model to the cluster, it became clear that this was just what was to be expected. Figure 3 shows the mass functions observed at the center and in an outlying field, and the fit of our theoretical model to them.

## 4. Dynamic Modeling of NGC 6397

For reasons that I have mentioned, we needed a dynamical model for NGC 6397. To create this I used a multi-mass embodiment of the algorithm that has given rise to the name "King models" (King 1966). In fitting the model, three conditions had to be satisfied. First, the model has to fit our observed mass function, when we project the model onto the plane of the sky at the radius at which our mass function was determined. Second, the central concentration of the model must match that of the cluster. Third, the projected luminosity profile of the model must agree in detail with that observed for the cluster.

The first two conditions were satisfied by trial and error. The third condition, however, introduced a separate aspect of cluster dynamics. The observed slope of the central part of the luminosity profile could be matched only by adding to the model a few hundred massive remnants. These stars have the effect of taking for themselves the $-1$ slope that belongs to a singular core of self-gravitating objects, so that in equipartition the red giants whose light we see have a flatter distribution. (This is a point that was made long ago by Lugger *et al.* [1987].) I arbitrarily assigned to these added objects masses of 1.4 $\mathcal{M}_\odot$ and called them neutron stars, but a considerable trade-off is possible between their individual mass and their number, so that one cannot categorically identify them as neutron stars. No doubt central binaries figure in the picture too. What is certain, though, is that there are some hundreds of objects near the cluster center whose mass appreciably exceeds that of the red giants.

One thing that the reader may have noticed with surprise is that I have used a King model here to represent a cluster with a post-collapse core. This is a new thing; it has always been supposed that core collapse takes a cluster off the sequence that is represented by the King models. What I have done is simply to use a model with a high central concentration—so high, in fact, that its exact value doesn't matter; the center is essentially singular. The mathematical reason that this works is that the high-concentration limit of the King models is the original collapsed-core model of Hénon (1961); the physical reason is that a cluster as highly evolved as a post-core-collapse one has velocity distributions that are very close to Gaussian, and velocity dispersions that are rather close to Maxwellian. These are the essential characteristics of a King model.

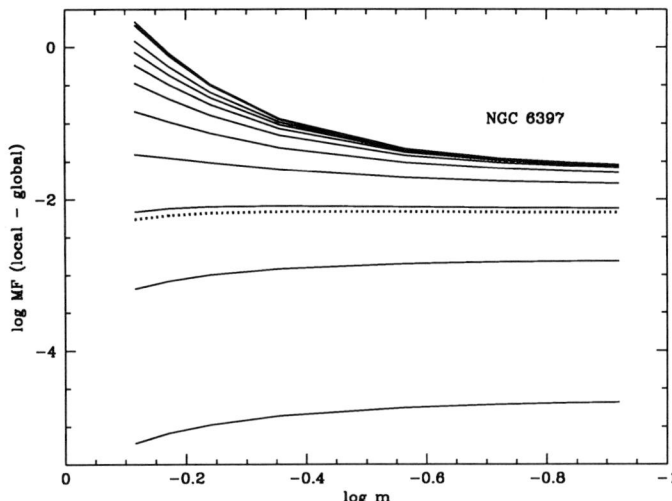

*Figure 4.* Differences between local mass functions at different radii (in projection) and the global mass function, in our model of NGC 6397. The dotted line represents the radius from which our observed MF came. The relative vertical positions of the curves reflect radial differences in density, but the zero point is arbitrary, since local and global MFs have different dimensions.

There are at least two reasons why King models of very high central concentration have not been explored. One is that beyond a central concentration of about 2.0 they have been shown to be unstable (Katz 1980, Wiyanto, Kato, & Inagaki 1985); the other is that energetically a model cannot evolve along the sequence of King models beyond this point, because the magnitude of the potential energy no longer increases with increasing central concentration (King 1966). It is likely that the effect of binaries at the cluster center solves both of these problems.

Still another objection may be made to the model that I have used here. The Fokker–Planck integrations of Murphy, Cohn & Hut (1990) showed that equipartition never quite catches up in a post-core-collapse cluster. The difference is quantitatively quite small, however; examination of one of their runs shows that a group with mass 0.14 $\mathcal{M}_\odot$ has a velocity dispersion that would be appropriate for 0.18 $\mathcal{M}_\odot$; I have not yet bothered to take this small difference into account, but easily could.

## 5. Local vs. Global Mass Functions

The availability of a cluster model makes it possible to convert from a mass function observed in one small part of the cluster to the global MF. Figure 4 illustrates the differences derived from our model of NGC 6397. The figure

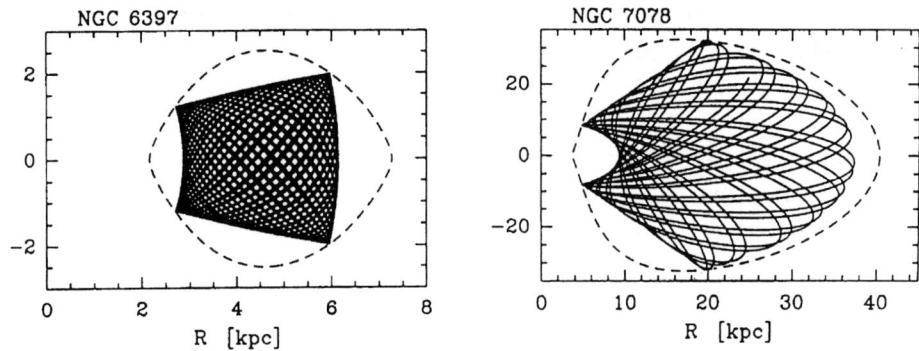

*Figure 5.* Comparison of the Galactic orbits calculated by Dauphole *et al.* 1995 for NGC 6397 and M15. NGC 6397's passages through the Galactic plane are much more frequent, and at lower inclination angles and smaller perigalacticon than those of M15, making it much more susceptible to star loss via tidal shocking.

again illustrates mass segregation, in that the higher-mass stars are greatly over-represented near the cluster center, relative to the less-massive stars, and they are under-represented near the edge. The figure also shows that the correction from local to global is relatively small over most of the outer part of the cluster.

## 6. The Effect of Tidal Shocks

Another interesting point about NGC 6397 is that the low-mass end of its luminosity function is deficient with respect to the LFs of other clusters of the same metallicity. The paper by Piotto, Cool, & King elsewhere in this volume shows that the LFs of M15 and M30 are practically identical to each other, whereas NGC 6397 is strikingly poorer in faint stars. I emphasize: this cannot be a metallicity effect, because all three clusters have nearly the same metallicity.

As many people have suggested, the difference could be in the differing susceptibility of these clusters to tidal shocks. In Figure 5 are the Galactic orbits of NGC 6397 and of M15, in the meridional plane of $z$-coordinate and distance from the axis of rotation, as calculated by Dauphole *et al.* (1995). (Unfortunately the orbit of M30 is unknown.) Notice the differing scales of the two graphs. NGC 6397 is clearly hit much harder and much more often by passages through the Galactic plane. It is thus important in two ways that NGC 6397 stays so close to the Galactic center. First, it goes rapidly back and forth through the plane—a passage every 40 or 50 Myr.

Second, the force exerted by the plane, which determines the strength of a shock, is much greater in there than farther out in the Galaxy.

But the effect of tidal shocks has never been quantified well enough, and Weinberg has shown (1994abc) that the traditional theories are not reliable. One project that we have under way is a numerical investigation in which we follow the orbits of a large number of representative stars in NGC 6397, as they move about in the cluster and the cluster moves in the Galactic gravitational potential.

## 7. The Structure of Collapsed Cores

Another problem that HST can illuminate is the structure of collapsed cores. Now for the first time we can see into these dense regions.

I will say rather little about this topic, however, because it is covered elsewhere in this volume by Sosin and King, who give the mass function and the radial density distribution near the center of M15.

One related point that I want to make is a somewhat sobering one, however. The FOC images that resolve the center of M15 are quite beautiful. But in some ways the situation is quite ugly. The core of M15 is so small that it does not have enough stars in it to trace its structure accurately. The trouble is that one cannot simply increase the statistical stability of the counts by including the numerous faint stars along with the brighter ones. The fainter stars have smaller masses, and therefore delineate a different density distribution. Hopefully this can be taken into account by more elaborate modeling, but the problem of the true nature of the core is clearly a more difficult one than can be solved by simply counting stars and drawing a profile.

There is one more matter about collapsed cores that I think deserves general attention. That is the question of stating the position of an object, such as an X-ray source, in terms of the number of core radii that it is from the center. This is perfectly proper for clusters that have a core, but it does not work for post-collapse clusters, most of which have nothing that can properly be called a core radius. The core radii that appear for these in the literature are meaningless force-fits of an inappropriate curve to some particular data set, and have little or nothing to do with the actual structure of the cluster center. (I myself confess to this sin [Trager, King, & Djorgovski 1995], which I now regret very much.)

But it is nevertheless useful to have measure of how far out in the cluster the object is. Without a core radius, what can we use? There is a simple and obvious solution: state what fraction of the cluster's light lies within the cluster-centered circle that goes through the object in question. Then, to avoid using one datum for collapsed clusters and a different one for uncollapsed ones, I would suggest that the locations of objects such as X-ray sources always be described by an "encircled fraction," and never by a number of core radii. The growth curves from which encircled fractions can be found are easily calculated from the cluster profiles given by Trager *et al.* (1995). (Their detailed data are indeed on the corresponding AAS CD-ROM, despite omission of the CD-ROM symbol in the table of contents.)

## 8. Conclusion

In considering the results that we and others are getting from HST, we should all remember that this is only a start. The results presented by several of us at this meeting all come from exposures taken in the first 15 months after the repair of the spherical aberration. As time goes on, HST will continue to contribute more and more to our understanding of the structure and dynamics of globular clusters.

## 9. Acknowledgments

It is a pleasure to recognize the collaboration of Adrienne Cool, Giampaolo Piotto, Craig Sosin, and Jay Anderson in the work that is described here. I also thank Andrea Lommen for orbital calculations and Charles Bartels for technical assistance. This work was supported by NASA grant NAG5-1607.

## References

Cool, A. C., & King, I. R. 1995, in Calibrating HST: Post Servicing Mission, ed. A. Koratkar & C. Leitherer (Baltimore: STScI), p. 290
Cool, A. C., Piotto, G., & King, I. R. 1996, ApJ, submitted
Dauphole, B., Geffert, M., Colin, J., Ducourant, C., Odenkirchen, M., & Tucholke, H.-J. 1995, preprint
Drukier, G. A., Fahlman, G. G., Richer, H. B., Searle, L., & Thompson, I. 1993, AJ, 106, 2335
Hénon, M. 1961, Ann.d'Ap., 24, 369
Katz, J. 1980, MN, 190, 497
King, I. R. 1966, AJ, 71, 64
King, I. R., Sosin, C., & Cool, A. M. 1995, ApJL, 452, L33
Lugger, P. M., Cohn, H., Grindlay, J. E., Bailyn, C. D., & Hertz, P. 1987, ApJ, 320, 482
Murphy, B. W., Cohn, H. N., & Hut, P. 1990, MN, 245, 335
Oort, J. H., & van Herk, G. 1959, BAN 14, 299 (No. 491)
Paresce, F., De Marchi, G., & Jedrzejewski, R. 1995, ApJL, 442, 57
Richer, H. B., & Fahlman, G. G. 1989, ApJ, 339, 178
Sandage, A. 1954, AJ, 59, 162
Shara, M. M., Drissen, L., Bergeron, L. E., & Paresce, F. 1995, ApJ, 441, 617
Trager, S. C., King, I. R., & Djorgovski, S. 1995, AJ, 109, 218
Weinberg, M. D. 1994a, AJ, 108, 1398
Weinberg, M. D. 1994b, AJ, 108, 1403
Weinberg, M. D. 1994c, AJ, 108, 1414
Wiyanto, P., Kato, S., and Inagaki, S. 1985, PASJ, 37, 715

# HST OBSERVATIONS OF THE GLOBULAR CLUSTER M4

G. G. FAHLMAN
*Department of Geophysics and Astronomy,*
*University of British Columbia,*
*Vancouver, B.C., Canada V6T 1Z4*

H.B. RICHER, R.I. IBATA, N.C. IVANANS, G. MANDUSHEV
*Department of Geophysics and Astronomy,*
*University of British Columbia.*

J.E. HESSER, P.B. STETSON
*Dominion Astrophysical Observatory,*
*Herzberg Institute of Astrophysics, National Research Council.*

M. A. WOOD
*Department of Physics and Space Science,*
*Florida Institute of Technology*

D.A. VANDENBERG
*Department of Physics and Astronomy, University of Victoria.*

C. PRYOR
*Department of Physics and Astronomy, Rutgers University.*

W.E. HARRIS
*Department of Physics and Astronomy, McMaster University.*

H.E. BOND
*Space telescope Science Institute.*

M. BOLTE
*Lick Observatory, University of California.*

AND

R.A. BELL
*Department of Astronomy, University of Maryland.*

**Abstract.**
   The WFPC2 aboard the Hubble Space Telescope has been used to obtain deep images in three fields at different radial positions in the nearest

Globular cluster, M4 (NGC 6121). In this paper, we discuss the white dwarf cooling sequence and show how the dynamical structure of the cluster will affect their cumulative distribution function. We also present the first discussion of our observations of the faint cluster main sequence stars.

## 1. Introduction

The high angular resolution achievable with the Hubble Space Telescope offers two significant advantages over ground based observations of globular star clusters: (1) accurate photometry to faint levels is feasible in crowded cluster cores, and (2) the improved contrast of the sharper stellar profile against a lower and much more stable background allows photometry of very faint stars. By exploiting both of these improvements, we are able to study two stellar populations, the cluster white dwarfs and the low mass main sequence stars, which have proven inaccessible from the ground. These stars play key roles in the dynamical evolution of the clusters and provide important data pertaining to stellar evolution and star formation in the early universe.

Messier 4 (NGC 6121), at a distance of only 2.0 kpc, is the globular cluster closest to the sun. In this paper, we will describe our multi–field observational program for M4. At the time of this meeting, we have completed and calibrated most of the photometry for our program fields but a full analysis of these results, which require extensive computations to determine the completeness corrections and photometric error distributions has not yet been done. Previous discussion of the white dwarfs can be found in Richer *et al.* (1995a; paper I) and Richer *et al.* (1995b; paper II) discuss the implications of our photometry on the issue of the binary fraction in M4.

## 2. Observations and Data Reduction

The data described here was acquired under the HST proposal number 5461 and the STScI electronic information service can be used to obtain full details of the program.

Images were obtained with the WFPC2 camera through $U$ (F336W), $V$ (F555W) and $I$ (F814W) filters in two inner fields whose position in the cluster was specified by the projected location of the PC chip. One of these fields was located within the cluster core (referred to as the core field) and the other was located at 1 core radius. (These positions are based on an adopted core radius of $r_c = 74''$ from Webbink 1985.) The principal goal for these observations was to study the white dwarf cooling sequence in M4.

The outer field was located at 4 core radii and only $V$ and $I$ images were obtained. The goal in this relatively uncrowded field was to get sufficiently deep to see the end of the cluster main-sequence.

The raw data frames had the standard HST pipeline processing applied to them as discussed in paper I. The photometry was carried out with the ALLFRAME package developed by Stetson (1994). The calibration of the data follows the methodology described by Holtzman et al. (1995) and is therefore placed on the ground-based $U, V, I$ system defined by the Landolt (1992) standards. We assume the reddening law specified by the following coefficients: $R = A_V/E(B-V) = 3.12$, $E(V-I) = 1.25\ E(B-V)$, $E(U-I) = 2.98\ E(B-V)$ and $A_U = 4.83\ E(B-V)$ (Bessell and Brett 1988, Mathis 1990). The apparent distance modulus and cluster reddening adopted are $(m-M)_V = 12.65$, and $E(B-V) = 0.37$, respectively (see paper I for justification of these choices).

We have completed the photometric reductions for all the data in the 1 core and 4 core fields. For the core field, we have reduced data for only a single WFC chip and the PC chip. The remaining data has a significant overlap with the 1 core field data and all will be simultaneously reduced with ALLFRAME. We will limit the discussion in this paper to the 1 core and 4 core data only.

## 3. White Dwarfs

The white dwarf cooling sequence is most clearly delineated on the $(U, U-I)$ plane, shown in figure 1. The PSF fitting radius was set to a relatively large value so that the brighter, saturated stars were measured by fitting to the profile wings. The errors for these stars are large, accounting for the evident spread seen in figure 1. The turn-off in this color-magnitude plane has been observed from the ground and this shows that the photometry of the saturated stars has a bias as well as large errors in that the stellar locus for $M_U \leq 7.0$ appears to be too red. The faintest stars have an apparent magnitude of $U \geq 26$, more than a magnitude below the very deep ground based data obtained by Richer and Fahlman (1988) at the CFHT for their study of the cluster M71. In paper I, we have shown that the white dwarf sequence is consistent with the cooling locus of 0.5 $M_\odot$ carbon white dwarfs and this is illustrated in figure 1. where this model cooling curve has been overlayed on the data without adjustment.

If it is assumed that stars are conserved through their post-main-sequence (pms) evolutionary phases, then the number of wds brighter than a given magnitude, $N(<M)$, should be proportional to the time taken to cool to that magnitude. In figure 2, we show the cumulative luminosity function (CLF), $N(<M_U)$, for the 1 core field. This is the only data for which

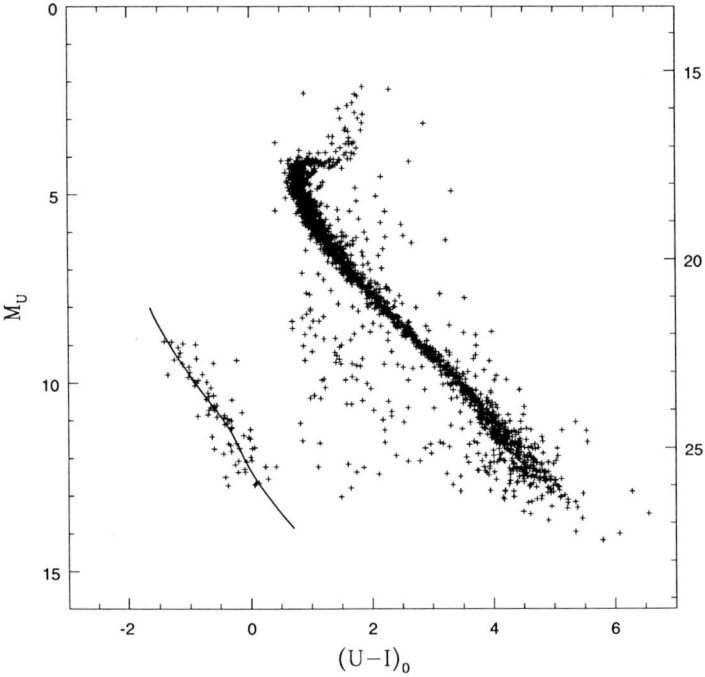

*Figure 1.* This CMD contains data from all four chips in the 1 core field of M4. The model curve is the cooling locus for pure carbon core white dwarfs with a mass $M = 0.5 M_\odot$. The right hand scale shows the apparent $U$ magnitude.

we have reasonable completeness corrections at present and the corrected counts, obtained by assigning a weight, equal to the completeness factor appropriate to the magnitude and CCD chip, to each observed star are also plotted. The three panels show theoretical curves for different interior compositions and are based on interior models developed by Matt Wood, and atmospheres from Bergeron *et al.* (1996). The theoretical counts are normalized to the indicated number of horizontal branch (HB) stars which were assumed to have a lifetime of $10^8$ yr. For reference, the number of HB stars expected in the HST field is 14.5 based on actual counts from ground based images of the M4 core.

The agreement between theory and observation is good at the bright end of the cooling sequence but, at the faint end there is a discrepancy which persists irrespective of the wd core composition. The deviation sets in at magnitudes for which the relevant stars are *easily* visible on the images and thus it would be surprising if this disagreement were due *solely* to poorly determined incompleteness corrections.

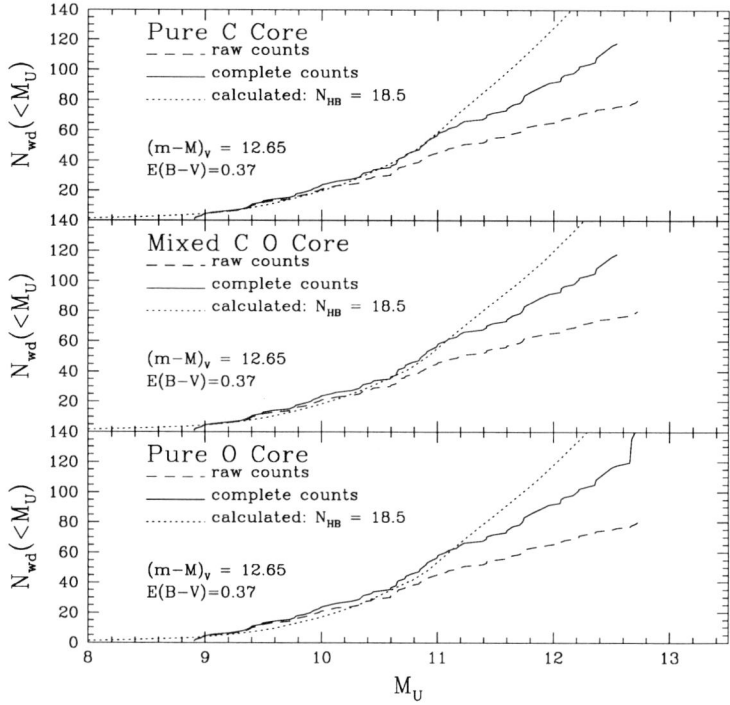

*Figure 2.* The observed cumulative luminosity function (CLF), both raw and completeness corrected, is compared to the theoretical CLF for white dwarf cooling sequences with three different interior compositions.

It appears that this disagreement could be due to mass segregation. The observed wds have a mass which is only 60% that of their progenitor stars and hence they will relax over time to a less concentrated spatial distribution than the cluster turn-off and pms stars we see today. The central and half-mass relaxation times in M4 are $4 \times 10^7$ yr and $4 \times 10^8$ yr respectively (Djorgovski 1993) whereas the white dwarfs with $M_U \geq 11.5$ have cooling times $\geq 5 \times 10^8$ yr. Hence the effects of relaxation should be apparent in the fainter and therefore older wds.

To illustrate the magnitude of the effect, we constructed an isotropic Michie-King model for M4. The present nuclear burning stars were assigned to 8 equal mass bins in the range 0.1–0.8 $M_\odot$ and the wd progenitors were assigned also to 8 equal mass bins in the range 0.8–8.0 $M_\odot$. An initial mass function (IMF) of the form $n(m) \propto m^{-(1+x)}$ was adopted and, with $x = 1.25$, the mass fraction in each of the 16 mass bins was calculated. The upper 8 mass bins were then converted to wds according to the rule

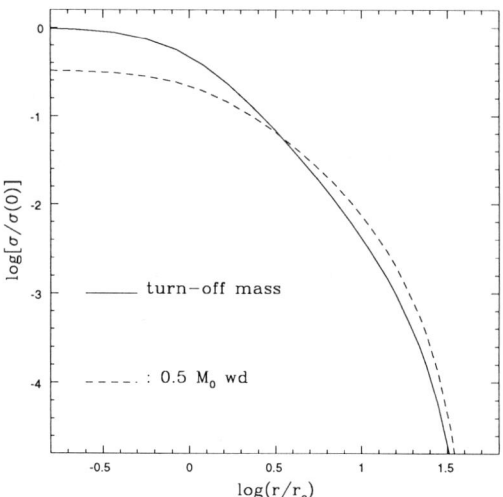

*Figure 3.* The normalized projected number density of turn-off stars is plotted as a function of cluster radius in a Michie-King model for M4. This profile is compared to the profile for 0.5 $M_\odot$ white dwarfs with the same mass fraction as the turn-off stars.

$m_{wd} = m_o \exp(m_i/8)$ as in Wood (1992). The constant $m_o$ was set to give $m_{wd} = 0.5$ for $m_i = 1.17$, which is the number weighted mean mass (in $M_\odot$ units) of the stars in the IMF bin just above the present turn-off. The central potential was then adjusted to give a model for which the stars in the turn-off mass bin gave a concentration parameter in good agreement with observation (Trager *et al.* 1995). In this model, stars identified with the turn-off constitute 2.6% of the cluster mass and the 0.5 $M_\odot$ wds contribute 15.2%.

The normalized, projected number density of the model turn-off stars is plotted in figure 3. This profile can be identified with the observed surface brightness profile and has a concentration parameter of c=1.67. We also show the profile of the 0.5 $M_\odot$ wds (dashed line) *scaled to the same mass fraction as the turn-off stars*. Essentially, the dashed line shows the effect of converting all the present turn-off stars to 0.5 $M_\odot$ wds and allowing them to fully relax to the quasi-equilibrium defined by the Michie-King model. As expected, the lower mass wds have a larger core radius than their progenitors. Close to the cluster core, the projected number density of wds is smaller than that of their progenitors and, therefore, smaller than the number which which would be predicted from the observed HB stars under the assumption of *local* star number conservation. The effect is reversed in the halo of the cluster.

For the model shown here, the projected number of wds at 1 core radius is only 47% that of an equivalent progenitor population, which is similar to the observed discrepancy in the CLF of figure 2. Now this particular King-Michie model is meant to be illustrative only since the IMF is unconstrained and, more fundamentally, the model ignores the fact that the formation and relaxation processes are occurring on comparable time scales (more or less) for the stars which we can observe. In spite of these shortcomings, the model does demonstrate that mass segregation can have a significant effect on a *locally measured* wd CLF. The dynamical evolution can be treated in Fokker-Planck models which include stellar evolution, e.g., those of Drukier (1995) and these would be preferable to the static Michie-King models.

## 4. The Main Sequence

In the left panel of figure 4, we show a very high quality ground based CMD derived from data obtained in 1990 at the du Pont 2.5m telescope of the Las Campanas Observatory (Thompson *et al.* 1990). This data was calibrated against Landolt standards and then transformed to the absolute CMD plane with the distance modulus and reddening quoted earlier. On the right, we have added the HST photometry for our 1 core and 4 core fields to give a CMD which spans the cluster m-s from the turn-off to within a few *thousandths* of a solar mass of the expected hydrogen burning cut-off. The excellent agreement between these data sets adds to our confidence in the Holtzman *et al.* calibration. The CMD for the HST data alone is shown in the left panel of figure 5.

Our line of sight to M4, at (l,b) = (351,16), intersects the galactic bulge and, for a solar galacto-centric distance of $R_o = 8.0$ kpc, passes the center of the Galaxy at a radius R = 2.5 kpc. Hence, the stars in bulge should define a m-s ridge line some 2.9 mag below that of M4 if they have roughly the same metallicity. This is in good agreement with the stellar sequence visible immediately below and to the left of the cluster m-s. The bulge turn-off stars appear to have a color just slightly to the red of that of M4, indicating that they indeed have a similar metallicity and a similar age.

In the panel on the right of figure 5, we have superimposed on the 4 core field CMD, a model isochrone from d'Antona and Mazzitelli (1995) that has a metallicity similar to that of M4. The stellar mass along the isochrone is also indicated with the last labeled point being the hydrogen burning limit. Clearly, this isochrone is a poor match to the lower m-s of M4. Although the gross morphology – the changes in slope along the m-s – do seem to be in rough agreement with the observations, the colors are too blue. The more limited models of Baraffe *et al.* (1995) show a similar discrepancy.

The physical distribution of most interest is the cluster mass function

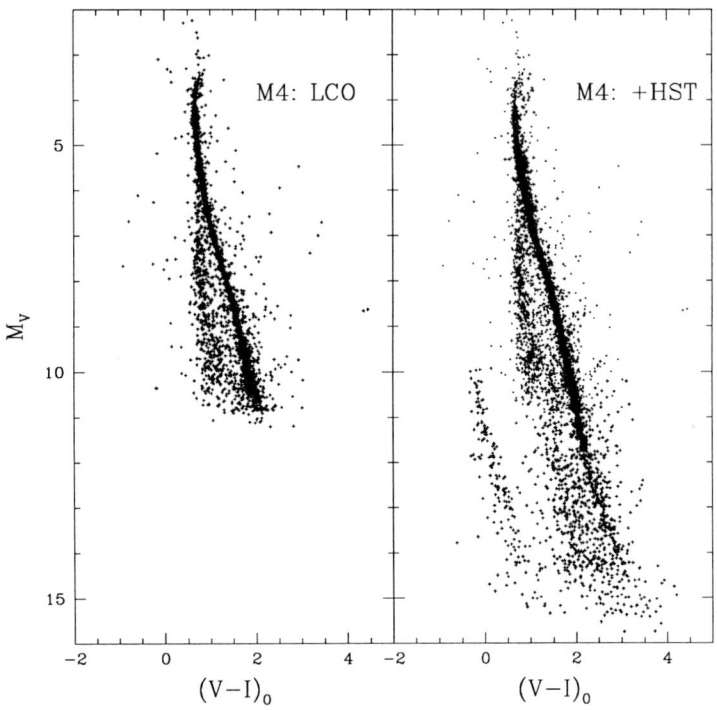

*Figure 4.* On the left is a ground based CMD obtained at Las Campanas Observatory (LCO) in 1990. On the right, we have added to the LCO data, our calibrated HST data from the 1 core and 4 core fields. See figure 5 for the HST data by itself on an apparent magnitude scale. The LCO field overlaps the 4 core HST field but covers about 3 times the area of the combined HST data accounting for the marked difference in the number of stars in the common magnitude region

(MF). This is obtained by multiplying the m-s LF by the slope of the luminosity-mass relationship which, in the absence of any empirical data, depends entirely on models for the cluster stars. Since adequate models are not yet in hand, we will defer a quantitative discussion of the cluster MF at this time. If the luminosity of the models is accepted, then figure 5 shows that our HST observations fall short by only a few *thousandths* of a solar mass from reaching the end of the end of the hydrogen burning sequence. As this limit is approached, the slope of the luminosity-mass relationship becomes extremely steep, especially in these visible photometric bands, and the m-s becomes very sparse at the faint end. Hence the scarcity of faint m-s stars in our fields does not necessarily imply a turn-over in the cluster MF before the end of the hydrogen burning (luminous) sequence.

We have also drawn on figure 5 the cooling locus for 0.5 $M_\odot$ wds with an

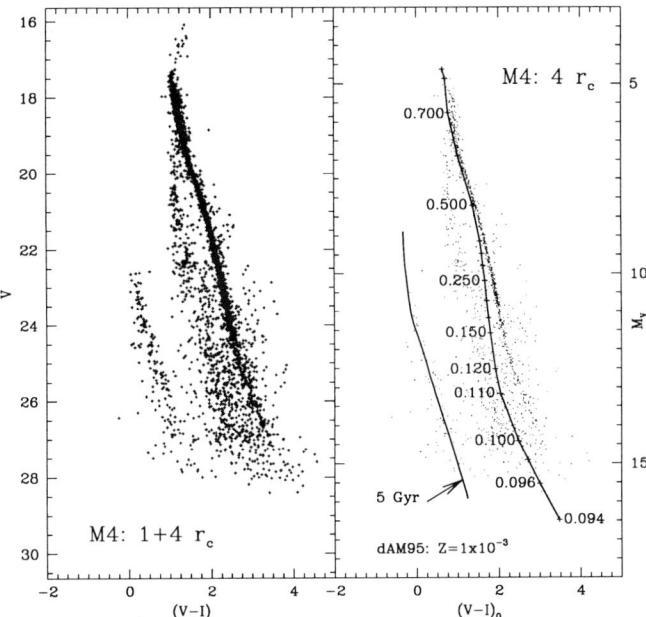

*Figure 5.* On the left, we show the CMD for our 1 core and 4 core fields in the apparent color-magnitude plane. The right hand panel shows the 4 core field data with a 10 Gyr. isochrone from d'Antona and Mazzitelli (1995) superimposed. We have labeled selected points along the isochrone with the mass of the model. The right hand ordinate is the absolute magnitude. Also shown is the cooling locus for a 0.5 $M_\odot$ C-O wd with the cooling time to reach the bottom of the observed sequence marked.

equal mix of Carbon and Oxygen in the interior. The faintest wds observed have a cooling age of about 5 Gyr – a result which is fairly insensitive to the details of the interior composition. This underscores a very important point: that the observed wds place constraints on the cluster age which are independent of the age assigned from the nuclear burning turn-off stars. With the present data, this age constraint on M4 is not particularly noteworthy. It is feasible to acquire deeper data and this could place much more interesting limits ( $\geq$ 10 Gyr, say) which would then bear on a number of fundamental problems in astrophysics.

## 5. Discussion

The refurbished HST has clearly lived up to its promise to reveal at last the globular cluster white dwarf cooling sequence and to extend our knowledge to encompass the entire hydrogen burning main-sequence. Our photometry in M4 has verified the basic theoretical tenets of white dwarf formation and evolution in the globular clusters. We have shown here that dynamical processes very likely affect the observed white dwarf cumulative luminosity function. Ultimately, this may help constrain models, like those of Drukier (1995), which incorporate stellar evolution into the Fokker-Planck framework.

It is unfortunate that the currently available model isochrones do not match the observed lower main-sequence. This problem, which may be largely due to deficiencies in the model atmospheres, will undoubtably be solved soon now that HST data is becoming available to challenge the theorists. The luminosity functions we observed in M4 still require completeness corrections but taking them as observed, and using the current, albeit imperfect, models as a guide, our data does suggest that the cluster mass function is rising to the limit of our data. There is no evidence available at this time that would suggest the cluster main-sequence is truncated before the hydrogen burning limit is reached. Hence it is very likely that the globular clusters contain a dark, brown dwarf population in addition to the cool, faint white dwarfs.

## References

Allard, F. & Hauschildt, P.M. (1995) *ApJ* **445**, 433.
Baraffe. I., Chabrier, G., Allard, F., & Hauschildt, P.M. (1995) *ApJ* **446**, L35.
Bergeron, P., Wesemael, F., & Beauchamp, A. (1996) *PASP*, in preparation.
Bessell, M. D., & Brett, J. M. *PASP* **100**, 1134.
d'Antona, F. and Mazzitelli, I (1995), preprint.
Drukier, G. A. (1995) *ApJS* **100**, 347.
Djorgovski, S. (1993) in *Structure and Dynamics of Globular Clusters*, eds. S.G. Djorgovski and G. Meylan, ASP, San Francisco, p. 373.
Holtzman, J.A., Burrows, C.J., Casertano, S., Hester, J.J., Trauger, J.T., Watson, A.M., & Worthey, G. (1995) *AJ*, in press.
Landolt, A.U. (1992) *AJ* **104**, 340.
Mathis, J. S. *ARA&A* **28**, 37.
Richer, H.B., and Fahlman, G.G. (1988) *ApJ* **325**, 218.
Richer, H.B., Fahlman, G.G., Ibata, R.I., Stetson, P.B., Bell, R.A., Bolte, M., Bond, H.E., Harris, W.E., Hesser, J.E., Mandushev, G., Pryor, C., & VandenBerg, D.A. (1995a) *ApJL* **451**, L17. (paper I)
Richer, H.B., Fahlman, G.G., Ibata, R.I., Ivanans, N.C., Manduschev, G., Hesser, J.E., Stetson, P.B., VandenBerg, D.A., Pryor, C., Harris, W.E., Bond, H.E., Bolte, M., & Bell, R.A. (1995b) in *The Origins, Evolution and Destinies of Binaries in Clusters*, ed. E.F. Milone, in press. (paper II)
Stetson, P.B. (1994) *PASP* **106**, 250.
Thompson, I.B., Sivaramakrishnan, A, Fahlman, G.G., & Richer, H.B. (1990) unpublished.
Trager, S.C., King, I.R., & Djorgovski, S. (1995) *AJ*, **109**, 218.
Webbink, R.F. (1985) in *Dynamics of Star Clusters, IAU Symposium No. 113*, eds. J. Goodman and P. Hut, Reidel, Dordrecht, p. 541.
Wood, M. A. (1992) *ApJ* **386**, 539.

# SEARCHES FOR CATACLYSMIC VARIABLES IN GLOBULAR CLUSTER CORES

MICHAEL M. SHARA
*Space Telescope Science Institute*
*3700 San Martin Dr.*
*Baltimore, MD 21218*
*USA*
*mshara@stsci.edu*

**Abstract.**
Close binaries are widely believed to exist in large numbers in the cores of globular clusters. If present, these binaries are critical sources and sinks of energy that drive the dynamical evolution of their host clusters. I report on HST searches for binaries (based on variability) in the outskirts and cores of several globular clusters; dwarf novae should be particularly easy to find. Dense and loose clusters have been thoroughly searched on timescales ranging from minutes to years. Detailed simulations demonstrate that virtually all binaries with $M < 8$, amplitudes $> 0.1$ mag and periods of $2 - 20$ hours should have been found. This includes virtually all known contact binaries. At least 1/3 of all dwarf novae present in several globulars should also have been seen (very easily!) in eruption at $M = 4 - 6$.

Simple tidal capture theory predicts that dozens of interacting binaries should have been found in our searches; the observed number is typically one or two objects per cluster. Unless tidal capture cataclysmic binaries are rapidly destroyed, ejected, or much fainter than most of their Galactic counterparts, we must conclude that very close binaries in globular cores are rare, and that their total influence on cluster dynamical evolution is less than currently claimed.

Tidal capture theory predicts that close binaries are formed by two- and three-body close encounters. The efficiency of the process (and the resulting numbers and types of interacting binaries) is currently a topic of much heated debate. The theoretical uncertainties in the physics of tidal capture are unlikely to be resolved soon. However, the Hubble Space Telescope has now made significant progress in observational searches for close binaries near globular cores. I briefly report on the results of three of these searches for cataclysmic binaries.

## 1. NGC 6752

The core of NGC 6752 has been imaged every 14 minutes over a span of 7 hours. 730 stars were monitored with the Hubble Space Telescope Faint Object Camera, working at 2200Å(Shara et al. 1995, ApJ, 441, 617). Artificial star tests demonstrate that we should have found virtually all variables brighter than $m_{220} = 20$ (i.e., $M_{220} \leq 6.75$) with variability amplitude $\Delta m \geq 0.25$ mag and most variables brighter than $m_{220} = 22$ (i.e., $M_{220} \leq 8.75$) with $\Delta m \geq 0.35$ mag. No variables were found; eight were expected based on tidal capture models in the one-fourth of the core that we observed. Our results strongly constrain the number of cataclysmic and contact binaries (with the above variablility and luminosity limits) in the core of NGC 6752 to be $\leq 10$ in total with 95% probability.

## 2. M80

A deep (U,B) color-magnitude diagram has been obtained of the core of the dense globular M80. A well-documented classical nova was seen to erupt in M80 in May 1860. Brightening to less than 7th apparent magnitude, the nova outshone the entire cluster for a week. Its decline was followed to 10th apparent magnitude 3 weeks later.

We have used an HST color-magnitude diagram to recover the quiescent classical nova, now known as T Sco (Shara & Drissen 1995, ApJ, 448, 203). It is almost an order of magnitude fainter than most decades-old novae. 8000 stars were examined in the M80 CMD. Only one other object (besides T Sco) occupies a position (in the CMD) bluewards and faintwards of the turnoff. The presence of only one other blue, faint star in the core of M80 is striking and highly significant. Dozens of bright, blue cataclysmic variables should easily have been detectable in our HST images if simple tidal capture theory is correct.

## 3. 47 Tuc

Finally, we have used archival HST images of the core of 47 Tuc (taken at 12 independent epochs) to search for erupting dwarf novae (Shara et al., 1996, submitted). We easily recovered the one known dwarf nova (V2) in 47 Tuc, and found it in eruption in 2 of our 12 epochs.

Our spatial and temporal coverage was sufficient to locate 1/3 of all dwarf novae (assuming that globular dwarf novae erupt every few weeks or months). Not a single dwarf nova (other than V2) was discovered. Our results strongly suggest that at most a few, and certainly not dozens, of cataclysmics inhabit the core of 47 Tuc. The one "escape hatch" that may still permit many interacting binaries to exist in globular cores is brightness - or lack thereof. If mass transfer rates are low (e.g. if most catalysmics are strongly magnetic or in hibernation most of the time) then simple tidal capture theory may still be vindicated.

## 4. Conclusions

Before HST resolved the cores of many Galactic globular clusters, theorists had predicted a plethora of cataclysmic variables in their cores. While a few CVs have now been found, the numbers appear to be one to two orders of magnitude less than the early expectations. Deeper searches with HST will soon show if tidal capture binaries, and in particular CVs, are truly rare, or if most CVs are intrinsically very faint.

# THE CENTERS OF GALAXIES

DOUGLAS RICHSTONE AND KARL GEBHARDT
*Department of Astronomy*
*University of Michigan*

ALAN DRESSLER
*Carnegie Observatories*

SANDRA FABER AND CARL GRILLMAIR
*Lick Board of Studies*
*University of California, Santa Cruz*

JOHN KORMENDY AND Y.-I. BYUN
*Institute for Astronomy*
*University of Hawaii*

TOD LAUER AND EDWARD AJHAR
*National Optical Astronomy Observatories*

AND

SCOTT TREMAINE
*Canadian Institute for Theoretical Astrophysics*

## 1. Introduction

This report has two major purposes. First, we summarize here work by our team on the determination of the density of stars near the centers of a large sample of galaxies observed with the Hubble Space Telescope. There appear to be two varieties of elliptical galaxies (and bulges). The stellar densities near the centers of small elliptical galaxies exceed those of globular clusters and the density of the universe at the recombination epoch. The radial dependence of density and implied gravitational force seems inconsistent (at least in the case of the smaller elliptical galaxies) with a long-lived triaxial configuration. It therefore seems likely that the central regions of less luminous elliptical galaxies are axisymmetric. For the more luminous ellipticals, the central densities are far more modest and the presence of a distinct core (defined below) is generally well established. Even in these cases, however, we find few if any galaxies with analytic (Taylor expandable) stellar densities near the center.

Second, we discuss some of the recent results on the detection of massive black holes in the centers of galaxies. In our view, the presence of *massive dark objects* with only upper limits to their radii and without visible emission, is now well established. In the case of NGC 4258 alternative models involving clusters of faint or degenerate stars appear to be ruled out. In other important cases that goal remains elusive.

## 2. Light Profiles of Centers of "Hot" Stellar Systems

We begin with an illustration of the emissivity distribution $\nu(r)$ (or density of stars) near the centers of $\sim 60$ ellipticals and S0 bulges based on Lucy-Richardson deconvolved images from the (optically uncorrected) Hubble Space Telescope (Figure 1). The reduction techniques are described in Lauer *et al.* 1995, Lauer *et al.* 1992a, and references cited therein. Most of the sample comes from Lauer *et al.*, although it has been enlarged as described in Gebhardt *et al.* 1996.

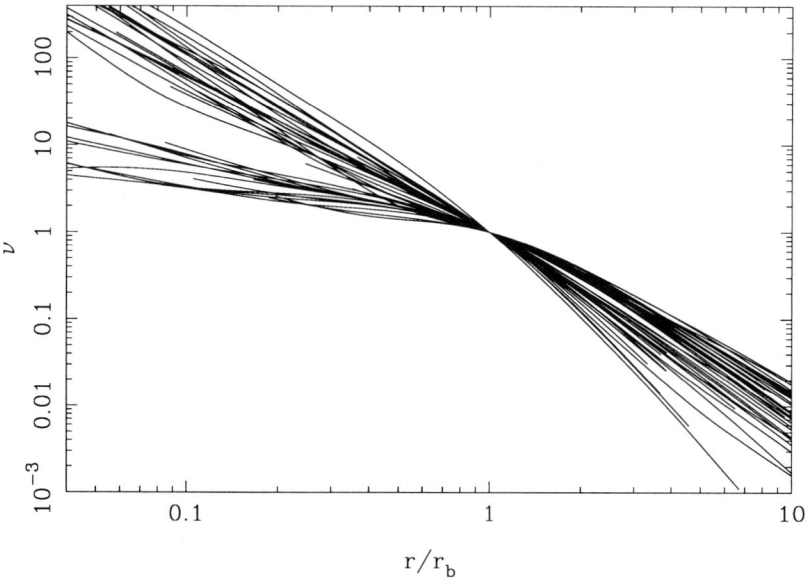

The stellar density $\nu$ shown in this figure has been deprojected from the observed surface brightness by the non-parametric technique of smoothing splines as described in Gebhardt *et al.* The galaxies have been superimposed by scaling both $\nu$ and $r$ to unit density and length at the radius of maximum logarithmic curvature $dS/d\log r$ where $S = d\log \nu/d\log r$.

Two interesting results can be instantly seen in Figure 1. First, in no case does the best estimate of $S$ reach zero at small radii (in all cases the density estimates are made only for $r \geq 0\rlap{.}''1$). Hence few if any of these objects can be approximated, even at HST resolution, by $\nu = A - Br^2$ near $r = 0$ and hence they are not *analytic cores* (Tremaine 1995). By contrast King models and nonsingular isothermal spheres (both sometimes used in the analysis of globular clusters — the primary subject of this book) do have analytic cores.

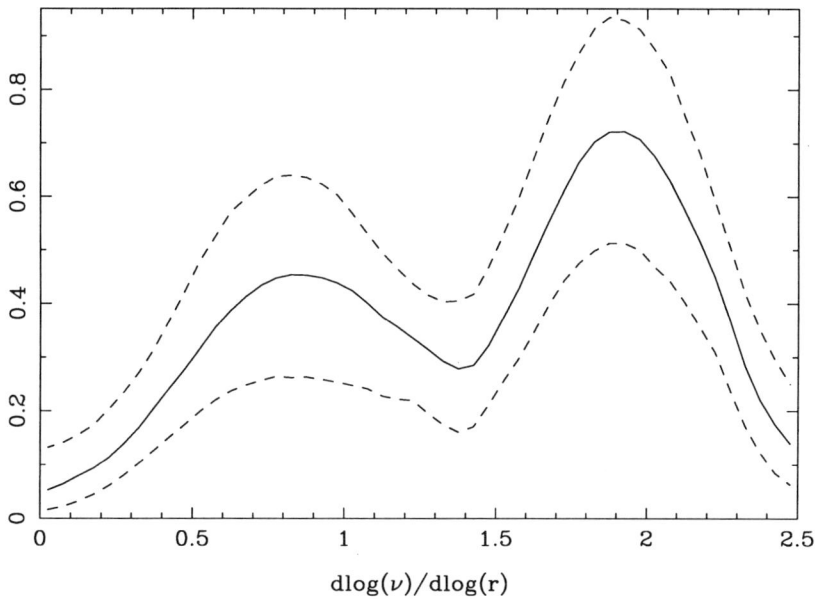

Second, considering only the light profiles inward from the position of maximum slope, these objects separate into two groups of galaxies. The distribution of logarithmic slopes at $0\rlap{.}''1$ from Gebhardt *et al.* supports this impression, at about the $2\sigma$ level (see Figure 2 above). The physical significance of this difference is buttressed by its correlation with properties of the main bodies of these galaxies including total luminosity, rotation $(v/\sigma)$, and departures from elliptical isophotes (boxiness or diskiness). These relationships are discussed in detail in Faber *et al.* 1996. As an illustration of this point we display the relationship between the logarithmic slope at $0\rlap{.}''1$ and the galaxy (or bulge) magnitude from Gebhardt *et al.* (see Figure 3 overleaf). We advocate the use of the terms "core" and "power law" to describe these two kinds of galactic centers. The term "core" (borrowed from its earlier usage in connection with isothermal spheres and King models) here refers to the presence of a clear break in the density distribution seen in the left-hand group in Figure 2. The term "power law" emphasizes the far less marked break in slope in the other group of galaxies. It remains to be seen, of course, whether the discovery of new objects will conform to this division into two clear groups, or will instead show that we have been overly impressed by extreme members of a continuous distribution.

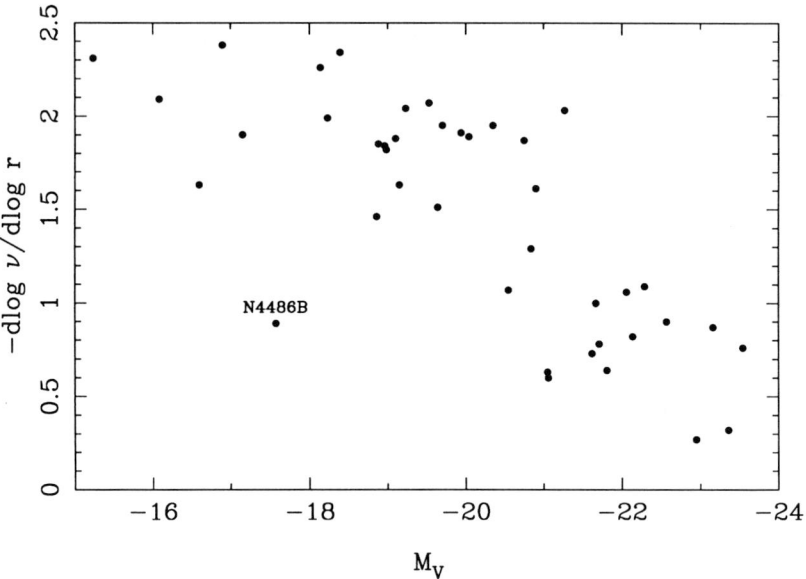

These logarithmic slopes can be immediately compared to a recent study by Merritt and Fridman (1995) of the regularity of orbits in triaxial mass distributions. They show that in profiles with $\rho \propto r^{-2}$ there are insufficient regular box orbits (flattened in the same direction as the mass distribution) to create an equilibrium configuration. With milder density cusps ($\rho \propto r^{-1}$), the critically necessary box-like orbits are still chaotic, but they mimic regular behavior for a long time before mixing in phase space. Such galaxies may be in equilibria that last a substantial fraction of a Hubble time. While Merritt and Fridman's work only investigated two density profiles and one particular set of axial ratios, the results seem very likely to be generic. Using these results as a guide, it seems very likely that the steep profile (low luminosity) elliptical galaxies in our sample are not triaxial anywhere near the center (although, since many are known to rotate they are probably oblate).

In the case of the shallow profile (and high luminosity) galaxies the situation is more complex. Even though these ellipticals and bulges are slightly more shallow then $\rho \propto r^{-1}$ case investigated by Merritt and Fridman, they do not approach the analytic case that would correspond to a Stäckel potential where regular box orbits are known to exist. The presence of a massive dark object would further steepen the potential and accelerate the stochastic mixing of orbits in such a system. There appear to be two possibilities in this case. The galaxies may be evolving slowly under the influence of orbit diffusion, and may still be triaxial. Or they may be axisymmetric near the center.

In either profile family, it seems doubtful that experience gained from the analysis of orbits in static Stäckel potentials or of triaxial objects with analytic cores is likely to have much connection to the real galaxies illustrated in figures 1, 2 and 3.

## 3. Some Thoughts on Formation and Survival of Galaxy Centers

The comments above were based only on the profile shapes of the observed galaxies, but now we consider also the densities at $0''.1$, as illustrated in figure 4 overleaf (only objects with filled symbols are resolved). The object with the largest density is M32, which is also plotted as it would appear if observed at the distance of the Virgo cluster (the distance of most of the galaxies in the sample). Its position in the figure shows that there is every reason to believe that all of the low luminosity ellipticals reach that same stellar density of nearly $10^6 L_\odot \mathrm{pc}^{-3}$. Moreover, this is a luminosity density. Using a reasonable $M/L$ of 2 and fitting a model to this object, Lauer et al. 1992b concluded that the M32 reaches densities of at least $10^7 M_\odot \mathrm{pc}^{-3}$. This mass density may be typical of low luminosity hot stellar systems and it exceeds the density of even the densest globular clusters (see Djorgovski 1993).

Taking $10^7 M_\odot \mathrm{pc}^{-3}$ as a useful fiducial number, we may compute the redshift at which the mean cosmic density of the universe was equal to this. That redshift is

$$1 + z = 3.3 \times 10^4 h^{-2/3} \Omega_0^{-1/3}, \qquad (1)$$

an epoch before recombination, and even before matter domination! The first moment of matter domination occurs at $1 + z_{eq} = 2.3 \times 10^4 \Omega_0 h^2$ (Padmanabhan 1995).

Since galaxy scale fluctuations could not have grown at that time they must have formed later and at lower densities. It therefore seems probable that dissipational processes played an important role in the formation of the densest parts of these low luminosity galaxies.

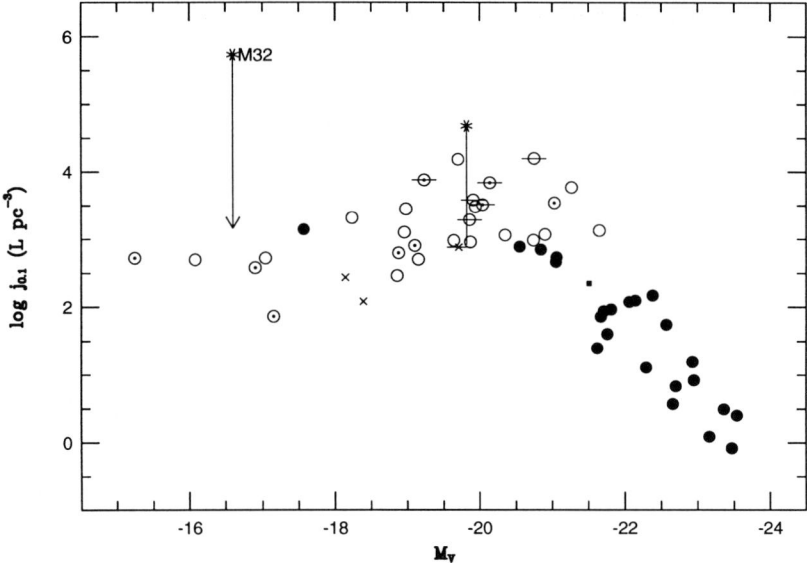

The low density centers of high luminosity galaxies offer a separate problem. It is well known that high luminosity galaxies accrete low luminosity galaxies, although the accretion rate is controversial. All of the low luminosity galaxies in our sample have high central densities and all of the quite high luminosity galaxies have very low densities, as much as six orders of magnitude smaller than M32. Since, during an accretion event the low luminosity objects never encounter tidal forces sufficient to destroy them, it is hard to understand the absence of M32-like nuclei in at least some of our luminous objects. Yet these are not seen. In cosmological simulations of formation of collisionless gravitating objects, (see, for example, Hernquist (1993) and Crone, Evrard and Richstone 1995), objects with well defined low density cores do not form because during the merger of small, dense objects with larger more diffuse ones there is quite incomplete energy redistribution among the particles. Hence, the most tightly bound subsystems retain their form, even when incorporated into much larger objects.

A possible resolution of this difficulty might be the presence of massive black holes in, at least, the more luminous ellipticals, which could provide the tidal forces necessary to disrupt the dense infalling dwarfs and scatter their stars over a large volume (and hence at low density). Although this explanation is consistent with all black hole detections in large ellipticals, it is nonetheless quite speculative.

## 4. The Case for Massive Black Holes

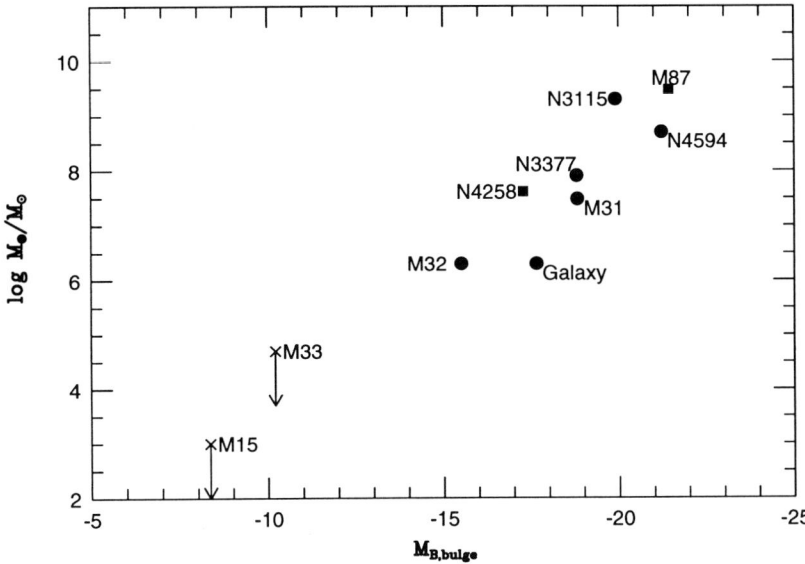

Finally, we turn to the question of evidence for massive black holes in galactic nuclei. As summarized by Kormendy and Richstone (KR, 1995), there are now several secure cases of objects in the literature, illustrated in Figure 5, in which there is strong dynamical evidence for unseen mass (or a sharp rise in M/L) at the center of the galaxy and on a scale whose upper limit corresponds to the spatial resolution of the data. As noted by KR, the mass of these objects appears to be proportional to the luminosity of the elliptical galaxy or spiral bulge in which they are located.

The demonstration that these mass concentrations are in fact black holes could really only be achieved through the observation of relativistic velocities. Failing that, the elimination of alternative possibilities, as attempted in the case of M32 by Goodman and Lee 1989, Richstone, Bower and Dressler 1990, and Lauer et al. 1993b, would be useful. By far the most spectacular example of this approach was achieved in the case of NGC 4258 by Maoz 1995. NGC 4258 is observed to have maser emission features over a range of radii from 0.13 pc to 0.25 pc, with an apparently Keplerian velocity dependence (Miyoshi et al. 1995). The mass enclosed within .13pc is $3.6 \times 10^7 M_\odot$. Maoz was able to show that if the enclosed mass had been a cluster of stars of individual masses greater than $.03 M_\odot$, the cluster would have evaporated in less than a Hubble time (even if the objects were degenerate). If, on the other hand, the cluster was composed of stars of still lower mass, the physical collision/merger timescale would be exceedingly short and more massive objects would build up rapidly, either leading to a single object or to evaporation. Hence, a dark cluster of stars appears to be ruled out in the case of NGC 4258 and we are left with the prospect of a massive black hole or of something not yet invented.

We acknowledge support from grants GO–02600.01–87A from STScI and NASA Theory grant NAG5-2758.

## References

Crone, M. M. , Evrard, A. E. , Richstone, D. 1994, ApJ, 434, 402.
Djorgovski, S. 1993, in Structure and Dynamics of Globular Clusters, ed. S. G. Djorgovski & G. Meylan (San Francisco: ASP), p. 373.
Dehnen, W. & Gerhard, O. 1993, MNRAS, 261, 311.
Faber, S.M. Gebhardt, K., Richstone, D. Lauer, T. R., Ajhar, E. A., Byun, Y. I., Dressler, A. Grillmair, C., Kormendy, J. & Tremaine, S. 1996, AJ, in preparation (Paper 4).
Gebhardt, K., Richstone, D. Lauer, T. R., Ajhar, E. A., Byun, Y. I., Dressler, A. Faber, S. M., Grillmair, C., Kormendy, J. & Tremaine, S. 1996, AJ, submitted (Paper 3).
Goodman, J. & Lee, HM. 1989, ApJ, 337, 84.
Hernquist, L. 1993, ApJ409, 460.
Kormendy, J. et al. 1995, in IAU Symposium 171, ed. R.L. Davies & R. Bender (Kluwer), to be published.
Kormendy, J. & Richstone, D. 1995, Inward Bound — The Search for Supermassive Black Holes in Galactic Nuclei, 1995, in Ann Rev Astron and Astroph, 33, 581.
Lauer, T. R., et al. 1992a, AJ, 79, 745.
Lauer, T. R., Faber, S. M. et al. 1992b, AJ, 104, 552.
Lauer, T. R., Ajhar, E. A., Byun, Y. I., Dressler, A. Faber, S. M., Grillmair, C., Kormendy, J. Richstone, D. & Tremaine, S. 1995, AJ, 110, in press (Paper 1).
Maoz, E. 1995, ApJ (Letters), 447, L91.
Merritt, D., & Fridman, T. 1995, ApJ, 456, in press.
Miyoshi, M., Moran, J., Hernstein, J. Greenhill, L., Nakai, N., Diamond, P.& Makato, I. 1995, Nature, 373, 127.
Padmanabhan, T. 1995 "Structure Formation in the Universe" (Cambridge University Press), p 95.
Richstone, D., Bower, G., and Dressler, A. 1990, ApJ, 353, 118.
Tremaine, S. et al. 1995, in Some Unsolved Problems in Astrophysics, ed. J. N. Bahcall & J. P. Ostriker (Princeton: Princeton U. Press) in press.
Tremaine, S. et al. 1994, AJ, 107, 634.

# STUDIES THROUGH RADIAL VELOCITY MEASUREMENTS OF THE PECULIAR MOTIONS OF STARS IN GALACTIC GLOBULAR CLUSTERS

G. MEYLAN
*European Southern Observatory*
*Karl-Schwarzschild-Strasse 2*
*D-85748 Garching bei Muenchen*
*Germany*

**Abstract.** In a brief introduction, we first compare, in the framework of globular cluster dynamics, the relative importance of the amounts of information acquired, during the past decades, through proper motion and radial velocity measurements. Next, we review the most recent and important studies based on measurements of radial velocities obtained with single-object and multi-object spectrometers and interpreted with the use of King-Michie and Fokker-Planck models or with non-parametric methods. Then, we present the results obtained from integrated-light spectra in the cores of a few collapsed globular clusters. We conclude with a summary of the most important scientific outputs secured from the radial velocity measurements of stars in globular clusters.

## 1. Introduction

As clearly stated by Oort & van Herk (1959) in their paper on M3, which was one of the first modern dynamical studies of a globular cluster (GC), *"the purpose of the investigation (...) is to see in how far the observed distribution of the stars corresponds to what we should expect theoretically, and to investigate what the comparison between observation and theory can teach us concerning the unobserved faint stars"*. Nowadays, the aim of their work is still completely valid.

Oort & van Herk (1959) were already mentioning the potential observational constraint to be obtained from the random motions of stars. Actually, at that time, so little was known concerning random motions in star clusters that this was not yet a sensitive test.

Random, or peculiar (u,v,w), motions were known only for stars in the solar neighbourhood. The velocity distributions of stars were known to be different along each axis and essentially gaussian (Blaauw & Schmidt 1965). This led to a distribution function in (u,v,w) space which has now acquired the name of "Schwarzschild velocity ellipsoid", as a reference to the elegant theoretical description of this problem (Schwarzschild 1907, 1908).

Proper motions and radial velocities of stars in GCs are significantly more difficult to measure than for field stars, since the typical projected velocity dispersion of a galactic GC is of the order of $\sigma_p = 5 - 10$ km s$^{-1}$.

## 1.1. PROPER MOTIONS: ERRORS $\sim \sigma_p$(GCS)

For a relatively nearby GC located at a distance of 5 kpc and with projected velocity dispersion $\sigma_p = 5$ km s$^{-1}$, the corresponding proper motion equals 20 marcsec/century, i.e., 1.5 $\mu m$ in 80 years on a Yerkes plate. This has made proper motions difficult to measure with the required precision. K.M. Cudworth has been the pioneer of this field (see Meylan & Pryor 1993 for some references). But even in the best studied GCs, the errors in the proper motions have been comparable in size to the motions themselves. Reijns et al. (1993) report on the largest proper motion work, concerning a few thousand stars in $\omega$ Centauri (see Section 4.3 below). Proper motions represent a gold mine of information which has still to be exploited.

## 1.2. RADIAL VELOCITIES: ERRORS $< \sigma_p$(GCS)

For the same nearby GC (with d = 5 kpc and $\sigma_p = 5$ km s$^{-1}$), cross-correlation techniques have provided, for nearly two decades, high-quality radial velocities $V_r$ with errors typically $\leq 1$ km s$^{-1}$. This has opened the way to essential observational constraints allowing numerous dynamical studies.

## 2. $V_r$ from cross-correlation techniques

There is only one GC — $\omega$ Centauri — which was studied, before the advent of cross-correlation techniques, by use of $V_r$ of individual stars. Harding (1965) measured the radial velocities of 40 individual stars, deriving the mean radial velocity of the cluster $\langle V_r \rangle = 238$ km s$^{-1}$, and its mean projected velocity dispersion $\sigma_p = 9.9 \pm 6.6$ km s$^{-1}$. Using the best 13 $V_r$ measurements, Harding (1965) detected the presence of rotation in this cluster, although in a more qualitative than quantitative way.

The basic idea of cross-correlation techniques — matching superimposed spectra — is not new: Evershed (1913) used the method for measuring solar spectra, but its application to stellar spectra has been suggested by Felget (1953) and Babcock (1955). Griffin (1967) was the pioneer of the field in being the first to build a modern instrument capable of using cross-correlation techniques in a high-performance mode. In this first study, Griffin quotes a standard deviation of one photoelectric observation equal to $1.06 \pm 0.07$ km s$^{-1}$. This started the beginning of a new era for dynamics of GCs.

Da Costa et al. (1977) measured, with a mean accuracy of 1.5 km s$^{-1}$, 11 stars in the cluster NGC 6397 and determined $\sigma_p = 3.1 \pm 0.7$ km s$^{-1}$. Such a low velocity dispersion would not have been measured without cross-correlation techniques. A milestone was reached, from both theoretical and observational points of view, with the work by Gunn & Griffin (1979) on M3. They measured, with a mean accuracy of 1.0 km s$^{-1}$, radial velocities for 111 member stars of the cluster. Density and velocity dispersion profiles were simultaneously fitted to multi-mass anisotropic dynamical models based on the King-Michie form of the phase-space distribution function $f_i(E, J) \propto [\exp(-A_i E) - 1] \exp(-\beta J^2)$. Their results were consistent with the absence of spectroscopic binaries in the cluster, a result which misled the astronomical community until such binaries were discovered about a decade later by Pryor et al. Gunn & Griffin (1979) also found two high-velocity stars with constant radial velocities at 3.5 and 4.5 times the mean cluster velocity dispersion. These two interlopers are difficult to explain from both a dynamical and a statistical point of view.

## 3. Single-object spectrometers and King-Michie models

The late 70's and early 80's saw the development of cross-correlation techniques in different astronomical institutes. All techniques were based on the acquisition of single-star $V_r$ in two different ways: (i) with spectrometers making online cross-correlation (CORAVEL type) and (ii) with spectrograph-CCD combinations recording the stellar spectra and allowing cross-correlation after the observations. Dynamical studies of galactic GCs, implying samples of $V_r$ for 68 to 469 stars, were published following the study of Gunn Griffin (1979) on M3: for 47 Tuc by Mayor et al. (1984) and Meylan (1988, 1989); for M92 by Lupton et al. (1985); for M2 by Pryor et al. (1986); for M13 by Lupton et al. (1987); for $\omega$ Cen by Meylan (1987a) and Meylan et al. (1995); for M15 by Peterson et al. (1989) and Grabhorn et al. (1992); for NGC 6397 by Meylan & Mayor (1991); for NGC 362 by Fischer et al. (1993) and for NGC 3201 by Da Costa et al. (1993).

Kinematical studies of a few other GCs were done with fewer ($\sim 20$) stars, e.g., by Peterson & Latham (1986) and by Pryor et al. (1989, 1991).

## 3.1. ω Centauri: THE MOST MASSIVE GALACTIC GLOBULAR CLUSTER

The most recent dynamical study of this type, containing also the largest number of such $V_r$ measurements, concerns ω Cen (Meylan et al. 1995). The mean radial velocities obtained with CORAVEL for 469 giant stars in this galactic GC, with a mean accuracy of 0.6 km s$^{-1}$, are used to derive the velocity dispersion profile. It increases significantly from the outer parts inwards: the 16 outermost stars, located between 19.2' and 22.4' from the center, have a velocity dispersion $\sigma_p = 5.1 \pm 1.6$ km s$^{-1}$, while the 16 innermost stars, located within 1' from the center, have a velocity dispersion $\sigma_p = 21.9 \pm 3.9$ km s$^{-1}$. This inner value of about $\sigma_p(0) = 22$ km s$^{-1}$ is the largest velocity dispersion value obtained in the core of any galactic GC.

TABLE 1. ω Cen: the four best (lowest $\chi^2$) K-M models

| $m_{hr}$ | x | $M_{hr}$ % | $M_{wd}$ % | $r_a/r_c$ | $M_{tot}$ [$10^6 m_\odot$] | $M/L_V$ |
|---|---|---|---|---|---|---|
| 1.4 | 1.50 | 2 | 15 | 2.0 | 5.23 | 4.32 |
| 1.4 | 1.25 | 4 | 21 | 3.0 | 4.55 | 3.74 |
| 2.0 | 1.75 | 1 | 10 | 2.0 | 5.63 | 4.21 |
| 2.0 | 1.50 | 3 | 15 | 3.0 | 4.88 | 3.90 |

The results of the four best (lowest $\chi^2$) simultaneous fits of these radial velocities and of the surface brightness profile to multi-mass King-Michie dynamical models are displayed in Table 1 (Meylan et al. 1995). Column (1) gives the stellar mass $m_{hr}$ of the heavy remnants, column (2) the mass function exponent x, columns (3) and (4) the fractions of the total mass in heavy remnants and in white dwarfs, column (5) the anisotropy radius $r_a$, column (6) the cluster total mass $M_{tot}$, and column (7) the cluster mass-luminosity ratio $M/L_V$.

The mean estimate of the total mass equals $M_{tot} = 5.1\ 10^6 m_\odot$, with a corresponding mean mass-to-light ratio $M/L_V = 4.1$. There is evidence for 0.6 - 1.3 $10^6 m_\odot$ of dark remnants in the form of white dwarfs and neutron stars. These results emphasize the fact that ω Cen is not only the brightest but also, by far, the most massive galactic GC.

## 3.2. PRIMORDIAL BINARIES IN ω Cen

The monitoring, over more than 10 years, of the radial velocities of 310 of the giant stars observed in ω Cen allows a search for primordial spectroscopic binaries (Mayor et al. 1996). The present period range for such binaries is limited, for short periods, by the onset of Roche-lobe overflow and, for long periods, by dynamical friction. Most of the primordial binaries among the red giants of $\omega$ Cen should have periods between 200 and 4,000 days. The two main important results of this study are as follows:

**Duplicity among the chemically peculiar giants of $\omega$ Cen**: The binary frequency observed among the 32 chemically peculiar giants in $\omega$ Cen is very low when compared to the binary frequency of similar stars in the field. This suggests either that enrichment mechanisms in $\omega$ Cen are quite different from those in the field, or that these stars may be the extreme tail of the abnormal abundances in $\omega$ Cen (Mayor et al. 1996).

**Global percentage of binaries in $\omega$ Cen**: Within the period range of 200 and 4,000 days, the binary frequency in $\omega$ Cen is about five times lower than for the field G dwarfs, and about 10 times lower than in the giants of the open cluster NGC 2477. Since about 20% of the nearby G dwarfs belong to binary systems with periods below $10^4$ days, the binary rate estimated in $\omega$ Cen is only a fifth of that rate, i.e., about 4% (Mayor et al. 1996).

## 3.3. ROTATION IN GLOBULAR CLUSTERS

Rotation, suspected for a long time to be at least partly responsible for the small flattening of GC, has now been actually observed. Non-cylindrical differential rotation has been measured in $\omega$ Cen, with $v_{rot}^{max} = 8.0$ km s$^{-1}$, between 3-4 $r_c$, and in 47 Tuc, with $v_{rot}^{max} = 6.5$ km s$^{-1}$, between 11-12 $r_c$ (Meylan & Mayor 1986). For $i = 90°$ and $60°$, the ratio of ordered to random motions $v_o/\sigma_o = 0.35$ and $0.39$ in $\omega$ Cen and $v_o/\sigma_o = 0.40$ and $0.46$ in 47 Tuc. Even with $i = 45°$, the importance of rotation remains weak compared to random motions: the ratio of rotational to random kinetic energies is $\simeq 0.1$, confirming the fact that GCs are hot stellar systems. Given their small ellipticities ($0.00 \lesssim \langle\varepsilon\rangle \lesssim 0.12$), GCs are located in the lower-left corner of the $v_o/\sigma_o$ vs. $\langle\varepsilon\rangle$ diagram, an area characterized by isotropy or mild anisotropy of the velocity-dispersion tensor (Pryor et al. 1986). A very clear relation between $\langle\varepsilon\rangle$ and $v_{rot}^{max}$ is presented in Fig. 5 in Meylan (1987b), pointing towards flattening by rotation.

Recently a non-parametric estimate of the mean line-of-sight velocity field on the plane of the sky has been obtained by Merritt et al. (in preparation) using the 469 stars from Meylan et al. (1995). Isorotation curves increase from 1 to 5 km s$^{-1}$ and confirm, both qualitatively and quantitatively, Fig. 3b in Meylan & Mayor (1986).

Models based on three integrals of the motion $(E, J_z, I_3)$ have been developed for galaxies (see, e.g., Dehnen & Gerhard 1993) and applied with a few subpopulations to GCs (Lupton & Gunn 1987, Lupton, Gunn & Griffin 1987). The third integral being still unknown, these two studies use guessed approximations for $I_3$. The advantage of the presence of rotation does not counterbalance totally, in the case of GCs, the difficulties of fitting processes with an increased number of free parameters.

## 4. Multi-object spectrometers and non-parametric studies

Acquiring even a few hundred stellar radial velocities one at a time is a slow and tedious job, even on 4-m class telescopes. But the number of stellar velocities in GCs has recently grown explosively because of new technologies made available. Fiber-fed, multi-object spectrographs like ARGUS at Cerro Tololo, HYDRA at Kitt Peak, and AUTOFIB at the Anglo Australian Observatory can obtain velocities for about 25 stars simultaneously. Similar gains result from using the Rutgers Fabry-Perot interferometer. The number of stars observed per GC may typically reach a few thousands in a few seasons instead of a few hundreds in more than a decade. Such large samples call for more subtle dynamical interpretation.

### 4.1. PARAMETRIC AND NON-PARAMETRIC METHODS

In order to build King-Michie type models, two essential steps have to be made: (i) the choice of the integrals of the motion, and (ii) the choice of the dependence of the phase-space distribution function on these integrals, i.e., the functional form of $f(E)$, $f(E, J)$, or $f(E, J_z, I_3)$. The results can be strongly biased by the above two assumptions: e.g., when going from a distribution function $f(E, J)$ to the gravitational potential $\Phi(r)$, the solution may not be unique, the presence or not of anisotropy and different potentials can mimic identical surface-brightness profiles.

For a few years, D. Merritt has been the pioneer of non-parametric methods in the framework of dynamical models (see Merritt & Tremblay 1994, and references therein). The general aim is to infer the gravitational potential $\Phi(r)$ and the phase-space distribution function $f(E)$, given the observations of the surface density and velocity dispersion profiles of a "tracer" population. Briefly, in the case of a GC, (i) the projected density $I(R)$ provides the space density $\nu(r)$, (ii) the projected velocity dispersion $\sigma^2(R)$ provides the space velocity dispersion $v^2(r)$, (iii) the Jeans equation provides the gravitational potential $\Phi(r)$, and (iv) the Eddington equation provides the phase-space distribution function $f(E)$. Nevertheless, a disadvantage of such techniques arises from the delicate process of deprojection using Abel integrals.

## 4.2. GCS STUDIED WITH NON-PARAMETRIC METHODS

Already for four GCs non-parametric studies have been published using samples from a few hundred up to a few thousand stars: 47 Tuc (Gebhardt & Fischer 1995), NGC 362 (Gebhardt & Fischer 1995), NGC 3201 (Gebhardt & Fischer 1995, Côté et al. 1995), and M15 (Gebhardt et al. 1994, Gebhardt & Fischer 1995). Non-parametric mass density and $M/L_V$ profiles are compared with theoretical slopes for core-collapse clusters. The two non-collapsed GCs, viz., NGC 362 and NGC 3201, seem to exhibit significant differences from the two possibly collapsed GCs, 47 Tuc and M15. The derived phase-space distribution functions are not consistent with King models: NGC 362 and NGC 3201 have significantly more tightly-bound stars than King models, and systematic differences appear between 47 Tuc and M15 and either the King models or the two less concentrated GCs. Côté et al. (1995), using King, King-Michie, and non-parametric models, present, for NGC 3201, an interesting comparison between the different results, a way to disentangle the consequences of the assumptions and disadvantages of each approach.

## 4.3. A MAJOR STUDY OF $\omega$ Centauri

The large collection of data announced by Reijns et al. (1993), yields radial velocities and proper motions for thousands of stars in $\omega$ Cen and illustrates what has become possible.

The proper motions measurements come from 50 early-epoch plates which are the best out of a collection of 443 plates taken of this cluster during the 1930's with the Yale–Columbia refractor while in South Africa. An equal number of $2^{nd}$-epoch plates have been obtained during the last decade at Mt. Stromlo. Nearly final proper motions with an internal 1-$\sigma$ accuracy of $\sim 7$ km s$^{-1}$ (0.3 marcsec/yr) have been determined for around 9800 stars brighter than V=16.5, of which an estimated 7500-8500 are members of $\omega$ Cen. Accuracies are well sufficient to investigate internal motions (Seitzer, private communication).

For the last four observing seasons, a major effort has been undertaken with ARGUS at Cerro Tololo to get accurate radial velocities ($\sim 1$ km s$^{-1}$ accuracy) for as many of these stars as possible. Some 4,500 velocities of over 3,500 stars have been obtained over the entire spatial extent of the cluster, which will permit investigation of the 3-D space velocity distribution. The same spectra are also being used to determine metallicity, to investigate the correlation between metallicity, radius, and kinematics (Seitzer, private communication).

## 5. Integrated-light spectra in the cores of GCs

In very high-concentration (collapsed) GCs, the measurement of the $V_r$ of individual stars becomes very difficult in the core because of crowding problems. A way to alleviate the problem consists of either obtaining integrated light spectra or of using a Fabry-Perot interferometer (e.g., Gebhardt et al. 1994). The former technique has been used by, e.g., Illingworth (1976), Peterson et al. (1989), Dubath et al. (1990), Zaggia et al. (1992), and Dubath et al. (1994a,b).

In the case of the extension of the CORAVEL technique to the integrated-light spectra, the cross-correlation technique produces a cross-correlation function (CCF) — relative light intensity as a function of radial velocity — which is nearly a perfect Gaussian whose $\sigma$ does not depend on the metallicity. The fit of such a function provides three physical quantities: (1) the abscissa of its minimum, equal to the radial velocity $V_r$, (2) its depth $D$, related to the metallicity, and (3) its standard deviation $\sigma_{CCF}$, related to line broadening mechanisms. Comparison of the cross-correlation function ($\sigma_{CCF}$) of a GC spectrum with the cross-correlation functions ($\sigma_{ref}$) of standard star spectra unveils the broadening of the cluster cross-correlation function ($\sigma_p$) produced by the Doppler line broadening present in the integrated-light spectra because of the random spatial motions of the stars along the line of sight. A precise estimate of the projected stellar velocity dispersion $\sigma_p$ in the integration area is then given by the following quadratic difference, $\sigma_p{}^2 = \sigma_{CCF}{}^2(\text{cluster}) - \sigma_{ref}{}^2$.

Depending on the relative numbers of bright and faint stars, such techniques present some potential sampling problems. When the light in the sampling area is dominated by one star, $\sigma_{CCF}$ is narrower and the derived velocity dispersion is too small; when the light is dominated by two stars with an unusually large radial velocity difference, $\sigma_{CCF}$ is wider and the derived velocity dispersion is too large.

### 5.1. THE CORE VELOCITY DISPERSION OF M15

M15 $\equiv$ NGC 7078 has been, for about two decades, the prototypical collapsed GC (see Peterson et al. 1989). Using the ESO New Technology Telescope with a $1'' \times 8''$ slit, Dubath et al. (1994) obtained five high-resolution integrated-light echelle spectra over the core of M15, covering a total central area of $5'' \times 8''$. By taking advantage of the spatial resolution along the slit, they extracted spectra at 120 different locations over apertures $\sim 1''$ square. The Doppler velocity broadening of the CCFs of these integrated-light spectra is always $\leq 17$ km s$^{-1}$, at all locations in the $5'' \times 8''$ area and with a mean velocity dispersion $\sigma_p = 11.7 \pm 2.6$ km s$^{-1}$. The individual radial velocities of the 14 best-resolved (spatially or spectroscopically) bright stars are also determined; they give $\sigma_p = 14.2 \pm 2.7$ km s$^{-1}$, a value consistent with the above determination. Two of the brightest central stars, separated by 2.5″, have radial-velocity values differing by 45.2 km s$^{-1}$. This study agrees with the recent work by Gebhardt et al. (1994), who measured the radial velocities of 216 stars located within 1.5′ of the cluster centre. From 0.1′ to 0.4′, their data suggest a constant velocity dispersion of about $\sigma_p = 11$ km s$^{-1}$. These measurements therefore provide no evidence for the velocity dispersion cusp observed by Peterson et al. (1989). Nevertheless, given the general shapes of its surface brightness and velocity dispersion profile, M15 remains the best GC candidate for being close to a state of deep core collapse.

5.2. CORE VELOCITY DISPERSION SURVEY

In order to study the globular cluster masses and mass-to-light ratios as a function of galaxy types and environments, Dubath et al. (1993 and in preparation) have, in the framework of their survey, obtained integrated-light spectra of the cores of about 60 Galactic, Magellanic, and Fornax globular clusters. Zaggia et al. (1992, 1993) have developed a similar technique which they applied to seven galactic globular clusters. The integrated absolute magnitudes, velocity dispersions, and core radii of the survey clusters are used to investigate the fundamental plane correlations for Galactic and Magellanic clusters (Dubath et al. in preparation). These correlations, which are analogous to the fundamental plane correlations for elliptical galaxies, have already been discussed (e.g., Djorgovski & Meylan 1994) and are consistent with the scaling law expected from the Virial Theorem. This suggests that globular clusters are virialized systems with a universal and constant M/L ratio to within the measurement errors.

## 6. Main scientific outputs from the radial velocities

The main scientific outputs from $V_r$ studies are as follows: (i) rotation is present in GCs (especially in $\omega$ Cen), but GCs remain hot dynamical systems; (ii) typical GC mass $\sim 10^5 M_\odot$ with $10^4 M_\odot \lesssim$ mass $\lesssim 10^6 M_\odot$ and $M/L_V \sim 2.5 \pm 1.0$; (iii) spectroscopic binaries do exist in GCs; (iv) velocity dispersion profiles imply no very massive haloes around GCs; ¡(v) study of global properties of GC systems, e.g., fundamental plane.

## References

Babcock H.W., 1955, Annual Report of the Director of the Mount Wilson and Palomar Observatories, 1954/1955, p. 27
Blaauw A., Schmidt M., eds, 1965, Galactic Structure, (Chicago: Chicago Univ. Press)
Côté P., Welch D.L., Fischer P., Gebhardt K., 1995, ApJ, in press

Da Costa G.S., Freeman K.C., Kalnajs A.J., Rodgers A.W., 1977, AJ, 82, 810
Da Costa G.S., Tamblyn P., Seitzer P., Cohn H.N., Lugger P.M., in Structure and Dynamics of Globular Clusters, ASP Conf. Series, Vol. 50, eds. Djorgovski S.G. & Meylan G., (San Francisco: ASP), p. 81
Dehnen W., Gerhard O.E., 1993, MNRAS, 261, 311
Djorgovski S.G., Meylan G., 1994, AJ, 108, 1292
Dubath P., Mayor M., Meylan G., 1993, in *The GC - Galaxy Connection*, ASP Conference Series, Vol. 48, eds. Smith G.H. & Brodie J.P., (San Francisco: ASP), p. 557
Dubath P., Meylan G., Mayor M., 1990, A&A, 239, 142
Dubath P., Meylan G., Mayor M., 1992, ApJ, 400, 510
Dubath P., Meylan G., Mayor M., 1994a, ApJ, 426, 192
Dubath P., Meylan G., Mayor M., 1994b, A&A, 290, 104
Evershed J., 1913, Kodaikanak Bull., No. 32
Felget P., 1953, Optica Acta, 2, 9
Fischer P., Welch D.L., Mateo M., Côté P., 1993, AJ, 106, 1508
Gebhardt K., Fischer P., 1995, AJ, 109, 209
Gebhardt K., Pryor C., Williams T.B., Hesser J.E., 1994, AJ, 107, 2067
Grabhorn R.P., Cohn H.N., Lugger P.M., Murphy B.W., 1992, ApJ, 392, 86
Griffin R.F., 1967, ApJ, 147, 465
Gunn J.E., Griffin R.F., 1979, AJ, 84, 752
Harding G.A., 1965, Royal Obs. Bull. No. 99, p. E65
Illingworth G., 1976, ApJ, 204, 73
Lupton R., Gunn J.E., Griffin R.F., 1985, in Dynamics of Star Clusters, IAU Symp. No. 113, eds. Goodman J. & Hut P., (Dordrecht: Reidel), p. 327
Lupton R., Gunn J.E., 1987, AJ, 93, 1106
Lupton R., Gunn J.E., Griffin R.F., 1987, AJ, 93, 1114
Mayor M. et al., 1984, A&A, 134, 118
Mayor M., Duquennoy A., Udry S., Andersen J., Nordström B., 1996, in The Origins, Evolution, and Destinies of Binary Stars in Clusters, ASP Conf. Series, Vol. ??, eds. Milone E.F. & Mermilliod J.-C., (San Francisco: ASP), in press
Merritt D., Tremblay B., 1994, AJ, 108, 514
Meylan G., 1987a, A&A, 184, 144
Meylan G., 1987b, in Stellar Evolution and Dynamics in the Outer Halo of the Galaxy, ESO Workshop Series, Vol. 27, ed. Azzopardi M., (Garching: ESO), p. 665
Meylan G., 1988, A&A, 191, 215
Meylan G., 1989, A&A, 214, 106
Meylan G., Mayor M., 1986, A&A, 166, 122
Meylan G., Mayor M., 1991, A&A, 250, 113
Meylan G., Mayor M., Duquennoy A., Dubath P., 1995, A&A, in press
Meylan G., Pryor C., 1993, in Structure and Dynamics of Globular Clusters, ASP Conf. Series, Vol. 50, eds. Djorgovski S.G. & Meylan G., (San Francisco: ASP), p. 31
Oort J.H., van Herk G., 1959, Bull. Astron. Inst. Netherlands, Vol. XIV, 299
Peterson R.C., Latham D.W., 1986, ApJ, 305, 645
Peterson R.C., Seitzer P., Cudworth K.M., 1989, ApJ, 347, 251
Pryor C., McClure R.D., Flechter J.M., Hartwick F., Kormendy J., 1986, AJ, 91, 546
Pryor C., McClure R.D., Flechter J.M., Hesser J.E., 1989, AJ, 98, 596
Pryor C., McClure R.D., Flechter J.M., Hesser J.E., 1991, AJ, 102, 1026
Reijns R., Le Poole R., de Zeeuw T., Seitzer P., Freeman K.C., 1993, in Structure and Dynamics of Globular Clusters, ASP Conf. Series, Vol. 50, eds. Djorgovski S.G. & Meylan G., (San Francisco: ASP), p. 79
Schwarzschild K., 1907, Göttingen Nachr. p. 614
Schwarzschild K., 1908, Göttingen Nachr. p. 191
Zaggia S., Capaccioli M., Piotto G., Stiavelli M., 1992, A&A 258, 302
Zaggia S., Capaccioli M., Piotto G., 1993, A&A, 278, 415

# STELLAR LUMINOSITY AND MASS FUNCTIONS OF GLOBULAR CLUSTERS[1]

GIAMPAOLO PIOTTO
*Dipartimento di Astronomia, Università di Padova*
*Vicolo dell'Osservatorio 5, I-35122 Padova, Italy*

AND

ADRIENNE M. COOL AND IVAN R. KING
*Astronomy Department, University of California*
*Berkeley, CA 94720-3411, U.S.A.*

**Abstract.** HST makes it possible for the first time to study nearly the entire mass range of globular-cluster main sequences, from the turnoff down almost to the theoretical limit for hydrogen ignition. We present main-sequence luminosity functions (LFs) for four clusters that include stars with $\mathcal{M} < 0.15 \mathcal{M}_\odot$ in all cases. We compare these and other LFs that have been obtained with HST for a total of five globulars to date. Two of the three clusters in the sample that have similar metallicities have nearly identical LFs, while the third is relatively deficient in low mass stars. Possible implications of this finding are briefly discussed. Inferred mass functions vary significantly depending on the mass–luminosity relations that are adopted.

## 1. Introduction

The *Hubble Space Telescope* (HST) allows photometry and counting of globular-cluster (GC) stars several magnitudes fainter than is possible with ground-based equipment. For the closest clusters, luminosity functions (LFs) and the resulting mass functions (MFs) can be extended down to $\sim 0.10 \mathcal{M}_\odot$, or nearly the hydrogen-burning limit. The MFs of GC stars provide impor-

---

[1] Based on observations with the NASA/ESA *Hubble Space Telescope*, obtained at the Space Telescope Science Institute, which is operated by AURA, Inc., under NASA contract NAS5-26555.

tant observational inputs into a wide variety of astrophysical problems, including (1) the realistic dynamical modeling of individual clusters; (2) the role of dynamical evolution in modifying globular-cluster MFs; (3) the amount of mass contained in very-low-mass (VLM) and brown-dwarf stars in globulars and, by extension, in the halo; (4) globular-cluster formation and initial MFs.

The first extensive comparison of Galactic GC MFs was presented by Scalo (1986), but was still based on star counts on photographic plates. Due to the limitation in mass range and to the big uncertainties in the incompleteness at the low-mass end, Scalo concluded there was no compelling evidence for differences among GC MFs. This result was questioned in the same year by McClure et al. (1986), who presented the first set of LFs based on CCD photometry. From a sample of 7 GCs, they noted a strong dependence of the slope of the MF on the metal content, the steepness increasing with decreasing metallicity. In a subsequent comparison of 17 GCs, Piotto (1991) suggested that dynamical evolution could also be important. A detailed multivariate statistical analysis (Djorgovski, Piotto, & Capaccioli 1993) confirmed both these results, and showed that the MFs depend mainly on a cluster's position in the Galaxy, and only secondarily on metallicity: clusters closer to the Galactic plane and center have flatter MFs. (The secondary dependence on metallicity is in the same direction as originally suggested by McClure et al.) The main limitation of all these ground-based studies is the small mass range ($0.5 < \mathcal{M}/\mathcal{M}_\odot < 0.8$) and the fact that the MFs were approximated by power laws. In the same period, Richer et al. (1991) tried to push the study of MFs to significantly fainter magnitudes in 6 globulars. Their results suggested that the MFs of at least some globulars are so steep at the faint end that VLM stars make a significant contribution to the total cluster mass. Extending this finding to the halo MF, the authors suggested that VLM stars could contribute to the missing mass of the Galactic halo.

Here we present LFs measured with the HST/WFPC2 for four globular clusters: M30 (NGC 7099), M15 (NGC 7078), 47 Tuc (NGC 104), and NGC 6397. We compare our results to existing HST studies of the latter three of these GCs, to the HST study of $\omega$ Cen by Elson, Gilmore, & Santiago (1995), and to earlier ground-based studies. We discuss the main properties of the LFs for the clusters of our sample, and infer the corresponding MFs using the most recently calculated mass–luminosity relations.

## 2. The Data Set

All the images were taken with the WFPC2 in parallel mode, at ~4.6 arcmin from the cluster centers. Filters and total exposure times are given

TABLE 1. Data Set

| Cluster | $(m-M)_I$ | [Fe/H] | Filter | Exp time [sec] | $M_{I,50\%}$ | $\mathcal{M}_{lim}$ [$\mathcal{M}_\odot$] |
|---------|-----------|--------|--------|----------------|--------------|-------------------------------------------|
| NGC 6397 | 12.05 | −1.9 | F555W | 14200 | 12.3 | 0.10 |
|          |       |      | F814W | 18700 |      |      |
| NGC 7078 | 15.26 | −2.2 | F606W | 6050  | 10.7 | 0.11 |
|          |       |      | F814W | 6050  |      |      |
| NGC 7099 | 14.48 | −2.1 | F555W | 15300 | 10.5 | 0.12 |
|          |       |      | F814W | 8700  |      |      |
| 47 Tuc   | 13.35 | −0.7 | F606W | 1600  | 10.0 | 0.13 |
|          |       |      | F814W | 2000  |      |      |

in Table 1, together with metallicities and adopted distance moduli. The reduction procedures are described by Cool & King (1995) and Cool, Piotto, & King (1995). Instrumental magnitudes were transformed into the WFPC2 "ground" system following Holtzman et al. (1995). Here we present only LFs from the $V_{555}$ and $I_{814}$ photometry, which are very similar to Johnson standard $V$ and $I$; the $V_{606}$ filter is significantly different from Johnson $V$.

## 3. The Luminosity Functions

Particular attention has been devoted to determination of the completeness and to corrections for field-star contamination (see Piotto, Cool, & King 1995). The LFs presented here include only points for which the completeness was found to be > 50%. With the exception of 47 Tuc, which has the most crowded images, the completeness drops below 80% only in the last $\simeq 1$ mag of any of the LFs. The last two columns in Table 1 give the absolute $I_{814}$ magnitude at which 50% completeness was reached, and the approximate corresponding mass. All the LFs have been obtained from color–magnitude diagrams (CMD), which allow the direct discrimination of cluster stars from background/foreground objects. Field-star contamination was significant only in the case of NGC 6397, which has the lowest Galactic latitude of the four clusters ($b = -12°$). (Compare the NGC 6397 CMD [Fig. 1 in Cool, Piotto, & King] with that of M30 [Fig. 2 in King], both elsewhere in these proceedings.)

In Figure 1 we present the $I_{814}$ LF for NGC 6397. The star numbers peak at $I_{814} \sim 20.5$ ($V_{555} \sim 22.2$), and then decline by a factor of 3 over the next three magnitudes. Also shown is the HST-based LF of Paresce, De Marchi, & Romaniello (1995). The two HST-based LFs span somewhat

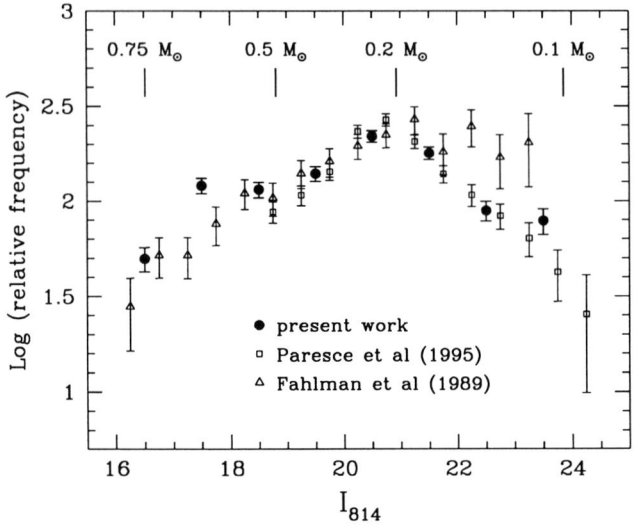

Figure 1. Two HST LFs for NGC 6397, along with the ground-based LF of Fahlman et al. (1989).

different magnitude ranges. Ours extends closer to the turnoff, owing to the inclusion of short exposures in our observing program, and ends with a bin at $I_{814} = 23$–$24$ ($V_{555} \simeq 25.3$–$26.5$), beyond which we consider the field-star corrections to be too uncertain to permit a reliable measurement of the LF (see the color–magnitude diagram in Fig. 1 of Cool, Piotto, & King elsewhere in this volume). Over the common range, the two HST-based LFs are in good agreement. Also shown is the ground-based $I$-band LF of Fahlman et al. (1989), which for $I_{814} \leq 21.5$ is in reasonable agreement with the HST-based LFs. At fainter magnitudes however, the ground-based LF rises significantly above both of the presumably more reliable HST-based LFs. Similar discrepancies between HST and ground-based results at the faint end of the LF were noted by Elson et al. (1995) for $\omega$ Cen.

In Figure 2 we compare the $I_{814}$ and $V_{555}$ HST LF with the ground-based $I$ LF (Drukier et al. 1993), and $V$ LF (Piotto, Ortolani, & Zoccali 1995) obtained at similar distance from the cluster center. The agreement is quite good, suggesting that while ground-based LFs should not be relied on at very faint magnitudes, they can be relied on at brighter magnitudes. This is important, as it means that the ground-based LFs can be usefully applied to extend the HST LFs to bright magnitudes (up to the turnoff), for which the use of HST would both be inefficient and unnecessary. In what

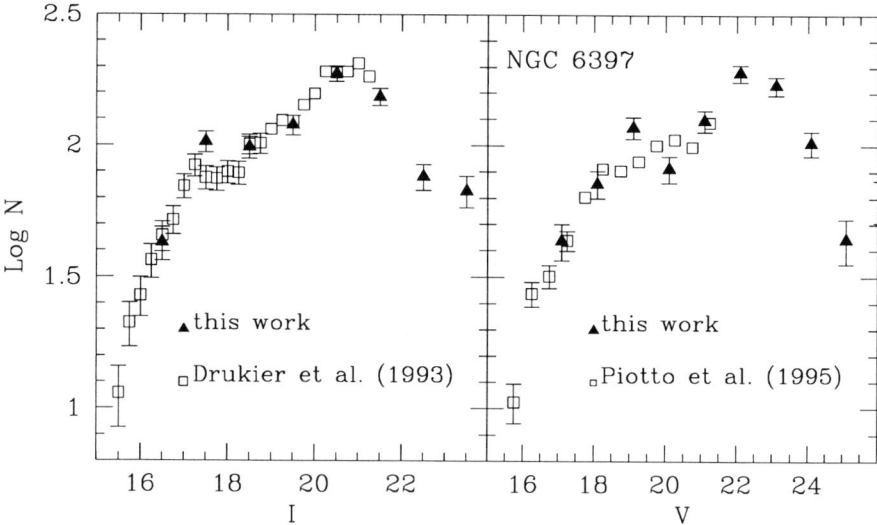

Figure 2. The $I_{814}$ and $V_{555}$ HST LFs for NGC 6397 are compared with the ground-based LFs by Drukier et al (1993) and Piotto et al (1995), obtained at similar distances from the cluster center. Where omitted, the error bars are smaller than the symbol size.

follows, the NGC 6397 HST LF, which is truncated at $M_{555} = 4.7$ ($M_{814} = 4$) due to the saturation of bright stars even in the short exposures, has been extended to the turnoff using the ground-based LFs shown in Fig. 2.

The $I_{814}$ and $V_{555}$ LFs for the three metal-poor clusters (NGC 6397, M15, and M30) are compared in the left and right panels of Figure 3. (M15 does not appear in the right panel, as the V data were taken with a different filter.) Vertical shifts were made to bring the LFs into alignment according to a least-squares algorithm, in the magnitude intervals $4.0 < M_{814} < 7.0$ and $4 < M_{555} < 7.0$. As shown by King elsewhere in this volume, the measured (local) LF of NGC 6397 closely resembles the cluster's global LF. As M30 and M15 are structurally very similar to NGC 6397, and the observations were similarly taken well outside of their cores, their observed LFs should also closely resemble the global LFs. What then makes these three clusters a particularly advantageous comparison is that their metal content is very similar (see Table 1, Col. 3), which implies that the mass-luminosity relations (MLR) for their stellar populations will also be very similar. Similarities or differences in their MFs can thus be discerned in a direct comparison of their observed LFs.

The broad features of the three $I_{814}$ LFs are similar. Each rises steadily to a peak near $M_{814} \simeq 8.5$–$9.0$, then bends over and drops significantly to the limit of the observations. But when the LFs are overlaid and lined up in the range $M_{814} = 4.0$–$7.0$ (left panel), it becomes clear that while the LFs

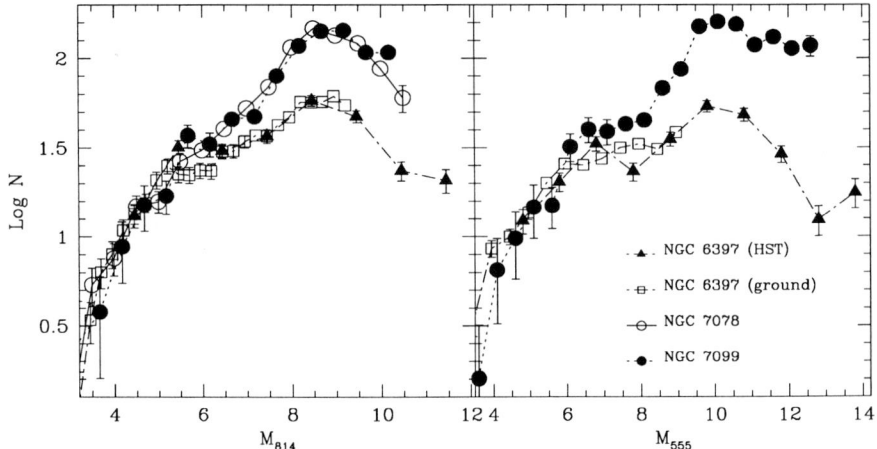

Figure 3. $I_{814}$ (left panel) and $V_{555}$ (right panel) LFs for the three metal-poor clusters, converted to absolute magnitudes. The LF for NGC 6397 has been extended to bright magnitudes (up to the TO), using ground-based LFs (cf. Fig. 2). Where omitted, the error bars are equal to or smaller than the symbol size.

of M30 and M15 are indeed very similar, NGC 6397 is markedly deficient in faint stars. A similar conclusion is reached by comparing the $V_{555}$ LF of NGC 6397 with that of M30 (right panel). The differences are statistically highly significant, being many times the uncertainties in the single points of the LFs.

Interestingly, the difference between NGC 6397 and the near-twin LFs of M30 and M15 becomes apparent only when the LFs are compared over a wide range of magnitudes. De Marchi and Paresce (1995a) analyzed the same M15 data set that we have analyzed, and obtained an LF that is quite similar to ours. However, when they compared their M15 LF to that of Paresce et al. (1995) for NGC 6397, the two appeared indistinguishable. The resolution of this apparent contradiction between our respective results is that their comparison was limited to magnitudes in the interval $M_{814} \simeq$ 6.5–10. The brighter magnitudes included in our NGC 6397 measurements allow a more stringent comparison, from $M_{814} = 4.5$–10.5. The relative paucity of faint stars in NGC 6397 then becomes unmistakable, and is further confirmed by the extension of the NGC 6397 LFs to even brighter magnitudes using ground-based data.

The similarity of the LFs of M15 and M30 over a span of more than 6 magnitudes is striking. It implies that their MFs are very similar in the range of approximately $0.12 < \mathcal{M}/\mathcal{M}_\odot < 0.8$. It would appear that either these two clusters were born with similar MFs and have changed little since, or that they were born similar and have changed similarly. A scenario in which they were born different and have evolved to become the same seems too contrived. In view of this similarity, it is natural to ask why NGC 6397, which has a similar metallicity and similar morphology (all three are post-core-collapse clusters), should have so many fewer low-mass stars. As further discussed by King elsewhere in these proceedings, the deficiency may be the result of tidal shocks, to which NGC 6397 is quite vulnerable, given its low-inclination orbit and small perigalacticon. Even without this dynamical explanation, the difference between these clusters is in qualitative agreement with the correlations between MF slopes and Galactic position identified by Djorgovski et al. (1993).

In Figure 4 we show the $I_{814}$ LF for 47 Tuc. The LFs of the metal-poor clusters from Fig. 3 are shown as lines, and the triangles represent the LF obtained by Elson et al. (1995) for $\omega$ Cen. The LFs have been aligned using a least-squares algorithm in the magnitude interval $5.0 < M_{814} < 7.0$. The 47 Tuc LF peaks at $M_{814} \sim 9$, at somewhat fainter magnitudes than the three metal-poor clusters, and then turns over. It agrees well with the LF obtained by De Marchi & Paresce (1995b) for a different field in 47 Tuc at a similar distance from the cluster center. Omega Cen is the only cluster of the five for which no drop-off has been detected to the limit of the existing HST observations.

Since $\omega$ Cen and 47 Tuc are intermediate and metal-rich clusters, respectively, the MLRs appropriate for their stellar populations will be different from each other and from the metal-poor clusters. In addition, a non-trivial correction of the observed LFs may be required to convert to global LFs for these clusters. For both these reasons, it is inadvisable to draw any conclusions from the LF comparison in Fig. 4. The dynamical modeling required to convert the local LFs to global ones is under way, and will be presented elsewhere.

## 4. The Mass Functions

If the results of star counts are to be used in modeling globular clusters, and if clusters with differing metallicities are to be compared, we must transform the observed LFs to MFs. This transformation requires an appropriate MLR, and it is particularly sensitive to the latter, as the transformation depends on its derivative. Unfortunately, the MLRs for low-mass ($\mathcal{M} < 0.5 \mathcal{M}_\odot$) stars are quite uncertain: there are still serious difficulties in both the calculation of the stellar structure (critically dependent on the difficult evaluation of both the opacity and the equation of state for a cool gas) and in the treatment of their atmospheres (Alexander et al. 1995).

Several MLRs for low-mass stars now exist in the literature, some of which have appeared quite recently. Consensus has yet to be reached, ow-

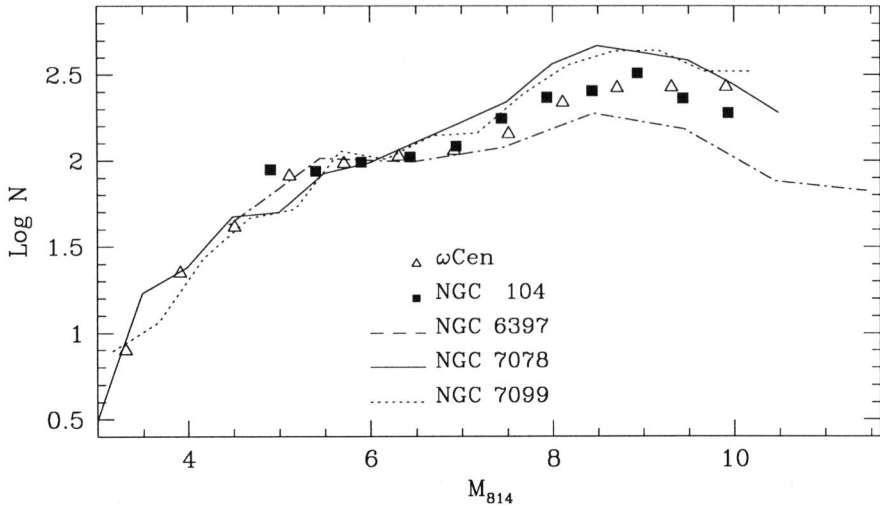

*Figure 4.* $M_{814}$ LFs for the 5 clusters in the sample discussed in this paper.

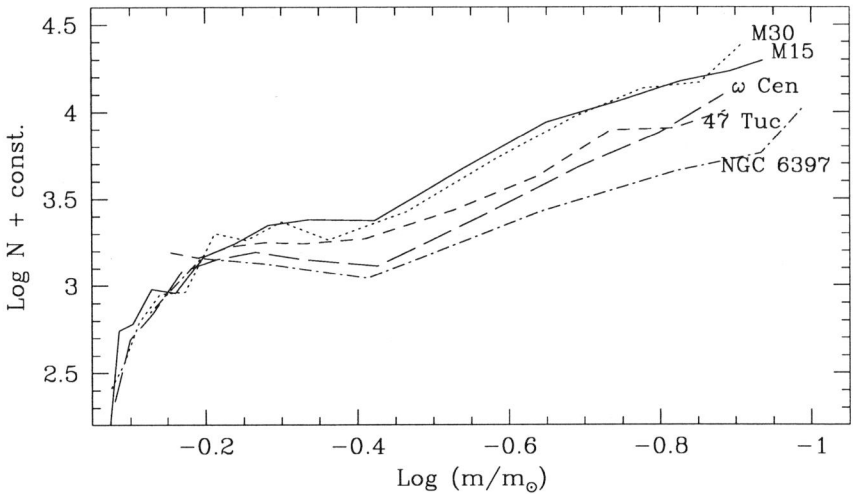

*Figure 5.* MFs derived from the LFs in Fig. 4, using the mass-luminosity relations of D'Antona & Mazzitelli (1995), which yield the steepest slopes at very low masses.

ing in part to a paucity of observational constraints. These MLRs differ from one another in ways that potentially can have a significant impact on the MFs derived from them. In particular, differences in the MLR slopes for very-low-mass stars are sufficient to produce significant differences in the slopes of the resulting MFs. In an effort to determine how large these variations can be, we have experimented with several of the available MLRs:

Baraffe et al. (1995), Bergbusch & VandenBerg (1992), D'Antona & Mazzitelli (1995), and Fahlman et al. (1989), as well as a set of models specifically calculated for the metallicity of these clusters by the Teramo group (Alexander et al. 1995).

In Figure 5 we show the MFs extracted from the $I_{814}$ LFs of Fig. 4, using the MLRs of D'Antona & Mazzitelli (1995). Similar results were obtained by converting the $V_{555}$ LF to mass. This set of MFs represents one extreme, since these MLRs have the steepest (and most rapidly steepening) slope of any of the MLRs we tried, and thus produce the steepest MFs. All the resulting MFs rise steadily at the low-mass end. In the interval $0.1 < \mathcal{M}/\mathcal{M}_\odot < 0.4$ the MFs can reasonably well be represented by power laws with slopes in the range $x = 0.6$–$1.0$ (where for the Salpeter law $x = 1.35$). The best fits to the MFs for NGC 6397 ($x = 0.6$) and $\omega$ Cen ($x = 0.9$) are both shallower than the MFs slopes found for these clusters in the ground-based study of Richer et al. (1991). None of the MFs obtained using D'Antona and Mazzitelli's (1995) MLR has a slope exceeding $x = 1.0$, though M30 and M15 approach this limiting case, which would correspond to a marginally divergent cluster mass *if* the MF slopes can be extrapolated into the brown dwarf regime.

At the other extreme are the MFs we obtain by using the MLRs of Alexander et al. (1995), which have a shallower slope for very-low-mass stars. With these MLRs, the MFs have slopes in the range $x = -0.1$ (NGC 6397) to $x = 0.7$ (M30), though the MFs deviate from power laws more than in the previous case. There is a tendency for the MFs to flatten out beyond $\log(\mathcal{M}/\mathcal{M}_\odot) = -0.8$ when these MLRs are used. In no case do we find evidence for a decline in the MF following the flattening.

## 5. Conclusions

With HST, main-sequence LFs can be readily measured from the turnoff down to less than $0.2\,\mathcal{M}_\odot$ in a large number of clusters, allowing for much more detailed comparisons of cluster LFs and MFs than have previously been possible. In the closest clusters, observations have already reached nearly to the hydrogen-burning limit, and future observations should reach even farther. While a sample of five clusters is too small to draw any broad conclusions, the initial results of comparisons between clusters with similar metallicities are intriguing, and provide support for the view that clusters may be born with similar MFs, which can then be altered by interactions with the Galaxy. As the sample of clusters observed with HST increases, much more will be learned about the various factors influencing cluster MFs.

The new capacity afforded by HST in the study of globular clusters will provide important new constraints on increasingly detailed dynamical modeling of globulars, which will interact with computations such as those we have admired at this meeting. At the same time, it is clearer than ever that the information contained in the observational data will be fully exploited only when the inner and outer workings of very-low-mass stars are more thoroughly understood.

## 6. Acknowledgments

This work was supported by NASA grant NAG5-1607. We thank F. D'Antona, E. Brocato, V. Castellani, and S. Cassisi for providing isochrones in advance of publication, and P. Stetson for his generosity with software.

## References

Alexander, D.R., Brocato, E., Cassisi, S., Castellani, V., Ciacio, F., & Degl'Innocenti, S. 1995, A&A, submitted
Baraffe, I., Chabrier, G., Allard, F., & Hauschildt, P.H. 1995, ApJ, 446, L35
Bergbusch, P. A., & VandenBerg, D. A. 1992, ApJS, 81, 163
Cool, A. C., & King, I. R. 1995, in Calibrating HST: Post Servicing Mission, ed. A. Koratkar & C. Leitherer (Baltimore: STScI), p. 290
Cool, A. M., Piotto, G., & King, I. R. 1995, submitted to ApJ
D'Antona, F., & Mazzitelli, I. 1995, ApJ, Dec. 1, in press
De Marchi, G., & Paresce, F., 1995a, A&A, in press
De Marchi, G., & Paresce, F., 1995b, A&A, in press
Drukier, G. A., Fahlman, G. G., Richer, H. B., Searle, L., & Thompson, I. 1993, AJ, 106, 2335
Djorgovski, S., Piotto, G., & Capaccioli, M. 1993, AJ, 105, 2148
Elson, R. A. W., Gilmore, G. F., & Santiago B. X. 1995, AJ, 110, 682
Fahlman, G. G., Richer, H. B., Searle, L., & Thompson, I. B. 1989, ApJ, 343, L49
Holtzman, J. A., Burrows, C. J., Casertano, S., Hester, J. J., Watson, A. M., & Worthey, G. S. 1995, preprint
McClure, R. D., VandenBerg, D. A., Smith, G. H., Fahlman, G. G., Richer H. B., Hesser, J. E., Harris, W. E., Stetson, P. B., & Bell, R. A. 1986, ApJL, 307, L49
Paresce, F., De Marchi, G., & Romaniello, M. 1995, ApJ, 440, 216
Piotto, G. 1991, in Formation and Evolution of Star Clusters, ed. K. Janes, (San Francisco: ASP), p. 200
Piotto, G., Cool, A.M., & King, I.R. 1995, preprint
Piotto, G., Ortolani, S., & Zoccali, M. 1995, preprint
Richer, H. B., Fahlman, G. G., Buonanno, R., Fusi Pecci, F., Searle, L., & Thompson, I. B. 1991, ApJ, 381, 147
Scalo J.M. 1986, *Fund. of Cosmic Phys.*, 11, 1

# THE NUCLEAR STAR CLUSTER OF THE MILKY WAY: STAR FORMATION, STELLAR COLLISIONS AND CENTRAL DARK MASS

R. GENZEL

*Max-Planck Institut fuer extraterrestrische Physik Garching, FRG*

**Abstract.**

High resolution near-infrared imaging and spectroscopy now gives detailed information about the structure, evolution and mass distribution in the nuclear star cluster of the Milky Way. The central parsec is powered by a cluster of luminous and helium rich, blue supergiants/Wolf-Rayet stars. The most likely scenario for the formation of the massive stars is a star formation burst a few million years ago at which time a dense gas cloud may have fallen into the center. The stellar density in the $\sim 0.3$ pc radius central core is high enough that collisions with main sequence stars destroy the largest late type giant stars. Radial velocity measurements for about 300 early and late type stars between 0.1 and 5pc radius from the dynamic center now strongly favor the existence of a central dark mass of $2.5 - 3.3 \times 10^6 M_\odot$ (density $(10^9 \ M_\odot \mathrm{pc}^{-3}, M/L_{2\mu m}) \sim 100 M_\odot/L_\odot)$ within 0.1pc of the dynamic center. This central dark mass cannot be a cluster of neutron stars. It is either a compact cluster of stellar black holes or, most likely, a single massive black hole.

## 1. INTRODUCING THE PHENOMENA

The nucleus of the Milky Way (adopted distance 8.5kpc) is one hundred times closer to the Earth than the nearest large external galaxy and more than a thousand times closer than the nearest active galactic nuclei. We can therefore study physical processes happening in our own Galactic Center at a level of detail that will never be reached in the more distant, but usually also more spectacular systems. What powers these nuclei and how do they evolve? What are the properties of the nuclear stellar clusters? Is star formation happening there? Do dormant massive black holes reside in their cores? In the present pap er I will describe the status of our present knowledge about these key questions. For a more extensive discussion I refer to Genzel,Hollenbach and Townes (1994).

The nuclear mass is dominated by stars, except probably in the innermost parsec. The density of stars increases with decreasing radius $R$ from the dynamic center approximately as $1/R^2$ and attains a value exceeding $10^6 M_\odot/pc^3$ in the central parsec. Infrared observations on a scale of 100 pc to 1 kpc show that these stars appear to be distributed in a rotating bar (Blitz and Spergel 1991). The gravitational torque of this bar may also explain the non-circular motions of interstellar gas clouds found by radio spectroscopy (Binney et al. 1991). The non-circular motions in turn may trigger gas infall into the nucleus. There is increasing evidence from gamma-ray spectroscopy of the 1.8 MeV 26Al line (Diehl et al. 1993) and from infrared stellar spectrophotometry (e.g. Cotera et al. 1994, Lebofsky and Rieke 1987) that (massive) star formation has occured throughout the Galactic Center region no longer than 10 million years ago.

Also on a scale of $\sim$ 100 pc several variable, spectacular hard X-ray and gamma-ray sources have been found (e.g. Skinner 1993). They may represent stellar black holes or neutron stars accreting gas from a companion or from nearby dense gas clouds. Throughout the central few hundred parsecs giant molecular clouds ($\sim 10^6 M_\odot$) are found whose gas density ($n_{H2} \sim 10^4$ to $10^6 cm^{-3}$ ) and temperature (40 to 200 K) are significantly greater than those of the clouds in the Galactic disk (e.g.Guesten 1989). The dynamics of this central molecular cloud layer is characterized by large internal random motions and unusual streaming velocities that can be partially explained by the presence of the central bar potential mentioned above.

Magnetic fields as large as $\sim$ 1 mGauss appear to permeate the central 50pc and are aligned approximately perpendicular to the Galactic plane (e.g.Morris 1993, Sofue 1994). Where they interact with neutral gas clouds remarkable filaments of nonthermal radio synchrotron emission are seen. The central radio rource, SgrA, can be separated into a thermal source, SgrA West and a non- thermal source, SgrA East. SgrA East may be evidence for one or several supernovae that have exploded in the central 10 parsecs within the last $10^5$ years (Mezger et al.1989).

While the velocities of gas and stars are approximately constant outside of a few parsec, the velocities are observed to increase within the inner core (e.g. Genzel and Townes 1987). The first evidence for this increase in gas velocities came from mid-infrared spectroscopy of [NeII] by Wollman (1976) and Lacy et al.(1979, 1980). These authors and others following interpreted the > 250 km/s gas velocities as signalling a concentration of non-stellar mass in the Galactic Center, possibly caused by a few million solar mass black hole at the dynamic center (Lacy et al. 1982, Serabyn and Lacy 1985). However, gas is affected by magnetic, frictional and wind forces, in addition to gravity so that stellar velocities are required to unambiguously determine the mass distribution.

At the dynamic center is a compact radio source, SgrA*, which is close to, but not coincident with a group of bright near-infrared sources (IRS16) of blue color (e.g. Backer 1994). Since its discovery 25 years ago SgrA* has been the most probable candidate for the central black hole. Surrounding the innermost ionized streamers one finds a system of dense orbiting molecular filaments approximately arranged in form of a circum-nuclear 'disk' (Genzel et al. 1985, Jackson et al. 1993). The circum-nuclear disk is probably fed by gas infall from dense molecular clouds at 10pc.

## 2. WHAT POWERS THE CENTRAL PARSEC ?

The observed broad band emission of the central few parsecs is dominated by intense mid- and far-infrared emission from 50 to 100 K dust grains originating in the circum-nuclear disk and in a cloud ridge associated with the ionized 'mini- spiral'. Taking into account the ($\sim$ 50%) fraction of the nuclear radiation not intercepted by the circum-nuclear gas the total UV and visible luminosity of the central parsec has been estimated to lie between 1 and $3 \times 10^7 L_\odot$ (Davidson et al.1992). The Lyman continuum flux is about $2 - 3 \times 10^{50} s^{-1}$ as determined from the thermal radio continuum (Lacy et al. 1980). Infrared spectroscopy of fine structure lines sampling a wide range of excitation stages implies that the effective temperature of the UV radiation field in the central parsec is only about 30,000 to 35,000 K. The line ratios also suggest a heavy element abundance of about twice that in the Sun (Lacy et al. 1980).

What powers this low excitation HII region and what are the properties of the central star cluster? Through the advent of sensitive, large format infrared detector arrays and speckle imaging it has become possible in the last few years to image the central parsec at the resolution to the diffraction limit of 4m class telescopes ($\sim$ 0.1" or 0.04 pc at 2 $\mu$m, Eckart et al. 1992, 1993, 1995). The best current images resolve the near-infrared emission of the central parsec into about 700 stars with K-band (2.2 $\mu$m) magnitudes < 16. Thus all red and most blue supergiants, all red giants/AGB stars later than K5 and all main sequence stars earlier than B0.5 should be detected on those images.The central IRS16 complex located within 1" of the compact radio source SgrA* consists of about two dozen single (or perhaps multiple) stars (see also Simons et al. 1990, Simon et al. 1990). From the number distribution of the near-infrared sources it appears that the centroid of the stellar cluster is more likely on SgrA* than on the IRS16 complex and that the core radius of the K$_i$15 stellar number density distribution is about 0.2 to 0.4 pc (Eckart et al. 1993, 1995, Genzel et al. 1996). If the stars with K$_i$15 are representative of the overall mass distribution of the cluster (an assumption that appears very plausible based on recent spectroscopic identification of the stars, Genzel et al. 1996) this core radius together with the mass of stars estimated to lie within a few parsecs indicates that the stellar density in the core about $4 - 8 \times 10^6 M_\odot \text{pc}^{-3}$. Recent imaging spectroscopy with the new MPE 3D-spectrometer shows that within the core radius bright late type stars (supergiants and the brightest AGB stars) are absent but that the core is surrounded by a ring of red supergiants/AGB stars (Genzel et al. 1996). Following earlier discussions of Lacy et al.(1982), Phinney (1989) and Sellgren et al.(1990), Genzel et al.(1996) interpret this finding in terms of destruction of the brightest (and hence largest) late type giant stars by collisions with main sequence stars. Assuming that such collisions in fact permanently destroy the outer atmosphere of the giants (see Davies et al. 1991) and that collisional destruction becomes observable whenever the collision time is less or equal than the lifetime of the red giant/supergiant phase, the observed lack of stars brighter than K∼ 10 in fact implies a core stellar density of about 5e6 $M_\odot \text{pc}^{-3}$, in excellent agreement with the density estimated from the number counts.

Another important ingredient of the near-infrared story has been the discovery of a HeI/Brgamma near-infrared emission line star (the AF-star, Forrest et al. 1987, Allen et al. 1990), followed by the discovery of an entire cluster of about 25 such stars in the central parsec and centered on the IRS16/IRS13 complex (Krabbe et al. 1991, 1995). Several of the brightest members of the IRS16 complex are HeI-stars, as is the nearby bright source IRS 13 (Eckart et al. 1995, Krabbe et al. 1995, Libonate et al. 1995, Blum et al. 1995b, Tamblyn et al. 1996). The IRS16 HeI "broad line region" discovered a decade ago by Hall et al. (1982) and Geballe et al. (1984) thus is now identified as a group of luminous mass losing, He-rich stars. Non- local thermodynamic equilibrium (NLTE) stellar atmosphere modeling of the observed emission characteristics of the AF-star (Najarro et al. 1994) confirms and quantifies earlier proposals (Allen et al. 1990, Krabbe et al. 1991) that the AF-star is a WN9/Ofpe star. WN9/Ofpe stars are a rare class of luminous blue supergiants related to luminous blue variables (LBVs), WNL Wolf-Rayet stars and Of/ON supergiants (Allen et al. 1990, Krabbe et al. 1991, Najarro et al. 1994, Libonate et al. 1994, Blum et al. 1995b, Tamblyn et al. 1996). These stars very likely represent the post-main sequence phase of massive stars (20 to 120 $M_\odot$) before they explode as supernovae. The AF-star has a luminosity of about $3 \times 10^5 L_\odot$, effective temperature near

20,000 K and main-sequence mass between 25 and 40 $M_\odot$ (Najarro et al. 1994). The surface He/H abundance ratio is near unity and the mass loss rate is $6 \times 10^{-5} M_\odot \text{yr}^{-1}$ at a velocity of 700 km/s (Najarro et al. 1994). Based on the most recent 3D spectroscopy and modelling the brightest HeI stars (IRS16NE,C,SW, IRS 13) also have effective temperatures between 20,000 and 30,000 K, are helium-rich and are about 5 to 10 times more luminous than the AF star (Krabbe et al. 1995). Their progenitor O stars likely had masses near $100 M_\odot$. In addition several stars display CIII/CIV/NIII emission lines, characteristic for late WC and WN Wolf-Rayet stars (Blum et al. 1995, Krabbe et al. 1995, Genzel et al. 1996. Combining the contributions from all its members, the HeI-star cluster can plausibly account for essentially all of the bolometric and Lyman-continuum luminosities of the central parsec (Krabbe et al. 1995). The HeI-star cluster also provides in excess of $10^{38}$ erg/s in mechanical wind luminosity which may have a significant impact on the gas dynamics in the central parsec (Genzel, Hollenbach and Townes 1994). Krabbe et al.(1995) successfully fitted the properties of the massive early type stars in the central parsec by a model of a star formation burst between 9 and $3 \times 10^6$ years ago in which a few hundred OB stars and perhaps a total of a few thousand stars were formed). This conclusion is in excellent agreement with earlier proposals by Lacy, Townes and Hollenbach (1982), Rieke and Lebofsky (1982) and Allen and Sanders (1986). In the model of Krabbe et al. the HeI stars are the most massive cluster members that in the mean time have evolved off the main sequence and the central parsec is now in the late, wind-dominated phase of the burst. The starburst model accounts naturally for the low excitation of the SgrA (West) HII region. Although there is also evidence for some very young, embedded OB stars the present star formation activity appears to be significantly less than during the peak of the burst. The present gas density in the central parsec is too low for gravitational collapse of gas clouds to stars in the presence of the strong tidal forces (Morris 1993). Perhaps the burst was triggered by infall of a dense gas cloud less than 10 million years ago, a scenario that is supported by an overall counter-rotation (in the sense of Galactic rotation) of the HeI star cluster (Genzel et al. 1996).

Based on earlier theoretical work by Lee (1987), Eckart et al.(1993) have proposed sequential merging by collisions as an alternative to the starburst scenario. A recent Fokker-Planck calculation of an evolving Galactic Center type, dense cluster with merging shows, however, that merging can account for only $\sim 10 - 20 M_\odot$ stars and no $> 30 M_\odot$ stars (Lee 1994). The basic reason is that in the calculations a sufficiently dense stellar core (density $10^7 M_\odot \text{pc}^{-3}$ or greater) cannot be maintained for a long enough time to build up very many massive stars. Morris (1993) has suggested that the HeI stars are not classical blue supergiants at all but transitory objects that have been created in collisions between ($\sim 10 M_\odot$) stellar black holes and solar mass, red giants. Both accounts of the HeI stars just cited are very specific to the high density environment of the central parsec. However, a number of stars similar to the SgrA HeI stars have now been found in several clusters 2 to 13' away from the central, high density SgrA region (Okuda et al. 1990, Moneti et al. 1991, Cotera et al. 1994, Harris et al. 1994, Figer 1995). In the case of the Morris scenario (1993) one probably would also expect a much larger X-ray emission than is observed. These facts and the requirement of having to account for $\sim 100\ M_\odot$ stars and the presence of heavy element nucleosynthesis products discussed above in my opinion now strongly favors the star formation model over the other scenarios.

There are less than a dozen red supergiants ($L > 10^4 L_\odot$) in the central starburst zone. Even after correction for collisional desctruction of late type stars mentioned above this suggests that there was relatively little star formation prior to 10-15 million years ago. In comparison there are a much great number of late type stars with luminosities $10^3$ to $10^4 L_\odot$ , both inside and outside (Haller and Rieke 1989) the central parsec. These medium luminosity stars are likely asymptotic giant branch stars of moderate mass (2 to 7 $M_\odot$). They may signify another starburst episode that happened $\sim 10^8$ years ago (Haller and Rieke 1989, Krabbe et al. 1995).

## 3. Is SgrA* a Massive Black Hole?

The next key issue that I want to discuss is the evidence for a central massive black hole. Ever since the original discovery of the nonthermal compact radio source SgrA* at the core of the nuclear star cluster (Ekers and Lynden-Bell 1971, Downes and Martin 1971, Balick and Brown 1974) that source has been the primary black hole candidate, in analogy to compact nuclear radio sources in other nearby normal galaxies (Lynden-Bell and Rees 1971). In fact ever more detailed radio observations have confirmed the unique nature of SgrA* in the Galaxy. Recent very long baseline interferometry (VLBI) observations at 7mm show its size to be less than a few AU (Backer 1994, Krichbaum et al. 1994). Its proper motion relative to a background quasar is now known to be less than about 38 km/s (Backer 1994), at least 6 times smaller than the (2d-) velocity dispersion of the stars. Hence SgrA* must have a mass in excess of about $150 M_\odot$. The source shows a mm/submm excess above the flat cm- spectral energy distribution (Zylka et al. 1992) probably indicative of the presence of a very compact ($\sim 10^{12}$ cm) radio core of stellar dimensions.

Yet observations at shorter wavelengths indicate nothing particularly impressive toward the radio position SgrA*. The high resolution maps of

Eckart et al. (1995) for the first time show that SgrA* is located near the centroid of a T-shaped concentration of $\sim 10$ compact near-infrared sources (=3DSgrA*(IR)). These sources are likely stars and one might speculate whether they represent a central stellar cusp around SgrA*. SgrA*(IR) also does not show intrinsic variability on scales of minutes or years, or significant line emission (Eckart et al. 1995). Depending on spectral type any possible infrared counterpart of SgrA* has a luminosity between a few $10^2$ and $10^4 L_\odot$. (Variable) hard X- ray emission is commonly considered a key signature of black holes. However, in contrast to the fairly bright infrared emission the present 1 to 30 keV X-ray luminosity of SgrA (West) and SgrA* is less than a a few hundred $L_\odot$ (Skinner 1993, Goldwurm et al. 1994). Recent observations with ASCA suggests that SgrA*'s X-ray luminosity may have been larger in the past few hundred years (a few $10^5 L_\odot$, Koyama et al. 1996) but still orders of magnitude smaller than the Eddington rate of a million solar mass black hole (Sunyaev et al. 1993).

The evidence for a (dark) central mass concentration in the Galactic Center thus is based entirely on the gas and stellar dynamics. As mentioned in the Introduction, evidence for a central mass concentration based on gas dynamics had already been growing in the 1980s but had not been considered compelling by most researchers in the field. However, ever better stellar velocities have become available during the past 8 years, fully vindicating the earlier measurements of gas velocities and substantially strengthening the evidence for a compact central dark mass in the Galactic center (Rieke and Rieke 1988, McGinn et al. 1989, Sellgren et al. 1990, Lindqvist et al. 1992, Krabbe et al. 1995, Haller et al. 1996, Genzel et al. 1996).

The most recent determinations by Sellgren et al. (1990), Krabbe et al.(1995), Haller et al.(1996) and Genzel et al.(1996) now are all in good agreement and show a highly significant increase of stellar radial velocity dispersion from about 55 km/s at 5 pc to about 180 km/s at 0.15 pc. From $\sim 1$"resolution 3D pectroscopy Genzel et al.(1996) have obtained velocities for 222 early and late type stars between 1" and 22" distance from SgrA*. After deprojection of the observed projected velocity dispersions and stellar surface densities Genzel et al.(1996) carried out a Jeans equation analysis. Assuming an isotropic stellar velocity field the new 3D data in combination with the other stellar measurements mentioned above require a combination of a M/L(2micron)$\sim 2 stellar$ cluster, and in addition a $2.5 - 3.3 \times 10^6 M_\odot$ dark mass. The dark compact mass is required at 6 to 8 sigma significance. It can be reduced but not fully removed even if highly anisotropic velocity fields are considered. For comparison, Haller et al.(1996) conclude that there must be a central mass of just under $2 \times 10^6 M_\odot$ and the most recent gas dynamics estimates find a central mass between 2 and $4 \times 10^6 M_\odot$ (Serabyn et al. 1988, Lacy et al. 1991, Herbst et al. 1993). The dark mass is not resolved ($R_{core} < 0.07$pc), has a M/L(2micron) ratio of at least 100 and a density of $> 10^9 M_\odot \text{pc}^{-3}$ (Genzel et al. 1996).

An experiment is now well underway to measure the proper motions of stars between 0.3" and 10" from SgrA* from repeated high resolution near-infrared imaging with the SHARP camera on the ESO NTT (Eckart and Genzel in prep). This experiment should give a clearcut answer on the anisotropy of the stellar orbits within a few years time.

As the dark mass has a core radius at least 5 times smaller and a core density at least 250 times greater than that of the visible (old) stellar cluster (average stellar mass $\sim 0.7 M_\odot$) it does not seem plausible that it consists of solar mass remnants (neutron stars or white dwarfs). Calculations of Chernoff and Weinberg (1990) indicate that such large density ratios between similar mass components cannot be attained even in core collapsed globular clusters. The dark mass concentration could either be a single massive black hole, or a very compact cluster of stellar mass ($\sim 10 M_\odot$) black holes (Morris 1993, Lee 1995) should such a cluster be stable. The most likely configuration is probably a single massive black hole.

If SgrA* is indeed a million solar mass black hole, the riddle is why it is presently so inactive. It is very interesting that the Galactic Center shares this 'luminosity deficiency' or 'blackness' problem with essentially all nearby nuclei for which there is substantial evidence for dark central masses (Kormendy and Richstone 1995), including the presently most convincing case, the 'mega' H2O maser source NGC4258 (Myoshi et al. 1995). It is possible that the tidal disruption and accretion of stars by the hole (happening in the Galactic center at a rate of $\sim 10^{-4} \text{yr}^{-1}$) occurs very efficiently albeit at low duty cycle (Rees 1988). Accretion of interstellar gas streamers by the hole may be prevented by the need to overcome the angular momentum problem, coupled with the outward force of the stellar winds as discussed above. Finally, the wind gas itself may be accreted largely spherically, with very low radiation efficiency (Melia 1992). A final and very interesting possibility is that most of the energy of the accreting material is advected into the hole and not radiated (Narayan et al. 1995). Nevertheless current models of black hole accretion have to be stretched to be comensurate with SgrA* being an underfed million solar mass black hole (Ozernoy and Genzel 1996).

## Acknowledgements

I thank the organizers of the conference for a very stimulating meeting and P.Hut and J.Makino for their patience for a very late proceedings contribution.

# References

Allen, D.A., in Genzel and Harris(1994), 293 (1994)
Allen, D.A., Hyland, A.R. and Hillier, D.J. , MNRAS 244, 706 (1990)
Allen, D.A. and Sanders, R.H., NATURE, 319, 191 (1986)
Backer, D., in Genzel and Harris(1994), 403 (1994)
Balick, B. and Brown, R.L., Ap.J. 194, 265 (1974)
Binney, J.J., Gerhard, O.E., Stark, A.A., Bally, J. and Uchida, K.A.,MNRAS 252, 210 (1991)
Blitz, L. and Spergel, D.N., Ap.J. 379, 631 (1991)
Blum, R.D., Sellgren, K. and dePoy, D.L.. ,Ap.J.440, L17(1995)
Blum, R.D., dePoy, D.L. and Sellgren, K., Ap.J.441, 603 (1995b)
Chernoff, D.F. and Weinberg, M.D., Ap.J. 351, 121 (1990)
Cotera, A.S., Erickson, E.F., Allen, D.A., Colgan, S.W.J., Simpson, J.P.and Burton, M.G., in Genzel and Harris(1994), 217 (1994)
Davidson, J.A., Werner, M.W., Wu, X., Lester, D.F., Harvey, P.M., Joy, M. and Morris, M., Ap.J. 387, 189 (1992)
Davies, M.B., Benz, W. and Hills, J.G., Ap.J. 381, 449 (1991)
Diehl, R. et al., Astr.Ap.(Suppl.) 97, 181 (1993)
Downes, D. and Martin, A., NATURE 233, 112 (1971)
Eckart, A., Genzel, R., Krabbe, A., Hofmann,R. van der Werf, P.P. and Drapatz, S. , NATURE 355, 526(1992)
Eckart, A., Genzel, R., Hofmann, R., Sams, B.J. and Tacconi-Garman, L.E. , Ap.J. 407, L77 (1993)
Eckart, A., Genzel, R., Hofmann, R., Sams, B. and Tacconi-Garman, L.E., Ap.J. 445, L26 (1995)
Ekers, R.D. and Lynden-Bell, D., Ap.Lett. 9, 189 (1971)
Figer, D. , PhD. Thesis, University of California, Los Angeles (1995)
Forrest, W.J., Shure, M.A., Pipher, J.L. and Woodward, C.A.in "The Galactic Center", ed.D.Backer, AIP Conf. Proc 155, 153 (1987)
Geballe, T.R. et al. , Ap.J. 284, 118 (1984)
Genzel, R., Watson, D.M., Crawford, M.K. and Townes, C.H.,Ap.J.297, 766 (1985)
Genzel, R. and Townes, C.H. , Ann.Rev.Astr.Ap. 25, 377(1987)
Genzel, R., Hollenbach, D. and Townes, C.H., Rep.Progr.Phys. 57, 417 (1994)
Genzel, R. and Harris, A.I., Nuclei of Normal Galaxies: Lessons from the Galactic Center, (Dordrecht:Kluwer) (1994)
Genzel, R., Thatte, N., Krabbe, A., Eckart, A., Kroker, H. and Tacconi-Garman, L.E. Ap.J.submitted (1996)
Goldwurm, A. et al., NATURE 371, 5889 (1994)
G—sten, R., in Morris(1989), 89 (1989)
G—sten, R. et al. Ap.J. 318, 124 (1987)
Hall, D.N.B., Kleinmann, S.G. and Scoville, N.Z., Ap.J. 262, L53 (1982)
Haller, J.W. and Rieke, M.J. in Morris(1989), 487 (1989)
Haller, J.W., Rieke, M.J. and Rieke, G.H., Tamblyn, P., Close, L. and Melia, F. , Ap.J. 456, 194 (1996)
Harris, A.I. et al., in Genzel and Harris (1994), 223 (1994)
Herbst, T.M., Beckwith, S.V.W., Forrest, W.J. and Pipher, J.L., A.J. 105, 956 (1993)
Jackson, J. et al., Ap.J. 402, 173 (1993)
Kormendy, J. and Richstone, D., Ann.Rev.Astr.Ap.33, 581 (1995)
Koyama, K.,et al. preprint (1995)
Krabbe, A., Genzel, R., Drapatz, S. and Rotaciuc, V., Ap.J. 382, L19 (1991)
Krabbe, A. Genzel, R., Eckart, A., Najarro, F., Lutz, D. et al., Ap.J.Lett. 447, L95 (1995)
Krichbaum, T.P., Schalinski, C.J., Witzel, A., Standke, K.J., Graham, D.A and Zensus, J.A.in Genzel and Harris(1994), 411 (1994)
Lacy, J.H., Baas, F., Townes, C.H. and Geballe, T.R. Ap.J. 227, L17 (1979)

Lacy, J.H., Townes, C.H., Geballe, T.R. and Hollenbach, D.J., Ap.J. 241, 132 (1980)
Lacy, J.H., Townes, C.H. and Hollenbach, D.J., Ap.J. 262, 120 (1982)
Lacy, J.H., Achtermann, J.M. and Serabyn, E., Ap.J. 380, L71 (1991)
Lebofsky, M.J. and Rieke, G.H., in The Galactic Center, ed. D.Backer, American Institute of Physics Conf.Proc.155, 79 (1987)
Lee, H.M. Ap.J. 319, 801 (1987)
Lee, H.M., in Genzel and Harris(1994), 335 (1994)
Lee, H.M., MNRAS, 272, 605 (1995)
Libonate, S., Pipher, J.L., Forrest, W.J. and Ashby, M.L.N., Ap.J. 439, 202 (1995)
Lindqvist, M., Habing, H. and Winnberg, A., Astr.Ap. 259, 118 (1992)
Lo, K.Y. and Claussen, M, J. , NATURE 306, 647 (1983)
Lynden-Bell, D. and Rees, M., MNRAS 152, 461 (1971)
Melia, F. ,Ap.J. 387, L25 (1992)
Mezger, P.G.et al. 1989, Astr.Ap. 209, 337
Mezger, P.G., in Genzel and Harris(1994), 415 (1994)
Miyoshi,M, Moran, J.M., Hernstein, J., Greenhill, L., Nakai, N., Diamond, P. and Inoue, M., NATURE 373, 127 (1995)
Moneti, A., Glass, I.,S. and Moorwood, A.F.M. , Mem.Soc.Astr.Ital.62, 4,755 (1991)
Morris, M. ,in Galactic and Extragalactic Magnetic Fields, eds. R.Beck, P.Kronberg and R.Wielebinski (Dordrecht:Kluwer), 361 (1990)
Morris, M., Ap.J . 408, 496 (1993)
Morris, M. (ed.), "The Center of the Galaxy" (Dordrecht:Kluwer) (1989)
Najarro, F. et al., Astr.Ap. 285, 573 (1994)
Narayan, R., Yi, I., and Mahadevan, R., NATURE 374, 623 (1995)
Okuda, H. et al. , Ap.J.351, 89 (1990)
Ozernoy, L. and Genzel, R., in "The Galaxy", ed.L.Blitz (Dordrecht:Kluwer), in press (1995)
Phinney, E.S., in Morris(1989), 543 (1989)
Rees, M., NATURE 333, 523 (1988)
Rieke, G. and Rieke, M. in Genzel and Harris(1994), 283 (1994)
Rieke, G.H. and Lebofsky, M.J., in "The Galactic Center", eds. G.Riegler and R.D.Blandford, AIP conf. proc. 83 (New York), 194 (1982)
Rieke, G.H. and Rieke, M.J., Ap.J. 330, L33 (1988)
Sellgren, K., McGinn, M.T., Becklin, E. and Hall, D.N.B., Ap.J. 359, 112 (1990)
Serabyn, E. and G—sten, R. , Astr.Ap. 184, 133 (1987)
Serabyn, E. and Lacy, J. , Ap.J. 293, 445 (1985)
Serabyn, E., Lacy, J., Townes, C.H. and Bharat,R., Ap.J. 326, 171 (1988)
Skinner, G.K., Astr.Ap.(Suppl.) 97, 149 (1993)
Simon, M. et al. Ap.J. 360, 95 (1990)
Simons, D.A., Hodapp, K.W. and Becklin, E.E. Ap.J. 360, 106 (1990)
Sofue, Y., in Genzel and Harris (1994), 43 (1994)
Sunyaev, R.A., Markevitch, M. and Pavlinsky, M., Ap.J. 407, 606 (1993)
Tamblyn, P., Rieke, G.H., Hanson, M.M., Close, L.M., McCarthy, D.W. and Rieke, M.J., Ap.J. 456, 206 (1996)
Wollman, E., Ph.D. Thesis ,Univ.of California, Berkeley, (1976)
Zylka, R., Mezger, P.G. and Lesch, J. , Astr.Ap.261, 119 (1992)

# TWO-DIMENSIONAL FOKKER-PLANCK MODELS

KOJI TAKAHASHI
*Department of Earth and Space Science,*
*Faculty of Science, Osaka University,*
*Toyonaka, Osaka 560, Japan*

**Abstract.**
The evolution of spherical single-mass star clusters was followed by numerically solving the orbit-averaged two-dimensional Fokker-Planck equation in energy–angular momentum space. Velocity anisotropy is allowed in the two-dimensional Fokker-Planck model. The development of the anisotropy is discussed in detail.

## 1. Introduction

Now, direct numerical integration of the orbit-averaged Fokker-Planck (hereafter FP) equation is a main tool to study the dynamical evolution of globular clusters. A direct-integration scheme was invented by Cohn (1979, 1980). Cohn (1979) first performed direct numerical integration of the time-dependent two-dimensional (hereafter 2D) FP equation in energy–angular momentum $(E, J)$ space with the calculation of a self-consistent potential. Although Cohn (1979) showed the potential power of the direct-integration scheme, he had to stop the calculation at a relatively early stage of gravothermal core collapse due to a numerical error concerning energy conservation; the central density of the cluster had increased by only three orders of magnitude, when the calculation was stopped. Later, Cohn (1980) assumed isotropy of the velocity distribution and treated the one-dimensional (hereafter 1D) energy-space FP equation. The use of the 1D FP equation has some advantages over the use of the 2D FP equation: one of these is that it saves a lot of computation time and computer memory space; another one is that the numerical error in energy conservation is greatly reduced (Cohn 1980). This reduction of the error is largely due to the adoption of Chang and Cooper's (1970) finite-differencing scheme for the 1D FP equation. Cohn (1980) reported that the secular energy drift rate was reduced by better than a factor of 100 by adopting of the Chang-Cooper scheme. In fact, he could follow the core collapse until the central density increased by twenty orders of magnitude.

Because of those advantages of the 1D FP equation, the 1D FP equation has been used in most FP studies concerning globular-cluster evolution, while the 2D FP equation has seldom been used. The adoption of the simpler 1D (isotropic) FP equation seems to be quite reasonable for studying the core evolution, which has been a main subject during the past two decades, because strong relaxation enforces the isotropy of the velocity distribution in the core.

On the other hand, it is true that the development of anisotropy in the halo is a natural consequence of cluster evolution driven by two-body relaxation. The relaxation is strong in the core because of its high density, and the strong relaxation continues to produce high-energy stars. Such high-energy stars have very radial orbits on the average, and they travel through the low-density halo almost without experiencing collisions. Therefore, radial orbits predominate in the halo and the velocity anisotropy increases as the halo grows. The penetration of anisotropy even into the inner region was pointed out in several early studies (e.g., Cohn 1985). This is closely related to gravothermal core collapse.

Thus, the development of velocity anisotropy is expected to occur throughout the whole cluster. In fact, so far, anisotropy has been considered in various simulations concerning the evolution of star clusters (see Takahashi 1995a). Recently, furthermore, more elaborate anisotropic gaseous and higher-order fluid-dynamical models of star clusters have been developed (e.g., Bettwieser and Spurzem 1986; Louis 1990; Louis and Spurzem 1991; Spurzem 1991; Giersz and Spurzem 1994; Spurzem and Takahashi 1995). They have made it possible to calculate in detail the evolution of anisotropic star clusters. On the other hand, more fundamental 2D FP simulations were carried out only for isolated single-mass clusters without binaries (Cohn 1979, 1985).

Considering the above situations, I think that it is now a good time to reconsider direct 2D FP calculations. Concerning the practical sides of computations, it is now possible to carry out the 2D calculations on standard workstations. A main obstacle to the 2D FP calculations is to develop computational schemes of high accuracy (especially in the energy conservation). The aim of this work is to develop reliable numerical schemes for the 2D orbit-averaged FP equation and to consider in detail the evolution of anisotropic star-clusters. More complete descriptions of the present topic are found in Takahashi (1995a, pre-collapse evolution; 1995b, post-collapse evolution).

## 2. The Orbit-Averaged Fokker-Planck Equation

We consider the evolution of spherical one-component star clusters. In a steady-state spherical system, the distribution function $f(\mathbf{r}, \mathbf{v}, t)$ is a function of only the energy $E$ and total angular momentum $J$ per unit mass. The evolution of $f$ due to two-body relaxation can be described by the orbit-averaged FP equation in $(E, J)$-space (Cohn 1979). The 2D FP equation under the fixed gravitational potential $\phi(r)$ can be written in a flux-conserving form,

$$A\frac{\partial f}{\partial t} = -\frac{\partial F_E}{\partial E} - \frac{\partial F_R}{\partial R}, \tag{1}$$

where

$$\begin{aligned}
-F_E &= D_{EE}\frac{\partial f}{\partial E} + D_{ER}\frac{\partial f}{\partial R} + D_E f, \\
-F_R &= D_{RE}\frac{\partial f}{\partial E} + D_{RR}\frac{\partial f}{\partial R} + D_R f.
\end{aligned} \tag{2}$$

Here, $R$ is the scaled angular-momentum: $R = J^2/J_c^2(E)$, where $J_c(E)$ is the angular momentum of a circular orbit of energy $E$. Thus, $R$ takes all values between 0 and 1, independent of $E$. The isotropized distribution function, $\bar{f}(E, r)$, introduced by Cohn (1979) was used to calculate the diffusion coefficients, $D_{EE}, D_E$, etc. The isotropization makes computing the coefficients much easier. It has generally been conceived that the use of the isotropized distribution function does not cause significant errors, because the coefficients only depend on the moments of $f(E, R)$, and because relaxation occurs mainly in the core where the distribution function is almost entirely isotropic. However, as shown below, since $f(E, R)$ strongly depends on $R$ for $E \sim 0$ as the halo develops, we may have to be more careful about the use of the isotropized distribution function.

## 3. The Method

The framework of our method is the same as that of Cohn's (1979) method. Cohn's method comprises two steps: the FP step and the Poisson step. In the FP step, the distribution function is advanced by solving the FP equation with the gravitational potential being held fixed. In the Poisson step, the potential is advanced by solving Poisson's equation with the distribution function being held fixed as a function of the adiabatic invariants.

An essential difference between our method and Cohn's method exists only concerning a discretization scheme of the FP equation. Two different discretization schemes have been developed: one is a finite-difference scheme where the Chang-Cooper scheme is simply applied for only the energy direction; the other is a finite-element scheme where the test and weight functions suggested by the generalized variational principle (Inagaki and Lynden-Bell 1990) are used. The details concerning the discretization schemes are described in Takahashi (1995a).

The FP equation is solved in a rectangular domain enclosed by boundary lines, $E = \phi(0)$, $E = E_{\min}$, $R = 0$, and $R = 1$, where $\phi(0)$ is the central potential and the value of $E_{\min}$ is chosen to be close to zero. We impose boundary conditions that $F_E = 0$ on boundaries $E = \phi(0), E_{\min}$, and $F_R = 0$ on boundaries $R = 0, 1$. We use variables $(X, Y)$ instead of $(E, R)$ in practical calculations. The variable $X(E)$ is defined by

$$X(E) = \ln\left[\frac{E}{2\phi(0) - E_0 - E}\right], \quad (3)$$

where $E_0$ is an adjustable parameter (Cohn 1979). The variable $Y(R)$ is defined by

$$Y(R) = \frac{\ln(1 + R/R_0)}{\ln(1 + 1/R_0)}, \quad (4)$$

where $R_0$ is an adjustable parameter, such that $0 < R_0 \ll 1$. We set $R_0 = 0.01$ in standard runs. This variable is introduced in order to give a good representation to radial orbits.

## 4. Results

Calculations were carried out using both the finite-difference and finite-element codes. The results obtained by the two (partially) different codes were generally in good agreement. For calculations of the pre-collapse evolution, the numerical accuracy of the two schemes is similar. However, for calculations of the post-collapse evolution, the accuracy (in particular, concerning total-energy conservation) of the finite-difference scheme is better. The reason why the accuracy of the finite-element scheme is not very good for the post-collapse calculations is not clear at present. The figures shown below were actually drawn from the results of calculations by the finite-difference code.

The initial condition of the calculations was Plummer's model, where the velocity distribution is isotropic everywhere. We use standard units such that $G = M = 1$ and $\mathcal{E}_i = -1/4$, where $G$ is the gravitational constant, $M$ is the total mass, and $\mathcal{E}_i$ is the initial total energy.

### 4.1. PRE-COLLAPSE EVOLUTION

Unless any heat sources are included in FP calculations, the core continues to contract and the core density continues to increase [gravothermal core collapse (Lynden-Bell and Eggleton 1980)]. A calculation was continued until the central density increased by about 14 orders of magnitude. During the calculation, the numerical error in the total mass was within 0.1%, and the error in the total energy was within 1%.

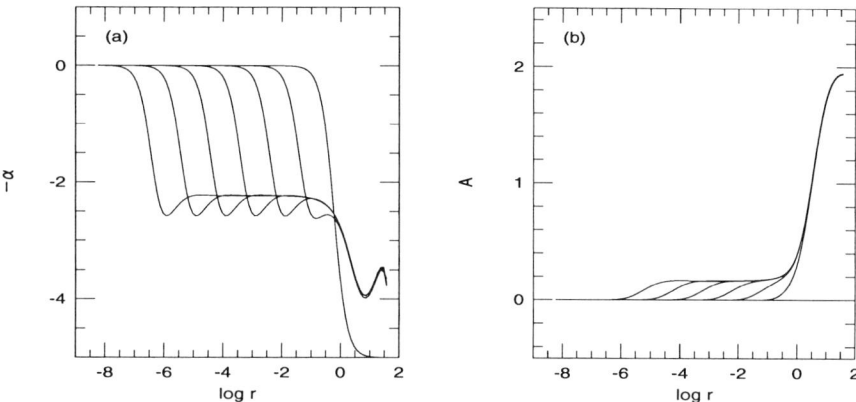

*Figure 1.* (a) Evolution of the radial profile of the logarithmic density gradient, $d\ln\rho/d\ln r = -\alpha$. The power-law region with $\alpha = 2.23$ extends into the inner region as the core collapse proceeds. (b) Evolution of the radial profile of the anisotropy parameter, $A \equiv 2 - 2\sigma_t^2/\sigma_r^2$. The self-similar region with $A = 0.16$ extends into the inner region as the core collapse proceeds.

Figure 1a shows the evolution of the radial profile of the logarithmic density gradient, $d\ln\rho/d\ln r = -\alpha$. The power-law region with $\alpha = 2.23$ extends into the inner region self-similarly as the core collapse proceeds. This value of $\alpha = 2.23$ coincides with that found in the 1D (isotropic) FP model (Cohn 1980; Heggie and Stevenson 1988). Figure 1b shows the evolution of the radial profile of the anisotropy parameter, $A \equiv 2 - 2\sigma_t^2/\sigma_r^2$, where $\sigma_r$ and $\sigma_t$ are the radial and (1D) tangential velocity dispersions, respectively. The self-similar region with $A = 0.16$ (or $\sigma_t^2/\sigma_r^2 = 0.92$) extends into the inner region as the core collapse proceeds. Cohn's (1985) calculation gave a similar value of $A \approx 0.15$. However, the degree of anisotropy in the halo is somewhat stronger in our calculation than in Cohn's calculation.

At late stages of the core collapse, the collapse rate $\xi \equiv t_r(0)d\ln\rho(0)/dt$, where $t_r(0)$ is the central relaxation time (Spitzer and Hart 1971a) and $\rho(0)$ is the central density, tends to an asymptotic constant value of $\xi = 2.9 \times 10^{-3}$. On the other hand, Cohn's (1979) 2D calculation gave $\xi = 6.0 \times 10^{-3}$ and Cohn's (1980) 1D calculation did $\xi = 3.6 \times 10^{-3}$. While the value of $\xi$ in our anisotropic model is smaller than that in the isotropic model, the value of $\xi$ in Cohn's (1979) anisotropic model is larger. Which tendency is true? I think that Cohn's value is less reliable than our value, because Cohn's calculation was accompanied by a large error in energy conservation ($\sim 11\%$) and did not follow the core collapse very deeply. Although I cannot give satisfactory proof of the slower collapse in anisotropic clusters, the following intuitive interpretation may be helpful: the reduction of the 2D FP equation to the 1D FP equation by averaging over angular momentum space causes artificial diffusion in addition to real diffusion, and, consequently, the core collapse proceeds faster in isotropic clusters. It is interesting that Louis (1990) also found the lower value of $\xi$ in anisotropic clusters using fluid-dynamical models.

Figure 2a shows the evolution of Lagrangian radii for the 2D and 1D models. The time is measured in units of the initial half-mass relaxation time $t_{rh,i}$ (Spitzer and Hart 1971a). We find again slower core collapse in the 2D calculation from figure 2a. The 2D calculation gives a core collapse time of $t_{coll} = 17.6 t_{rh,i}$ and the 1D calculation gives $t_{coll} = 15.6 t_{rh,i}$. Figure 2b also shows the evolution of Lagrangian radii, but the time for the 1D calculation is scaled so that the collapse time in the 1D calculation should coincide with that in the 2D calculation. In this figure, we do not find any significant differences between the two models for the 1-75% Lagrangian radii. However, the 90% radius of the 2D model expands further than that of the 1D model; that is, the 2D model has a more extended halo.

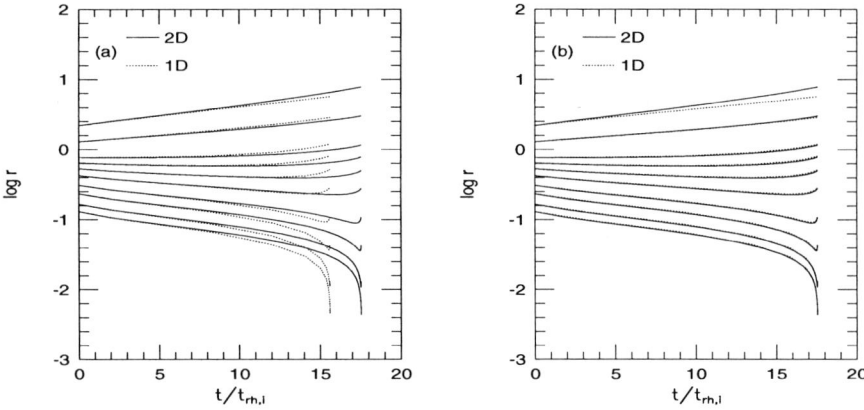

*Figure 2.* (a) Evolution of Lagrangian radii containing inner 1, 2, 5, 10, 20, 30, 40, 50, 75, and 90% of the cluster mass. The solid curves are the result of the 2D FP calculation, while the dotted curves are that of 1D calculation. The time is measured in units of the initial half-mass relaxation time $t_{rh,i}$. (b) Same as (a), but the time axis of 1D calculation is multiplied by a constant factor so that the collapse time in the 1D calculation should coincide with that in the 2D calculation.

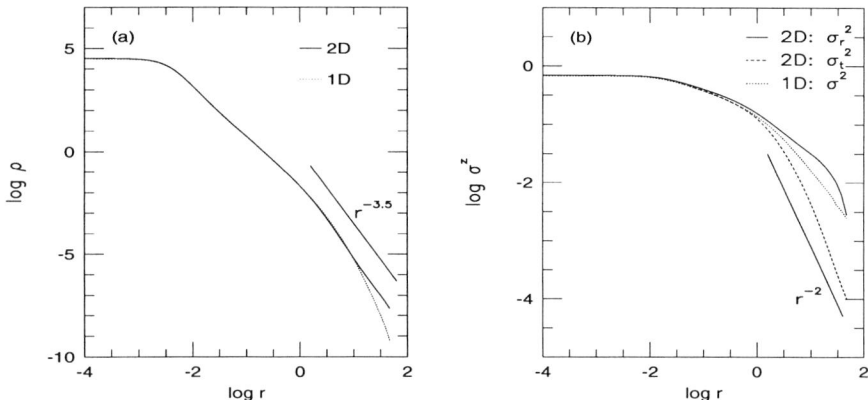

*Figure 3.* (a) Density profile at an epoch when the central density increases by about 4.5 orders of magnitude for the 2D model (the solid curve), and for the 1D model (the dotted curve). The asymptotic line $\rho \propto r^{-3.5}$ is shown for a comparison. (b) Velocity dispersion profiles at the same epochs as in (a). The solid and dashed curves are the radial and 1D tangential velocity dispersions for the 2D model, respectively. The dotted curve is the 1D velocity dispersion profile for the 1D model. The asymptotic line $\sigma^2 \propto r^{-2}$ is shown for a comparison.

Figure 3a shows the density profiles at epochs when the central density increases by about 4.5 orders of magnitude for the 2D and 1D models. The density profile in the outer halo is approximated by a power law, $\rho \propto r^{-3.5}$, rather well for the 2D model (cf. Spitzer and Shapiro 1972). In figure 3b, the velocity dispersion profiles at the same epochs as in figure 3a are shown. In the halo the radial velocity dispersion exceeds the tangential velocity dispersion considerably. This is because the halo is dominated by eccentric orbits. A power law, $\sigma_t^2 \propto r^{-2}$, gives a reasonable fit to the result of the 2D FP calculation in the halo. This power law corresponds to the constant mean squared angular momentum (cf. Spitzer and Hart 1971b; Hénon 1971).

## 4.2. POST-COLLAPSE EVOLUTION

Now we take account of heating effects by three-body binaries. The three-body binary heating rate per unit mass is given by

$$\dot{E}_b = C_b G^5 m^3 \rho^2 \sigma^{-7}, \qquad (5)$$

where $C_b$ is a numerical coefficient (Hut 1985). A standard value of $C_b = 90$ was chosen in the present calculations. The local heating rate (5) is orbit-averaged (Cohn 1979), and then the orbit-averaged heating rate, $<\dot{E}_b>_{\rm orb}$, is added to the usual first-order diffusion coefficient $< \Delta E >_{\rm orb}$. Furthermore, we assume that binary scatterings do not produce a net change of the scaled angular momentum $R$, i.e. $< \dot{R}_{\rm b} >_{\rm orb} = 0$.

For calculations of the post-collapse evolution, we must specify the number of stars in the cluster, $N$, and the numerical constant, $\mu$, in the Coulomb logarithm $\ln(\mu N)$. For all the present calculations, the value of $\mu = 0.11$ was chosen, which was suggested by Giersz and Heggie (1994a) for the pre-collapse evolution of single-mass clusters. Calculations were performed for $N = 5000$, 10000, and 20000. Concerning computation time, for example, the 2D FP calculation for $N = 20000$, where 151 $X$-mesh, 35 $Y$-mesh and 91 $r$-mesh points were used, required about 95 hours of CPU time on HP 9000/715 (50 MHz).

The core expansion is stable for $N = 5000$, marginally stable (overstable) for $N = 10000$. For $N = 20000$, the core expansion is unstable: the central density oscillates chaotically with the large amplitude [gravothermal oscillation (Bettwieser and Sugimoto 1984)]. There are no qualitative difference concerning the features of the gravothermal oscillations between 1D and 2D calculations. Figures 4a and 4b show the evolution of the anisotropy $A$ averaged over 0–1%, 1–2%,..., and 75–90% Lagrangian radii, for the case of $N = 20000$. The anisotropy at inner Lagrangian radii oscillates with the core oscillation. We can see the anisotropy oscillation even at the half-mass radius.

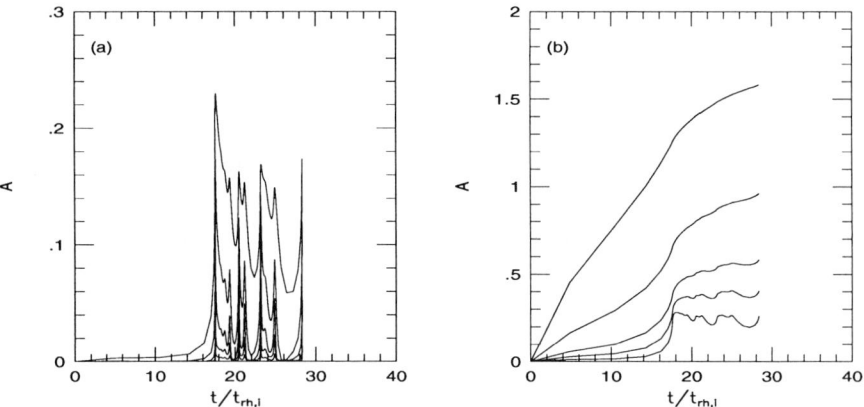

*Figure 4.* (a) Evolution of the anisotropy $A$ averaged over 0–1%, 1–2%, 2–5%, 5–10%, and 10–20% Lagrangian radii, for the case of $N = 20000$. (b) Same as (a), but for the anisotropy $A$ averaged over 20–30%, 30–40%, 40–50%, 50–75%, and 75–90% Lagrangian radii. $A$ increases as the radius increases.

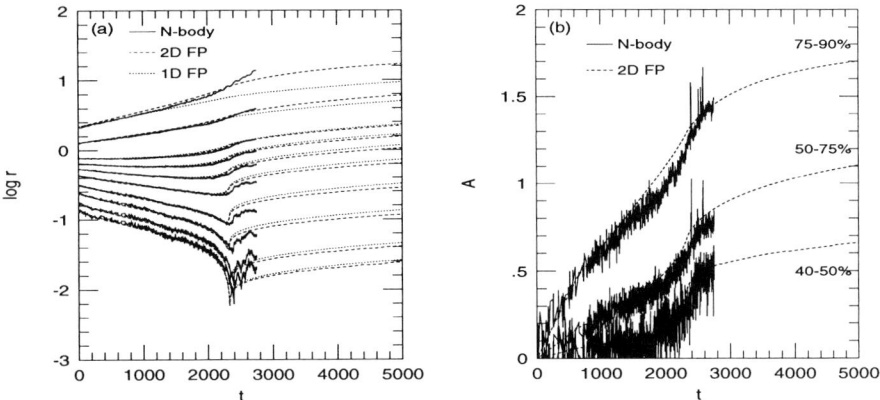

*Figure 5.* (a) Evolution of the 1, 2, 5, 10, 20, 30, 40, 50, 75, and 90% Lagrangian radii for $N = 10000$. The solid curves are the result of the $N$-body calculation (by R. Spurzem), the dashed curves are that of the 2D FP calculation, and the dotted curves are that of the 1D FP calculation. The time for the 1D calculation is scaled as in figure 2b. (b) Evolution of the anisotropy $A$ averaged over the 40–50%, 50–75%, and 75–90% Lagrangian radii. The solid curves are the result of the $N$-body calculation, and the dashed curves are that of the 2D FP calculation.

Lastly, we compare the FP models and an $N$-body model for $N = 10000$. The result of the 10000-body calculation has kindly been made available by Rainer Spurzem (cf. Spurzem and Aarseth 1995). Figure 5a shows a comparison of the evolution of Lagrangian radii. The time for the 1D calculation is scaled as in figure 2b. The result is generally in good agreement between the 2D FP and $N$-body models. In particular, the collapse times for the two models are very close. However, we should remember that the collapse time in physical time units in the FP model depends on $\mu$, and that there is a scatter of collapse times for each individual $N$-body simulation (Giersz and Heggie 1994a, b; Spurzem and Aarseth 1995). The good agreement of the collapse time in this case may be accidental. In the $N$-body model, the inner Lagrangian radii oscillate with the small amplitude after the core bounce. This may be due to stochastic binary activities. Figure 5b shows the evolution of the anisotropy $A$ in the outer regions. The agreement between the 2D FP and $N$-body models is good. This fact supports the reliability of the present 2D FP model.

## 5. Conclusions and Discussion

I have developed numerical codes to solve the orbit-averaged $(E, J)$-space FP equation with high accuracy, and have investigated in detail the pre- and post-collapse evolution of single-mass spherical star clusters where velocity anisotropy is allowed. Two different integration schemes have been developed: they are the finite-difference scheme and the finite-element scheme. Using these schemes, we could follow the core collapse until the central density increased by about 14 orders of magnitude, and follow long-term evolution past the core collapse.

We have seen that the relaxation process is always accompanied by the velocity anisotropy production. We still do not know very much the effects of anisotropy on the evolution of more realistic clusters (e.g., multi-mass clusters, tidally limited clusters, etc.). Since the present work has shown that 2D FP calculations can be performed with reasonable numerical accuracy, we do not have to hesitate doing 2D FP calculations from now on. The evolution of more realistic cluster models will be studied in the future.

The author is a research fellow of the Japan Society for the Promotion of Science, and this work was supported in part by the Grand-in-Aid for Encouragement of Young Scientists by the Ministry of Education, Science and Culture of Japan (No. 1338).

# References

Bettwieser E., Spurzem R. 1986, A&A 161, 102
Bettwieser E., Sugimoto D. 1984, MNRAS 208, 493
Chang J.S., Cooper G. 1970, J. Comp. Phys. 6, 1
Cohn H. 1979, ApJ 234, 1036
Cohn H. 1980, ApJ 242, 765
Cohn H. 1985, in Dynamics of Star Clusters, IAU Symp No.113, ed J. Goodman, P. Hut (D. Reidel Publishing Company, Dordrecht) p161
Giersz M., Heggie D.C. 1994a, MNRAS 268, 257
Giersz M., Heggie D.C. 1994b, MNRAS 270, 298
Giersz M., Spurzem R. 1994, MNRAS 269, 241
Heggie D.C., Stevenson D. 1988, MNRAS 230, 223
Hénon M. 1971, Ap&SS 13, 284
Hut P. 1985, in Dynamics of Star Clusters, IAU Symp No.113, ed J. Goodman, P. Hut (D. Reidel Publishing Company, Dordrecht) p231
Inagaki S., Lynden-Bell D. 1990, MNRAS 244, 254
Louis P.D. 1990, MNRAS 244, 478
Louis P.D., Spurzem R. 1991, MNRAS 251, 408
Lynden-Bell D., Eggleton P.P. 1980, MNRAS 191, 483
Spitzer L.Jr., Hart M.H. 1971a, ApJ 164, 399
Spitzer L.Jr., Hart M.H. 1971b, ApJ 166, 483
Spitzer L.Jr., Shapiro S.L. 1972, ApJ 173, 529
Spurzem R. 1991, MNRAS 252, 177
Spurzem R., Aarseth S.J., 1995, in preparation
Spurzem R., Takahashi K. 1995, MNRAS 272, 772
Takahashi K. 1995a, PASJ 47, in press
Takahashi K. 1995b, in preparation

# MONTE-CARLO SIMULATIONS

MIREK GIERSZ
*Nicolaus Copernicus Astronomical Centre,*
*Polish Academy of Sciences,*
*Bartycka 18, 00-716 Warsaw,*
*Poland*

**Abstract.** The revision of the Stodółkiewicz's Monte-Carlo code is presented. It treats each *superstar* as a single star and follows the evolution and motion of all individual stellar objects. The first calculations, for equal-mass $N$-body systems with three-body energy generation accordingly to Spitzer's formulae, show good agreement with the direct $N$-body calculations for $N = 2000$ and $10000$ particles.

## 1. INTRODUCTION.

Our knowledge about the stellar content, kinematics and the influence of the environment on observational features of globular clusters and even richer stellar systems are increasing dramatically (as we could learn, for example, from talks presented on this conference). First, observations are reaching the point where segregation of mass within globular clusters can be observed directly and quantitatively. Second, observations have revealed that clusters with dense (collapsed) cores are relatively more concentrated to the galactic center than uncollapsed ones. Thus the influences of the environment and mass spectrum are crucial for cluster evolution. Third, recent observations show that many different and fascinating types of binaries and binary remnants are present in abundance in globular clusters. Binaries, in addition to being a diagnostic of the evolutionary status of clusters, are directly involved in the physical processes of energy generation, providing the energy source necessary to stop the core collapse and then drive the core expansion. So, to model the evolution of real stellar systems and make meaningful comparison with observation one has to take into account the complex interactions between stellar evolution and stellar dynamics. Of course all these demands can be easily fulfilled by direct $N$-body models. But they are very time-consuming and they need a special-purpose hardware to be run efficiently (Makino 1995, Taiji 1995). Another possibility is to use a code, which is very fast and can properly reproduce the standard relaxation process, and at the same time provides a clear and unambiguous way of introducing all the physical processes, which are important during globular cluster evolution. Monte-Carlo codes, which use a statistical method of solving the Fokker-Planck equation, provide all the necessary flexibility. They were developed by Spitzer (1975, and references therein) and Hénon (1975, and references therein) in the early seventies, and substantially improved by Marchant & Shapiro (1980, and references therein) and Stodółkiewicz (1986, and references therein). Unfortunately, lack of fast computers at that time and development of the direct Fokker-Planck and gaseous models contributed to the abandonment of this method. But recent developments in computer hardware now make it possible to run a Monte-Carlo code efficiently. The great advantages of this method, besides of its simplicity and speed, are connected with the inclusion of anisotropy, and with the fact that added realism does not slow it down.

The Monte-Carlo code can have another possible use. Despite the simplified nature of continuum models (Fokker-Planck and gaseous models) they will continue for a while to be the most commonly used codes for stellar dynamical evolution. The Monte-Carlo models can be used to optimize free parameters of continuum models and to check their validity as it was done in the comparison between small $N$-body simulations and continuum ones (Giersz & Heggie 1994ab, Giersz & Spurzem 1994). This procedure should further increase our confidence of the results obtained in Fokker-Planck or gaseous simulations.

## 2. MONTE-CARLO METHOD.

### 2.1. BASIC IDEAS.

The Monte-Carlo method can be regarded as a statistical way of solving the Fokker-Planck equation. The basic idea behind the Monte-Carlo method is as follows. During the time interval $\Delta t$, much smaller than the relaxation time, the fluctuating gravitational field (connected with distant two-body interactions between stars) can be neglected in a first approximation and the system can be regarded as being in a steady state. However, the fluctuating gravitational field causes slow and random changes of the particle orbit parameters. This effect is small over $\Delta t$ but it builds up and becomes significant over the relaxation time scale and it has to be taken into account. To compute it, the standard Monte-Carlo tricks can be applied. The perturbation of a test star orbit is a random quantity, so only its statistical properties matter. The exact value of each perturbation is unimportant. The procedure to calculate perturbations is as follows: **(1)** instead of integrating the perturbations along the orbit, the perturbation is computed at a randomly selected point, **(2)** instead of considering the effect of all stars in the system, the perturbation is computed from a randomly chosen star, **(3)** the computed perturbation is multiplied by an appropriate factor in order to account for all the time points and all the system stars which have not been considered. If the procedure is correctly set up the evolution of the artificial system will be statistically the same as the evolution of the real one. For technical reasons (available computer memory and speed of computations) in all Monte-Carlo methods the whole system was divided into a certain number of *superstars* each consisting of a certain number of stars with the same mass, distance from the cluster center, radial and tangential velocities.

The way of implementing this basic strategy divides Monte-Carlo codes into three different groups; referred to as "Princeton", "Hénon" and "Cornell" methods (Spitzer 1987, and references therein). Each of these implementation was in the past successfully used in simulations of evolution of globular clusters and galactic nuclei.

## 2.2. NEW IMPLEMENTATION OF THE STODÓLKIEWICZ'S MONTE-CARLO SCHEME.

The real power of Monte-Carlo codes was demonstrated by Stodółkiewicz (1982, 1985, 1986). He substantially improved Hénon's version of Monte-Carlo code by adding individual time-step scheme and a special procedure, which makes the total energy conservation much more strict. He used the code to model the evolution of globular clusters influenced by the following processes: formation of binaries by dynamical and tidal interactions, interactions between binaries and field stars and between binaries themselves, collisions between stars, stellar evolution, the tidal field of the Galaxy and tidal shocks. These were unique calculations, and have never been repeated or superseded by anybody. Excellent and very detailed description of the Stodółkiewicz's code can be found in Stodółkiewicz (1982, 1986).

Unfortunately the Stodółkiewicz's method is not suitable to correctly represent the very center of the system. In the core, as a result of the collapse, the density in a small and ununiform region reaches high values. This area is represented by only a few *superstars*. Therefore the statistical properties of this region are very poorly described. Moreover, *superstars* which constitute the core take part in processes which are responsible for energy generation and creation, in direct star interactions, many different and fascinating types of binaries, binary remnants and coalesced stars. In the new code, in order to properly describe this region and these processes, each *superstar* is treated as a single star and evolution and motion of all individual objects are followed. This improvement is possible only due to enormous increase of speed and memory in present day general-purpose computers.

Note, that individual treatment of all objects in the system enables, for example, to investigate influence of primordial binaries on the system evolution. In the Stodółkiewicz's method all binaries or coalesced stars take part only in relaxation process. They are neglected in computation of gravitational potential, so the process of mass segregation is not properly described.

Basically, the improvement mentioned above is the only major change to the Stodółkiewicz's original code. Other changes are rather cosmetic and do not have any influence on code flow or implementation of physical processes.

## 3. FIRST RESULTS

The Monte-Carlo method contains several free parameters, which have to be adjusted in order to get the proper description of the system evolution. The best way of adjusting them is to compare the results of Monte-Carlo and direct $N$-body simulations. The same strategy was used to optimize free parameters of the continuum models (Giersz & Heggie 1994ab, Giersz & Spurzem 1995).

The good statistical quality data for single mass $N$-body simulations are available only for $N = 250, 500, 1000, 2000$ and $10000$ (Giersz & Heggie 1994ab, Giersz & Spurzem 1994, Aarseth & Spurzem 1995). Simulations with $N = 2000$ are the best for our purposes. They cover the system evolution up to twelve collapse times and consist of sixteen separate runs.

Most of the Monte-Carlo simulations were run for systems consisting of 2000 equal mass particles, but additionally a few simulations were performed with $N = 10000$ and $30000$. Pilot runs have shown that the best choice of free parameters is practically the same as chosen by Stodółkiewicz (1982). The results discussed bellow ($N = 2000$) were averaged over 25 simulations, each having the same initial parameters but with different sequence of random numbers used to initialize the positions and velocities of the stars.

During the phase of core collapse the $N$-body and Monte-Carlo models follow each other very closely (Fig. 1). The first differences start to build up around the time of the core bounce. This is, particularly well, visible for the middle and outer Lagrangian radii. In the Monte-Carlo simulations the

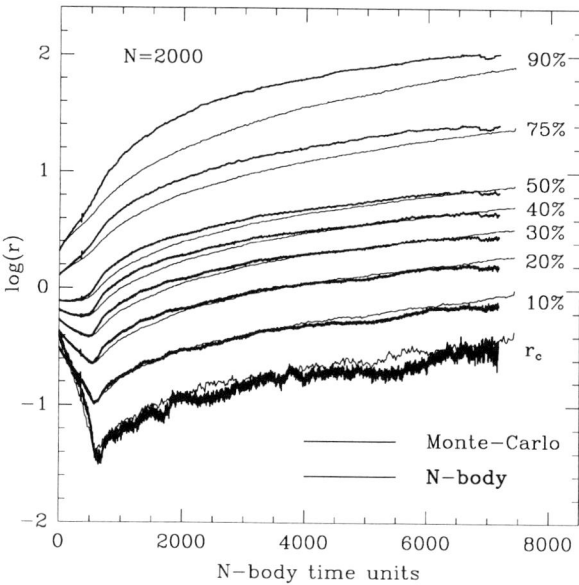

*Figure 1.* Evolution of Lagrangian radii for $N = 2000$ for Monte-Carlo model (solid line) and $N$-body model (thick solid line). The core radii for both model are also shown.

rate of system expansion is slower just after the core bounce and slightly faster for the advanced post-collapse evolution. For the inner parts of the system, up to the Lagrangian radius containing 10% of the total mass, the evolution for both models is very similar. Only the core collapse is slightly deeper in the $N$-body model. It is worth to note that the collapse time for both models is practically the same. This further confirms the value of $\gamma = 0.11$, in the Coulomb logarithm, obtained by comparison of $N$-body and continuum models (Giersz & Heggie 1994a).

Similarly as for the inner Lagrangian radii the anisotropy for the inner half of the system is very well reproduced by the Monte-Carlo model (Fig. 2). For the outermost part of the system (Lagrangian radii containing more than 90% of the total mass) the anisotropy in the $N$-body simulations is larger from the very beginning of the core expansion. This suggests that in the $N$-body simulations halo is developed faster than in the Monte-Carlo ones. At least part of the differences in the anisotropy labeled by 75% can be explained by the way of anisotropy computation. For the $N$-body simulations the displayed anisotropy is computed for shell between 50% and 75% Lagrangian radii, while for the Monte-Carlo simulations for shell between 70% and 75% Lagrangian radii. Because the anisotropy increases with radius, so it should be slightly larger for the Monte-Carlo model than for the $N$-body one.

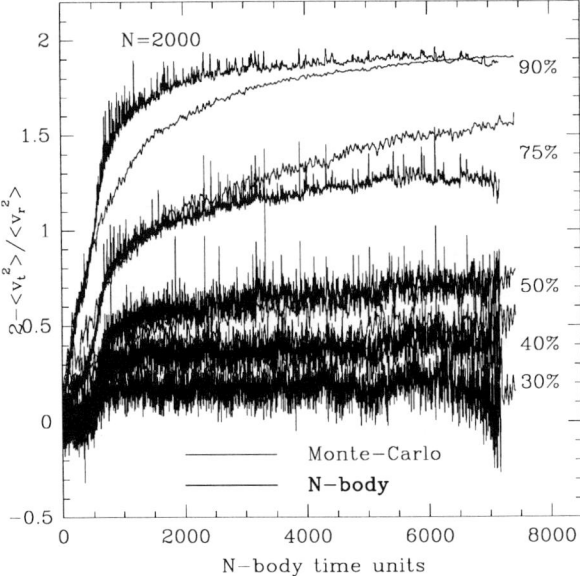

*Figure 2.* Evolution of the anisotropy for $N = 2000$ for Monte-Carlo model (solid line) and $N$-body model (thick solid line).

The clue to explain the differences between the $N$-body and Monte-Carlo models gives Fig. 3, which shows the energy balance. There are three main differences between models visible in this figure. Firstly, the energy carried out by single star escapers, $E_{ext}^{es}$, is much larger in the Monte-Carlo simulations. This is connected with the fact that the number of escapers is larger by about 30% and that the escapers connected with interactions between three-body binaries and field stars are more numerous in the Monte-Carlo model. The larger number of escapers in the Monte-Carlo simulations is mainly connected with the fact that stars are immediately removed from the system, while in the direct $N$-body simulations they are only removed when they are further than ten times the half-mass radius. Secondly, binaries start to form slightly earlier and the number and the total internal energy of escaping binaries, $E_{int}^{eb}$, are larger in the Monte-Carlo simulations. Thirdly, around the time of core bounce, the total internal binding energy of the three body binaries bound to the system, $E_{int}^{b}$ is smaller in the Monte-Carlo simulations. This is despite that the number of binaries bound to the system is practically the same in both models. Too early energy generation by three-body binaries makes the core collapse less deep and as well less abrupt expansion of the outer parts of the system. On the other hand in the $N$-body simulations binaries, which stay in the core, harden to higher binding energies than in the Monte-Carlo simulations. So they can produce

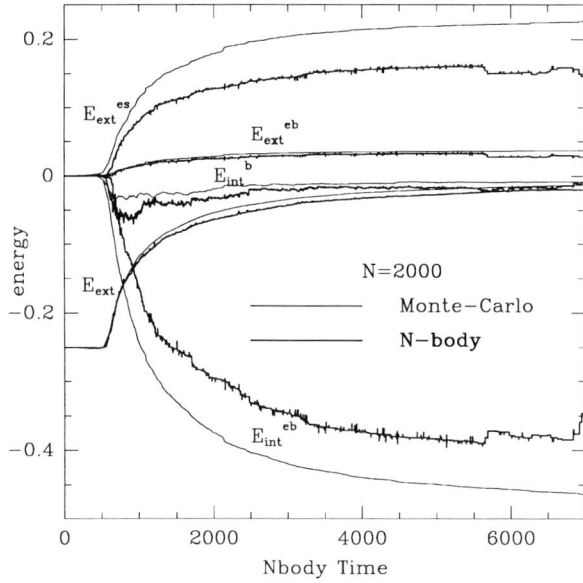

*Figure 3.* Energy balance for $N = 2000$ for Monte-Carlo model (solid line) and $N$-body model (thick solid line). The meaning of the symbols is given in the text.

more stars, which penetrate, on nearly radial orbits, the outermost parts of the system partially contributing to larger anisotropy. Smaller binding energy of the system, $E_{ext}$ for Monte-Carlo simulations is connected with the fact that there are substantially more escapers than in the $N$-body simulations.

Comparison between the Monte-Carlo simulations and the $N$-body ones for $N = 10000$ shows, basically, the same features as in the case of $N = 2000$.

The gravothermal oscillations are the most pronounced feature of the post-collapse evolution of $N$-body systems with number of stars greater than a few thousands. They were observed in gas (Bettwieser & Sugimoto 1984, Goodman 1987, Heggie & Ramamani 1989), Fokker-Planck (Cohn et al 1986, Cohn, Hut & Wise 1989, Gao et al 1991) and recently in $N$-body simulations (Makino 1995). The lowest value of $N$ for which the gravothermal oscillations begin to show up is uncertain. But recent pilot $N$-body simulation of system consisting of 16000 particles ( Makino 1995) shows clear oscillations. All this suggests that gravothermal oscillations should, as well, be present in the Monte-Carlo simulations, at least for $N = 30000$.

In Fig. 4 the evolution of the logarithm of the central density for $N = 2000$, 10000 and 30000, respectively is shown. It seems that gravothermal oscillations are visible in system with 30000 particles and there are some

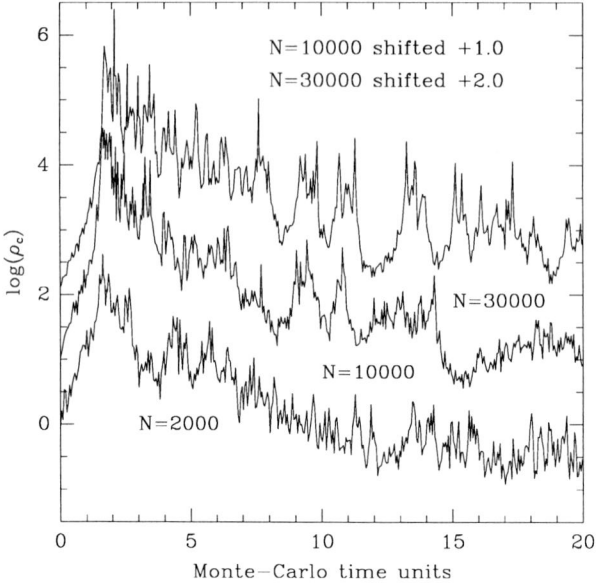

*Figure 4.* Evolution of the central density for $N = 2000$, 10000 and 30000. Data for $N = 10000$ is shifted up by 1 in the logarithm of the central density. Data for $N = 30000$ is shifted up by 2 in the logarithm of the central density.

signs of gravothermal oscillations for $N = 10000$. This conclusion is somewhat ambiguous, because binary activities can give similar behaviour of the central density. But amplitudes of oscillations connected with binary activities are usually smaller than observed in 10000 and 30000 body simulations. Evidently more simulations with broader range of $N$ are needed to clarify this problem and to get a better agreement with the direct $N$-body simulations.

At the end I would like to show an unpublished (in international astronomical literature) result of the Stodółkiewicz Monte-Carlo simulations of two-component system with $10^5$ stars, performed in 1986 (Fig. 5) The evolution of central density for unevolved and evolved stars is shown together with the evolution of the total central density. The total central density (dominated by evolved stars) behaves in manner characteristic for gravothermal oscillations. I think that this is the remarkable result obtained well before the gravothermal oscillations were confirmed and widely accepted by the astronomers.

Finally, a few words about the efficiency of the new code. The calculation of $N = 2000$, 10000 and 30000 models takes about 2, 20 and 130 hours, respectively. The theory predicts a linear increase of computing time with $N$. This is connected with the fact that the most time consuming part of the code is the computation of the potential, which is proportional to $N$

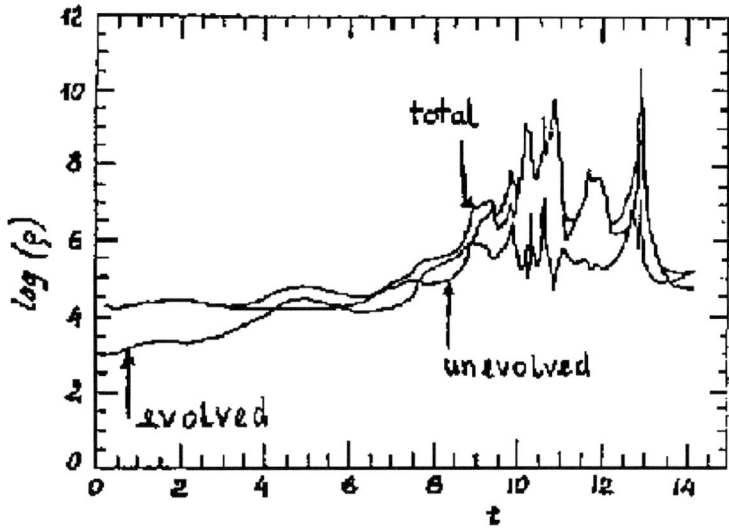

*Figure 5.* Evolution of the central density for Stodółkiewicz Monte-Carlo model for $N = 10^5$ two component system. Arrows show the densities for evolved, unevolved stars and the total density.

due to spherical symmtery of the system. The real calculations give slightly steeper dependence on $N$. It seems that this is connected with the fact that larger systems have more density peaks, which implies smaller time steps in order to properly resolve the system evolution. The high speed of the code makes it possible to run several different models to improve statistical quality of the data and investigate influence of different physical processes on cluster evolution.

## 4. FUTURE DEVELOPMENTS OF THE NEW MONTE-CARLO CODE

The new Monte-Carlo code presented in the previous section can be treated only as a first approximations. Several important physical processes have to be included to make the simulations of stellar systems more realistic. The final code will contain the following physical processes:
- formation of binaries due to dynamical and tidal interactions,
- primordial binaries,
- stellar evolution,
- tidal field of Galaxy, and tidal shocks connected with crossing of the galactic plane and with large molecular clouds,
- collisions between stars,

- interactions between binaries and stars and between binaries themselves.

In the first stage of the code development (nearly completed) all processes connected with interactions between objects will be modeled using analytical cross sections available in the literature. This will allow the code to be tested and make possible comparison with continuum models.

In the next stage interactions between groups of three and four (finite size) stars will be modeled by numerical integrations of their orbits (the first attempts are tested now). This more realistic approach ensures that processes of energy generation (the most important factor in the dynamical evolution of globular clusters) will be modeled more closely.

The final stage will be the inclusion of a detailed 3-D hydrodynamical modeling of collisions between stars. This will be done by use of Smooth Particle Hydrodynamics (SPH) for a limited number of particles per star (a few hundred). This will allow for the first time close comparison between numerical models and observations of real globular clusters. I refer here to observations of various, peculiar objects like blue stragglers and milliseconds pulsars, which can be formed during collisions and encounters between stars.

## References

Aarseth S.J. & Spurzem R., 1995, in preparation
Bettwieser E. & Sugimoto D., 1984, MNRAS, 208, 498
Cohn H., Wise, M.W., Yoon T.S., Statler T.S., Ostriker J., Hut P., 1986, in Hut P., McMillan S., eds., The Use of Supercomputers in Stellar Dynamics, New York: Springer, p.206
Cohn H., Hut P., 1989, ApJ, 226, 1087
Gao B., Goodman J., Murphy B., Cohn H., 1991, ApJ, 370, 567
Giersz M. & Heggie D.C., 1994a, MNRAS, 268, 257
Giersz M. & Heggie D.C., 1994b, MNRAS, 270, 298
Giersz M. & Spurzem R., 1994, MNRAS, 269, 241
Goodman J., 1987, ApJ, 313, 576
Hénon M., 1971, Astrophys. Sp. Sci., 14, 151
Hénon M., 1975, in Hayli A., ed, Dynamics of Stellar Systems, Reidel: Dordrecht, p.133
Heggie D.C., & Ramamani N., 1989, MNRAS, 237, 757
Makino J., 1995, this volume
Marchant A.B, & Shapiro S.L., 1980, ApJ, 239, 685
Spitzer L., Jr., 1975, in Hayli A., ed, Dynamics of Stellar Systems, Reidel: Dordrecht, p.3
Spitzer L., Jr., !987, Dynamical Evolution of Globular Clusters. Princeton Univ. Press, Princeton, p.77
Stodółkiewicz J.S., 1982, Acta Astr., 32, 63
Stodółkiewicz J.S., 1985, in Goodman J. & Hut P., eds, Dynamics of Star Clusters, Reidel:Dordrecht, p.361
Stodółkiewicz J.S., 1986, Acta Astr., 36, 19
Taiji M., 1995, this volume

# FLUID TECHNIQUES AND EVOLUTION OF ANISOTROPY

RAINER SPURZEM
*Institut für Astronomie und Astrophysik der Universität,
Olshausenstraße 40, D-24098 Kiel*
*e-mail: supas028@astrophysik.uni-kiel.d400.de*

**Abstract.**
Fluid dynamical techniques to model the dynamical evolution of star clusters, and their successors, gaseous models using an equation of heat conductivity to model relaxation effects, including anisotropy, are presented. The historical merits of such models are reviewed as well as the current status of their credibility, based on quantitative comparisons with other methods, like orbit-averaged Fokker-Planck solutions and direct $N$-body simulations.

## 1. Introduction with some historical remarks

Fluid or gasdynamical models of star clusters have successfully been used since many years. As Sugimoto (1985) expressed it ten years ago in the last IAU symposium on the dynamics of star clusters "we can understand physics of self-gravitating systems in terms of gaseous models in so far that their global nature and effects of self-gravity are concerned. ... They include gravothermal collapse/expansion ... and post-collapse evolution with gravothermal oscillation." In this review I want to discuss the progress, which has been made in the understanding of the relevance of fluid or gas dynamical models during the past ten years and convince the reader that there are prospects to create realistic models of stellar systems, including effects of particle-particle interactions, on the basis of such models.

I want to use this occasion first to stress some of the historical merits of the models, which have served for many important discoveries in the past. Thereafter one of them, the anisotropic gaseous model is presented in some more detail, its comparison with direct $N$-body simulations and direct solutions of the orbit-averaged Fokker-Planck equation is discussed, and finally first results on the path towards real models of star clusters are presented.

The term fluid dynamical model was used by Larson in his seminal series of papers on star clusters (Larson 1969, 1970a, b). He derived dynamical equations from the fundamental kinetic Boltzmann equation, including a Fokker-Planck collisional term based on the derivation of Rosenbluth, McDonald & Judd (1958). He assumed that some higher order moments of the velocity distribution function can be derived as in the case of a Maxwell-Boltzmann distribution. Using a series expansion of the distribution function in Legendre polynomials up to second order he could derive the local collisional anisotropy decay timescale (note, however, that his derivation was not fully self-consistent, since he used an isotropized background distribution function). In a direct comparison between Larson's models, direct $N$-body simulations, and Monte-Carlo models of star clusters Aarseth, Hènon & Wielen (1974) showed, however, that Larson's models deviated from the other models in late core collapse.

The physics of global instabilities of self-gravitating gas spheres in the linear approximation has been studied by Hachisu & Sugimoto (1978), who showed that the gravothermal catastrophe, detected by Antonov (1962) and Lynden-Bell & Wood (1968) could be understood as a global instability against the redistribution of heat in a self-gravitating isothermal gas cloud. It was the effect of the negative specific heat in the core of self-gravitating systems, which caused the runaway. In a similar approach Hachisu (1979) studied the so-called "gravo-gyro" catastrophe, caused by a negative specific momentum of inertia in self-gravitating systems. However, until recently there has not been paid much attention to models of rotating star clusters (but see C. Einsel & R. Spurzem, and P.Y. Longaretti, this volume). Spurzem (1991) showed that thermodynamic arguments (maximizing an entropy functional) could also be used to understand the linear response of anisotropy to a redistribution of heat.

Hachisu et al. (1978) first utilized the gas dynamical equations (which can be seen as isotropic version of the moment equations of Larson) with an equation of heat transfer and various scalings of the heat conductivity $\Lambda \propto \rho^\alpha T^\beta$ to model a star cluster. This is a phenomenological closure of the moment equations, instead of a specialization of the distribution function as in Larson's case. Lynden-Bell & Eggleton (1980) found that $\Lambda \propto \rho T^{-1/2}$ is the physical case in which the conductivity scales with the standard stellar dynamical two-body relaxation rate. With such a model fair agreement (e.g. of the self-similar solution for gravothermal collapse) could be reached with the at that time recent competitive models based on the numerical solution of the 1D orbit-averaged Fokker-Planck equation (Cohn 1980). But it was much less clear, to what extent these models were really appropriate for real star clusters like globulars.

Although the process of core collapse can be understood without inclusion of anisotropy, the prospect of modelling real star clusters, which exhibit anisotropy (e.g. Lupton et al. 1987), would require anisotropic models. Here anisotropy is understood as a difference between the radial and tangential velocity dispersions ("temperatures") in a spherically symmetric system. Unfortunately the very efficient scheme of Cohn (1980) to numerically solve the 1D orbit-averaged Fokker-Planck equation could not easily be generalized to the 2D case. So until very recently (Takahashi 1995, K. Takahashi, this volume) anisotropic Fokker-Planck models were not available (with the exception of an early attempt by Cohn (1979), who did not continue this work because of problems with numerical accuracy). Remarkably early, however, Stodółkiewicz (1982, 1986) developed a Monte-Carlo method based on Hènon's work, which was able to numerically simulate large $N$

star clusters including anisotropy. Such a model has been revisited recently by M. Giersz (this volume).

The moment equations, as they were used e.g. by Larson (1970a) include anisotropy; however, it is less obvious how to generalize the closure equation of heat transport in an anisotropic case. So-called one-flux and two-flux closures have been examined, (Bettwieser & Spurzem 1986). In a more systematic study (Louis & Spurzem 1991) it could be shown, that at least the self-similar solutions in both cases were very similar, and even close to the solutions of a higher order moment model of Louis (1990).

Soon the question what happens to globular clusters after core collapse was raised. After Hènon (1975) and Stodółkiewicz (1982) first extended their Monte-Carlo models to the post-collapse phase Inagaki & Lynden-Bell (1983) showed that there is a self-similar post-collapse solution with a central pointlike energy source by using a gaseous sphere model. Bettwieser & Sugimoto (1984), and Heggie (1984) put into their gas sphere models a distributed heating term tailored to describe the energy generation in the core due to formation and hardening of three-body binaries. Here a post-collapse model with a non-singular core was reached, in the case of the first two papers large amplitude gravothermal oscillations were found. Until such oscillations were also detected in the solutions of the Fokker-Planck equation (Cohn, Hut & Wise 1989) it was widely believed that they are an artifact of the gaseous model or, even worse, of particular codes to solve the model equations. Goodman (1987) proved that post-collapse oscillations can be understood as an instability of a self-similar solution. Both gaseous and Fokker-Planck models exhibit a rich dynamical behaviour of their oscillating solutions, with period doublings and possibly chaotic behaviour, similar to the non-linear dynamics in the case of the Rössler- and Lorentz attractors (see e.g. Jackson 1990), which originate from heat conduction problems as well (Breeden et al. 1994, Spurzem 1994).

The question whether large amplitude gravothermal oscillations occur in real, discrete $N$-body systems made the need clear for quantitative, detailed comparisons between direct simulations and models based on the Fokker-Planck equation (gaseous models as well as models based on orbit-averaging, henceforth denoted as statistical models). It has long been argued (cf. e.g. Inagaki 1986) that stochastic fluctuations at core bounce (which always occurs at a very small core particle number, provided three-body binary formation dominates and there were no primordial binaries) destroy the characteristic temperature inversion, which creates the steady gravothermal expansion in the statistical models. To start with the most recent result, J. Makino (this volume) has shown that such a temperature inversion indeed occurs in a real $N$-body system of 32k particles near core bounce and thus should trigger a gravothermal reexpansion.

But there are more questions than that of gravothermal oscillations. The validity of the Fokker-Planck approximation (uncorrelated small angle two-body encounters dominate the evolution) and the possibility to model energy transport by two-body encounters in a nearly collisionless (mean free path long compared to systems dimensions) stellar system by a phenomenological heat transport equation, which can be strictly derived only in the case of a collisional Boltzmann gas, are open theoretical questions. Surprisingly there have not been many quantitative comparative studies between direct $N$-body and other models (despite of considerable development of software and hardware. except of the pioneering study of Aarseth, Hènon & Wielen (1974). Recently, however, Giersz & Heggie (1994a,

b) and Giersz & Spurzem (1994, henceforth GS) have shown that for the equal point mass case there is fair agreement between all models for $N \leq 2000$, provided the stochastic $N$-body fluctuations are overcome by ensemble averaging of several statistically independent $N$-body simulations.

## 2. The anisotropic gaseous model

To be definite and clear for the reader unfamiliar with gaseous models I would like to give in the following a short, but complete description of the variables and equations used for a standard model.

Let the dependent variables be the mass $M_r$ contained in a sphere of radius $r$, the local mass density $\rho$, radial and tangential pressure $p_r$, $p_t$, bulk mass transport velocity $u$, and transport velocities $v_r$, $v_t$ of the radial and tangential energy, respectively. As auxiliary quantities the radial and tangential 1-D velocity dispersions $\sigma_r^2 = p_r/\rho$, $\sigma_t^2 = p_t/\rho$, the average velocity dispersion $\sigma^2 = (\sigma_r^2 + 2\sigma_t^2)/3$, the anisotropy $A = 2 - 2\sigma_t^2/\sigma_r^2$ and the relaxation time

$$T = \frac{9}{16\sqrt{\pi}} \frac{\sigma^3}{G^2 m \rho \log(\gamma N)} \qquad (1)$$

in the definition of Larson (1970a) are used, where $N$ is the total particle number of the star cluster, $m$ the individual stellar mass and $\gamma$ a numerical constant whose value will be discussed below. The equations are

$$\frac{\partial M_r}{\partial r} = 4\pi r^2 \rho \ ; \quad \frac{\partial \rho}{\partial t} + \frac{1}{r^2}\frac{\partial}{\partial r}(\rho u r^2) = 0 \qquad (2)$$

$$\frac{\partial u}{\partial t} + u\frac{\partial u}{\partial r} + \frac{GM_r}{r^2} + \frac{1}{\rho}\frac{\partial p_r}{\partial r} + 2\frac{p_r - p_t}{\rho r} = 0 \qquad (3)$$

$$\frac{\partial p_r}{\partial t} + \frac{1}{r^2}\frac{\partial}{\partial r}(p_r u r^2) + 2p_r \frac{\partial u}{\partial r} + \frac{3}{r^2}\frac{\partial}{\partial r}(p_r(v_r - u)r^2)$$

$$- 4\frac{p_t(v_t - u)}{r} = -\frac{2}{3}\frac{p_r - p_t}{\lambda_A T_A} + \left(\frac{\delta p_r}{\delta t}\right)_{bin3} \qquad (4)$$

$$\frac{\partial p_t}{\partial t} + \frac{1}{r^2}\frac{\partial}{\partial r}(p_t u r^2) + 2\frac{p_t u}{r} + \frac{1}{r^2}\frac{\partial}{\partial r}(p_t(v_t - u)r^2)$$

$$+ 2\frac{p_t(v_t - u)}{r} = \frac{1}{3}\frac{p_r - p_t}{\lambda_A T_A} + \left(\frac{\delta p_t}{\delta t}\right)_{bin3} \qquad (5)$$

$$v_r - u + \frac{\lambda}{4\pi G\rho T}\frac{\partial \sigma^2}{\partial r} = 0 \ ; \quad v_r = v_t \qquad (6)$$

The net transport velocities for radial and tangential energy $(v_r - u)$ and $(v_t - u)$ can be derived from the energy fluxes $F_r$ and $F_t$ (which are identified with the third order moments of the velocity distribution) by dividing out a convenient multiple of the relevant pressure ($2p_t$ for $(v_t - u)$, $3p_r$ for $(v_r - u)$ ). The reader interested in more details about this and the connection of the variables to moments of the stellar velocity distribution is referred to Louis & Spurzem (1991).

The numerical constants $\lambda_A$, $\lambda$ and $\gamma$ occurring in Eqs. (4) to (6) are related to the timescales of collisional anisotropy decay and heat transport, and to the Coulomb logarithm, respectively. $\lambda$ is related to the standard $C$ constant in isotropic gaseous models (see e.g. Heggie 1984) by $\lambda = 2.7\sqrt{\pi}C$. $T_A$ is the anisotropy decay timescale for an anisotropic local velocity distribution function; for a generalization of Larson's (1970a) distribution function (series of Legendre polynomials) including anisotropy it is $T_A = 10\,T/9$ (Louis & Spurzem 1991). $\lambda_A$ is discussed in Sect. 4. Additional terms due to the average heating by formation and hardening of three body binaries (see e.g. Cohn 1985) are

$$\left(\frac{\delta p_r}{\delta t}\right)_{bin3} = \frac{2}{3}C_b\frac{\rho^3}{m^2\sigma^2}\left(\frac{Gm}{\sigma}\right)^5 \;;\; \left(\frac{\delta p_t}{\delta t}\right)_{bin3} = \left(\frac{\delta p_r}{\delta t}\right)_{bin3}. \quad (7)$$

This is an isotropic energy input. It is shown in GS and Giersz and Heggie (1994a, b) that for particle numbers between $N = 1000$ and $N = 10.000$ the best agreement between direct $N$-body calculations, direct solutions of the orbit-averaged Fokker-Planck equation and this anisotropic gaseous models is achieved for one set of parameters, namely $\lambda = 0.4977$ (i.e. $C = 0.104$), $\gamma = 0.11$, $\lambda_A = 0.1$, and $C_b = 90$. The latter value used to be the standard value derived from theoretical arguments, based on a numerical factor of $\tilde{C} = 0.9$ in the formula for the formation rate of three-body binaries (Hut 1985). Recently, Goodman & Hut (1993) argue, that $\tilde{C} = 0.75$ is a better value, but still within some uncertainty. The results of comparisons with $N$-body simulations show that $C_b = 90$ is a fairly reasonable value, but within the uncertainty $C_b = 75$ (which would ensue with the new formation rate) cannot be ruled out. Note that the value of $\gamma$ found empirically is somewhat smaller than e.g. Spitzer's (1987) standard value ($\gamma = 0.4$).

As for multi-component models the simplest approach is to take dynamical equations like those above (including the closure equation) for each component separately, then coupling them by gravity (via Euler's equation) only and self-consistent collisional terms for the decay of anisotropy and the exchange of energy. The results for the collisional terms have been reported in the Appendix of Spurzem & Takahashi (1995, henceforth ST). Therein we also argue, that such a model is in much better agreement with direct Fokker-Planck solutions than previously argued (Bettwieser & Inagaki 1985). The main reason is a much more complicated additional coupling between the components within the conductivity equation adopted by Bettwieser & Inagaki (1985) as compared with the new model (Spurzem 1992).

## 3. Comparisons

Figs. 1 to 3 are taken from GS and visualize the quality of agreement between $N$-body an standard gaseous model for an equal mass system in pre- and post-collapse (low $N$, averaged $N$-body simulation). It also illustrates the influence of a possible variation of $C_b$. Note that the agreement is non-trivial, because the main free parameter in the gaseous model ($\lambda$ has been fixed already by comparison with the orbit-averaged Fokker-Planck model.

Fig. 4 shows a similar comparison, but now for an *individual* $N = 10000$-body simulation, compared with the standard anisotropic gaseous model ($C_b = 90$). The

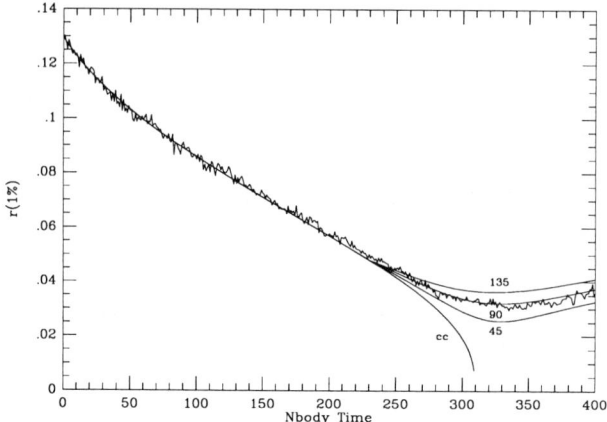

*Figure 1.* Evolution of the 1% Lagrangian radius in an averaged $N = 1000$ $N$-body model in comparison to the standard anisotropic gaseous model for varying strengths of the binary energy generation (curves labelled by $C_b$-value, see main text). The curve labelled $cc$ is the one for $C_b = 0$, i.e. pure core collapse.

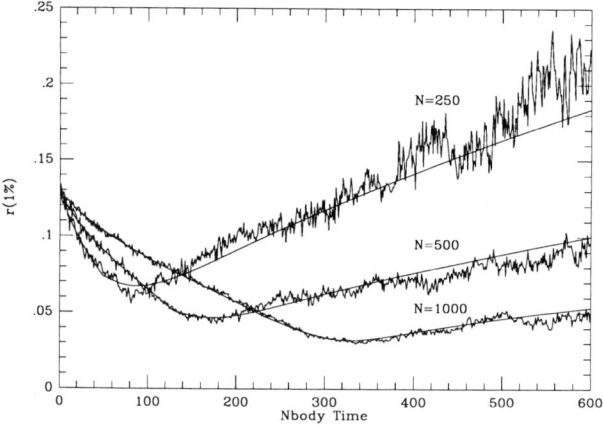

*Figure 2.* Evolution of the 1% Lagrangian radius in averaged $N$-body models for $N = 250, 500, 1000$ (here taken $C_b = 55$, $C_b = 70$, $C_b = 90$, respectively). For the reasons why to take smaller $C_b$ for small $N$ see discussion in GS.

good agreement in pre-collapse again underlines that the Fokker-Planck approximation is valid in this evolutionary phase, however, we now note a significant discrepancy in collapse times and oscillations of the $N$-body central density which do not show up in the gaseous model. The discrepancy in collapse time is seen as a result of poor statistics. One can estimate an expected spread in collapse times for an $N = 10000$ system of about 130 time units, so the actual collapse time is just 1.4 $\sigma$ apart from the average (Spurzem & Aarseth 1996). How about the oscillations of the $N$-body model? Are they gravothermal? The author has spent considerable effort in looking at these data for inversions of the temperature gradients, or its signature in the cumulative distribution function to find any trace of this necessary feature of gravothermal oscillations, as it was already clearly outlined in the original paper by Bettwieser & Sugimoto (1984). Since they could *not* be found the oscillations are interpreted as binary-driven, generated by stochastic binary

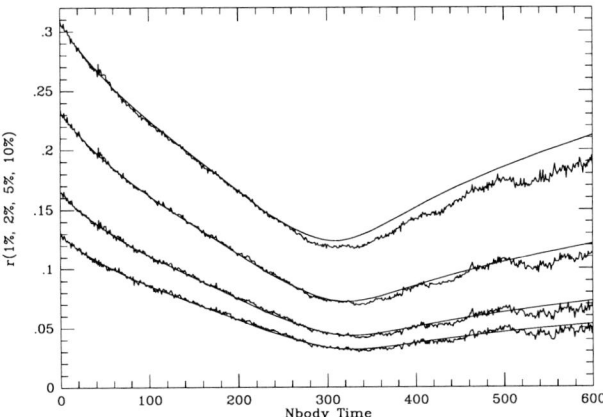

*Figure 3.* Evolution of the 1 to 10% Lagrangian radius in an averaged $N = 1000$ $N$-body model in comparison to the standard anisotropic gaseous model ($C_b = 90$).

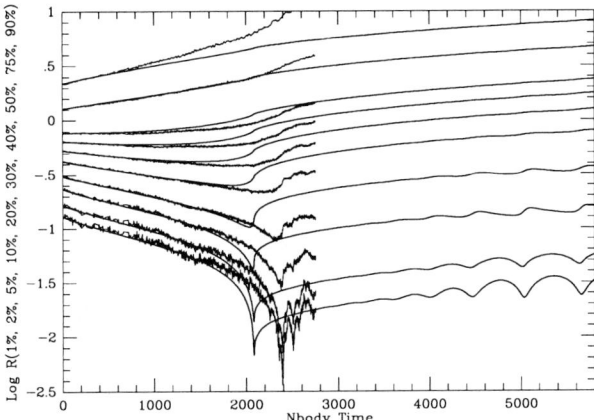

*Figure 4.* Evolution of the 1 to 10% Lagrangian radius in a single $N = 10000$ $N$-body model in comparison to the standard anisotropic gaseous model ($C_b = 90$).

encounter events, which do not persist for long times *after* the binary activity has ceased (see also the critical assessment of what real gravothermal oscillations are by S.L.W. McMillan, this volume). Such interpretation is supported by observations in the $N$-body simulation that at the beginning of an expansion phase there is a strong binary scattering event and that the expansion ceases after an active binary has been lost by escape or becomes inactive (e.g. by ejection into the halo) (Spurzem & Aarseth 1996).

It is interesting to note that Takahashi (1995, and this volume) finds in his new 2D Fokker-Planck models a collapse time for an $N = 10000$ model, which agrees much better with the here presented $N$-body data. He claims that the collapse time is considerably longer in the anisotropic case, which coincides with a result published by Louis (1990), based on a fifth-order moment model. But presently it is difficult to judge about this conjecture from the viewpoint of $N$-body simulations, because the variations of collapse times in $N = 10000$ $N$-body simulations is of the same order as the postulated difference between isotropic and anisotropic

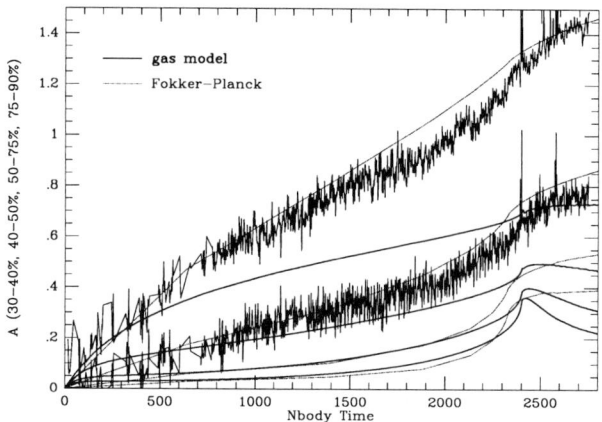

*Figure 5.* Anisotropy averaged between the indicated Lagrangian radii for direct $N = 10000$ body simulation (fluctuating curves, only for two outermost zones) and for comparison 2D Fokker-Planck model by Takahashi (1995) and standard anisotropic gaseous model. Times were rescaled such that core bounce occurs at the same point of the abscissa for all models in order to allow a better comparison.

models. So we have to wait for better statistics and more $N$-body models with larger $N$. Since the scatter in collapse times in relation to the collapse time itself becomes smaller with increasing $N$ a few larger $N$ simulations will give much more significant results on the average collapse time.

## 4. The anisotropy

It turns out that a canonical value of $\lambda_A$ in the gaseous model equations, as it would turn out for some standard anisotropic distribution function, yields much larger anisotropy than in the $N$-body models. Reasonable agreement inside the the half-mass radius can be achieved for $\lambda_A = 0.1$ for all cases of $N$, single and two-mass models (see GS and ST). Such a result is theoretically not well understood, could be related, however, to the fact, that the local approximation used in the gaseous model becomes obsolete in the halo regions, where anisotropy prevails. Stars on radial orbits suffer encounters during there passage through the core, where there is a much higher density, thus the collisional decay of anisotropy ought to be shorter than by a local estimate in the halo, which is consistent with the above findings. Since the orbit average of the direct Fokker-Planck solution includes for a given orbit contributions to the diffusion coefficients from different radii such a discrepancy should not occur in 2D orbit-averaged Fokker-Planck models. Indeed first results by Takahashi (1995, this volume) point this out. In Fig. 5 we show the anisotropy for the outer mass shells ($N = 10000$) in comparison between his new 2D Fokker-Planck results, standard anisotropic gaseous model and the direct $N$-body simulations. The 2D Fokker-Planck model agrees in the outermost shells much better with the $N$-body system. For the intermediate shells (Lagrangian radii containing 30 to 50 % of total mass) fluctuations of the measured $N$-body anisotropy were too large to allow for a reasonable plot.

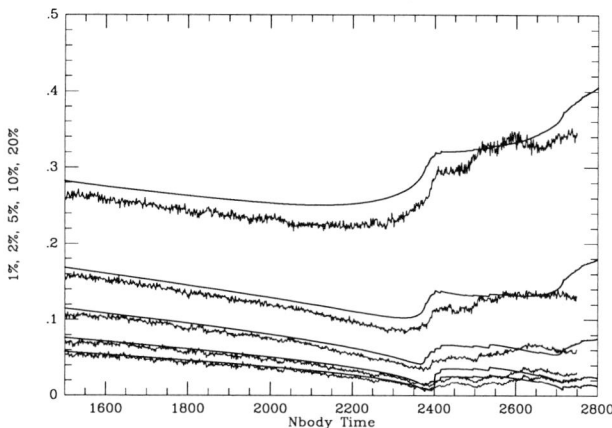

*Figure 6.* Lagrangian radii containing 1 to 20 % of the total mass for an anisotropic gaseous model with stochastic binaries compared to the direct $N = 10000$ $N$-body simulation. To compare the characteristic features of the post-collapse evolution the data have been scaled in time such that both models have core bounce at the same point of the abscissa, as in Fig. 5.

## 5. Outlook and Conclusion

It has been shown that the anisotropic gaseous model of star clusters matches for a wide range of $N$ and single and two-mass star clusters, and for a standard parameter choice, very well with expectations from direct solutions of the orbit-averaged Fokker-Planck equation and averaged direct $N$-body simulations. The underlying assumptions (Fokker-Planck approximation, small angle two-body encounters dominate evolution in core collapse, and their energy transport properties can be modelled by a heat transfer equation analogous to gas dynamics, but with a specially tailored conductivity to account for the stellar dynamical relaxation timescale) and a statistical treatment of the average heating rate by formation and hardening of three-body binaries are supported by this result.

Differences between gaseous models and other solutions still remain in the case of the anisotropy in the outer halo regions and and in the stochastic behaviour of *individual* $N$-body models compared with the statistical gaseous model. The first problem might be overcome in higher order moment models, because it is related to the anisotropy decay in the collisional terms, which rely on certain assumptions on the functional form of the velocity distribution. The second problem is tackled by a stochastic treatment of binaries as in a Monte-Carlo model (cooperation with M. Giersz in progress, see a very similar approach for isotropic Fokker-Planck models by Takahashi & Inagaki 1991). Fig. 6 shows a gaseous model with stochastic binaries as compared with the $N$-body simulation; the characteristic behaviour of the real $N$-body system is matched very well. Including stochastic binaries and other effects like stellar evolutionary mass loss into the model in the near future will generate a very efficient realistic model of a star cluster.

## Acknowledgements

Many thanks to M. Giersz and K. Takahashi for fruitful cooperation and data exchange. Computations for the $N$-body model made at HLRZ Jülich, Germany.

# References

Aarseth S.J., Hénon M., Wielen R., A& A, 37, 183
Antonov V.A., 1962, Vest. Leningrad Gos. Univ., 7, 135
Bettwieser E., Inagaki S.J., 1985, MNRAS, 213, 473
Bettwieser E., Spurzem R., 1986, A& A, 161, 102
Bettwieser E., Sugimoto D., 1984, MNRAS, 208, 493
Breeden J.L., Cohn H.N., Hut P., 1994, ApJ, 421, 195
Cohn H., 1979, ApJ, 234, 1036
Cohn H., 1980, ApJ, 242, 765
Cohn H., 1985, in J. Goodman, P. Hut, eds, Proc. IAU Symp. 113, Dynamics of Star Clusters, Reidel, Dordrecht, p. 161
Cohn H., Hut P., Wise M., 1989, ApJ, 342, 814
Goodman J., 1987, ApJ, 313, 576
Goodman J., Hut P., 1993, ApJ, 403, 271
Giersz M., Heggie D.C., 1994a, MNRAS, 268, 257
Giersz M., Heggie D.C., 1994b, MNRAS, 270, 298
Giersz M., Spurzem R., 1994, MNRAS, 269, 241
Hachisu I., 1979, PASJ, 31, 523
Hachisu I., Sugimoto D., 1978, Prog. Theor. Phys., 60, 123
Hachisu I., Nakada Y., Nomoto K., Sugimoto D., 1978, Prog. Theor. Phys., 60, 393
Heggie D.C., 1984, MNRAS, 206, 179
Hénon M., 1975, in A. Hayli, ed, Proc. IAU Symp. 69, Dynamics of Stellar Systems, Reidel, Dordrecht, p. 133
Hut P., 1985, in J. Goodman, P. Hut, eds, Proc. IAU Symp. 113, Dynamics of Star Clusters, Reidel, Dordrecht, p. 231
Jackson E.A., 1990, Perspectives of nonlinear dynamics, Cambridge Univ. Press, Cambridge, GB
Inagaki S., 1986, PASJ, 38, 853
Inagaki S., Lynden-Bell D., 1990, MNRAS, 244, 254
Larson R.B., 1969, MNRAS, 145, 405
Larson R.B., 1970a, MNRAS, 147, 323
Larson R.B., 1970b, MNRAS, 150, 93
Louis P.D., 1990, MNRAS, 244, 478
Louis P.D., Spurzem R., 1991, MNRAS, 251, 408
Lupton R.H., Gunn J.E., Griffin R.F., 1987, AJ, 93, 1114
Lynden-Bell D., Wood R., 1968, MNRAS, 138, 495
Lynden-Bell D., Eggleton P.P., 1980, MNRAS, 191, 483
Rosenbluth M.N., MacDonald W.M., Judd D.L., 1957, Phys. Rev. 2nd Ser., 107, 1
Spitzer L., 1987, Dynamical Evolution of Globular Clusters, Princeton Series in Astrophysics, Princeton University Press, New Jersey, USA
Spurzem R., 1991, MNRAS, 252, 177
Spurzem R., 1992, in Rev. Mod. Astr., 5, ed. G. Klare, Springer-Verlag, p. 161
Spurzem R., 1994, in Pfenniger D., Gurzadyan V.G., eds, Ergodic Concepts in Stellar Dynamics, Springer-Vlg., Berlin, Heidelberg, p. 170
Spurzem R., Aarseth S.J., 1996, MNRAS, preprint, subm.
Spurzem R., Takahashi K., MNRAS, 272, 772
Stodółkiewicz J.S., 1982, Acta Astronomica, 32, 63
Stodółkiewicz J.S., 1986, Acta Astronomica, 36, 19
Sugimoto D., 1985, in J. Goodman, P. Hut, eds, Proc. IAU Symp. 113, Dynamics of Star Clusters, Reidel, Dordrecht, p. 207
Takahashi K., Inagaki S., 1991, PASJ, 43, 589
Takahashi K., 1995, PASJ, 47, 561

# THE ROLE OF BINARIES IN THE DYNAMICAL EVOLUTION OF THE CORE OF A GLOBULAR CLUSTER

PIET HUT
*Institute for Advanced Study, Princeton, NJ 08540, U.S.A.*

**Abstract.**
    The size of the core is one of the main diagnostics of the evolutionary state of a globular cluster. Much has been learned over the last few years about the behavior of the core radius during and after core collapse, under a variety of different conditions related to the presence or absence of large numbers of binaries. An overview is presented of the basic physical principles that can be used to estimate the core radius. Four different situations are discussed, and expressions are presented for the ratio $r_c/r_h$ of core radius to half mass radius. The regimes are: deep collapse in the absence of primordial binaries; steady post-collapse evolution after primordial binaries have been burned up; chaotic post-collapse evolution under the same conditions; and post-collapse evolution in the presence of primordial binaries. In addition, modifications to all of these cases are indicated for the more realistic situation where effects of the galactic tidal field are taken into account.

## 1. Introduction

Core collapse is the most dramatic phenomenon in the evolution of globular clusters. It corresponds to the pre-main-sequence stage in stellar evolution, when the energy losses from the photosphere are not yet balanced by nuclear energy generation in the center. In contrast, during the post-collapse evolution of a star cluster some form of central energy source turns on, which makes up for the energy lost through conduction and evaporation of stars from the central regions (for a review of different mechanisms, see Goodman 1993).

The details of the halt of core collapse only emerged some dozen years ago, and were reported in IAU symposium 113 (Goodman & Hut, 1985). A few years later, the plot thickened again through the discovery that a significant fraction of stars in globular clusters were contained in binaries. These primordial binaries modify the picture of core collapse considerably. Rather than proceeding down to the miniscule core and enormous densities required in the earlier single-star models, core collapse is halted at moderate central densities of order $\rho_c \sim 10^6 M_\odot \text{pc}^{-3}$, and a typical core radius of $r_c \sim 0.1\text{pc}$. The corresponding core contains a few $\times 10^3 M_\odot$, roughly one percent of the inner half mass of the cluster enclosed in a radius $r_h \sim 10\text{pc}$ (Goodman & Hut, 1989).

A few years ago, we have given a detailed review of primordial binaries in globular clusters, and their effects (Hut et al. 1992). I refer to this paper for many original references, as well as summaries of the history of this field. Many of the more recent theoretical developments concerning the role of binaries in star clusters are discussed by others in these proceedings, e. g. in the contributions by Aarseth, Clarke, Heggie et al., Kiseleva, Leonard, Mardling, McMillan, Phinney, and Rasio & Heggie. In order to avoid too much overlap with these other papers, I will concentrate here on a single question that is perhaps most relevant to observations: how to estimate the core size $r_c$ of a post-collapse cluster. In four separate situations, I will describe the basic physics from which the ratio $r_c/r_h$ can be estimated, where $r_h$ is the half-mass radius. These are given in §§2-5 below. §6 discusses the consequences of adding a galactic tidal field, and §7 sums up.

## 2. Deep Single-Star Core Collapse

In the late stages of core collapse, the core dynamically decouples from the bulk of the cluster, in a process called 'gravothermal catastrophe' (Lynden-Bell & Wood 1968; Spitzer 1987). The reason for this instability is the fact that dynamical evolution proceeds on a relaxation time scale. The relaxation rate is proportional to the number of encounters per star, which in turn is proportional to the density, which is far higher in the core of an evolved cluster than in the bulk of the cluster. Therefore, when the core is losing energy by local 'evaporation' of stars into the surrounding regions, as well as by heat conduction into those regions, the rest of the material has no time to keep up with the shrinking core. Collapse will occur, leading to an infinite density in finite time, unless some other physical process will switch on.

Let us make a rough estimate of the critical size for the core of a star cluster, when core collapse is halted. In the equal-mass point-mass approximation, heat production is proportional to the rate of binary formation in three-body encounters, a process that is proportional to the third power of the density:

$$\dot{E}_+ = C_3 \rho^3. \quad (1)$$

Energy losses, caused by two-body relaxation, are proportional to the number of encounters per unit volume, which in turn is proportional to the square of the density $\rho$:

$$\dot{E}_- = C_2 \rho^2. \quad (2)$$

A natural local choice of physical units is one in which the gravitational constant $G$, the mass of a single star $m$, and the root-mean-square velocity of the single stars $v$ are all set equal to unity: $G = m = v = 1$. Since the core is close to being self-gravitating, we can use the virial theorem:

$$N_c(\frac{1}{2}mv^2) = \frac{G(mN_c)^2}{4r_c}, \quad (3)$$

where $N_c$ is the number of particles in the core and $r_c$ the core size. We thus find $N_c = 2r_c$, and $\rho = N/((4/3)\pi r_c^3) \simeq 2N^{-2}$. Requiring now that $\dot{E}_+ = \dot{E}_-$, we find

$$N_c = \sqrt{\frac{2C_3}{C_2}} = \sqrt{\frac{20}{0.003}} \simeq 80, \quad (4)$$

where the estimates for the numerical values for the constants are taken from Hut & Inagaki (1985). A similar estimate leading to $N_c \simeq 50$ was made by Goodman (1984). It is clear that deep core collapse leads to the momentary appearance of a tiny core with less than one hundred stars, with $r_c \ll 0.01$pc corresponding to a huge density, of order $10^9 M_\odot \text{pc}^{-3}$, for typical globular cluster parameters.

## 3. Single-Star Post-Collapse Evolution: Smooth Core Reexpansion

Long after core collapse, on a time scale of several half-mass relaxation times, the outer regions of the cluster are finally able to catch up with what has happened in the center. For small to moderate numbers of stars ($N \lesssim 10^4$; Goodman 1987), the core is able to adjust itself to the demands of these outer regions. Given the rate of energy loss from these regions, the core will find an equilibrium size and corresponding density such that it will form new binaries at the correct rate to make up for the outer energy losses.

Goodman (1984) has made a detailed analysis of this situation, and derived a relation $N_c \propto N^{1/3}$, which implies $r_c/r_h \propto N^{-2/3}$. For realistic globular cluster parameters, he found $N_c \sim 3 \times 10^2$, six times larger than the deep collapse figure of $N_c \sim 50$ that he found. The reason that the core radius becomes about six times smaller during deep collapse, compared to subsequent smooth expansion, is that the latter takes place in near-isothermal equilibrium. As a result, the conduction rate of energy through two-body relaxation is much lower than during the original collapse, which leaves a more substantial temperature gradient in its wake. Lowering the conduction rate is equivalent to lowering the value $C_2$, which has the effect of increasing $N_c$, according to eq. 4.

## 4. Single-Star Post-Collapse Evolution: Chaotic Core Reexpansion

For large numbers of stars ($N \gg 10^4$), post-collapse expansion of the core of a star cluster does not proceed smoothly. As discovered by Sugimoto and Bettwieser (1983; Bettwieser and Sugimoto 1984), chaotic fluctuations occur in the size of the core radius. These can be explained as a consequence of the gravothermal instability, and were therefore called 'gravothermal oscillations'.

The underlying physical mechanism can be characterized as follows. For a large number of stars in the system, the inner relaxation timescale is much larger than the half-mass relaxation timescale, which determines the overall rate of expansion. Therefore, the inner regions have the tendency to evolve on a timescale much smaller than the bulk expansion timescale. As a result, the inner regions tend to get impatient, and a small fluctuation can trigger a local re-collapse, followed by a local re-expansion. The larger the number of stars, the more the central and outer timescales are decoupled, and the more chaotic the oscillations become. The dynamical behavior of these oscillations can be shown to be characterized by a low-dimensional chaotic attractor (Breeden et al. 1990; Cohn et al. 1991; Breeden & Cohn 1995). The gravothermal character of the core oscillations was confirmed explicitly by Goodman (1987), who performed a linear stability analysis of a new regular self-similar model for post-collapse evolution, and classified the different modes of behavior according to the type of linear instability they exhibit.

Like gravothermal collapse, gravothermal oscillations appear to be a ubiquitous phenomenon, at least in the models which treat the stars and all physical processes as continuous quantities. Inagaki (1986) and McMillan (1986, 1989) have expressed doubts as to whether the oscillations persist in real clusters, where the stars and the physical processes are discrete, and statistical fluctuations may be large. This issue has now been resolved by direct N-body simulations of systems containing $> 10^4$ stars, and the conclusion is that oscillations do occur, and indeed have a clearly gravothermal nature (Makino, this volume).

Although the details of the gravothermal oscillations are strongly dependent on the total number of stars, as well as their mass spectrum, the maximum value of the core radius is relatively insensitive to those details. Since the evolution slows down most around the time of maximum expansion, observations of a cluster core will find the core to be near maximum expansion in the overwhelming majority of cases (just as a binary star in a very eccentric orbit will almost always have a separation close to twice the semimajor axis). Typical values, from Fokker-Planck calculations including a mass spectrum, are $r_c/r_h \sim 10^{-2}$ (Murphy et al. 1990).

## 5. Post-Collapse Evolution with Primordial Binaries

Among globular cluster stars, a fair fraction are born as members of binaries. The fraction may not be as large as that in the galactic disk, but it is large enough to have a significant dynamical effect (see Hut et al. 1992 for a detailed review and references). The overall cluster binary population is conveniently parametrized by the binary fraction $f_B$, defined as the number of "objects" in a cluster that are actually binaries (so, if binary components are representative of the cluster as a whole, the binary mass fraction is $\sim 2f_B/[1+f_B]$). Because of the presence of many observational selection effects, this quantity is not known accurately, but it probably lies in the range 3–30%, and a value of $f_B \sim 10\%$ is widely taken to be "typical."

The presence of primordial binaries radically changes the picture sketched above, in which clusters undergo deep core collapse, before re-expanding, with or without core oscillations. As was pointed out by Goodman & Hut (1989), gravitational 'burning' of binaries will cause the core collapse to be halted at a far larger core radius of $r_c \approx 0.02 r_h$, compared to $r_c \sim 10^{-4} r_h$ for deep core collapse.

This phenomenon has its analogue in stellar evolution, where a protostar does not land directly on the hydrogen main sequence, but instead spends a brief time on the deuterium main sequence. While primordial deuterium is being burned in the center of the star, its radius remains significantly larger than its eventual value. The reason is the high efficiency of deuterium burning at relatively low temperature, compared to hydrogen burning. As a result, central heat production and surface heat loss are balanced at a lower central temperature, and hence larger stellar radius.

Similarly, the high efficiency of gravitational energy extraction from an existing binary population, compared to binary production from single stars only, allows the cluster core to remain relatively large, until the primordial binaries have been depleted. Numerical conformation of the semi-quantitative estimate given above, for $N$-body simulations with $N \leq 2 \times 10^3$

(McMillan et al. 1990, 1991; McMillan and Hut 1994) showed that for $N = (1 \sim 2) \times 10^3$, the core radius $r_c \approx (0.10 \sim 0.15)r_h$. This is larger than the value predicted by Goodman and Hut (for this $N$ range, their eq. (6) would give $r_c \approx 0.05r_h$). Most likely, this discrepancy is caused by the low $N$ values used in direct $N$-body simulations so far; it is not clear whether values of $N \sim 10^3$, with total numbers of core stars $N_c < 10^2$, are high enough to reach the asymptotic scaling regime. Fokker-Planck calculations by Gao et al. (1991) resulted in values in the range $r_c \approx (0.01 \sim 0.04)r_h$, in broad agreement with the earlier analytical estimate. A more definite answer will soon be provided, when we will run primordial binary simulations on the GRAPE-4, for $N \geq 2 \times 10^4$.

## 6. Effects of the Galactic Tidal Field on Post-Collapse Evolution

Let us now take into account the fact that globular clusters do not evolve in isolation, but are tidally limited by the presence of the gravitational field of the galaxy. This means that the average density of the cluster is forced to remain at a fixed value, comparable to the average density of the matter inside its orbit (for a circular orbit; the situation is qualitatively similar for orbits of moderate eccentricity). As a result, $r_h$ is forced to shrink upon mass loss through evaporation of cluster stars, in stark contrast to the evolution of isolated clusters, for which $r_h$ grows steadily after core collapse.

Since mass segregation tends to concentrate the heavier binaries towards the core, most of the evaporating stars are single. Therefore, the depletion of binaries in the core by gravitational 'burning' may or may not be offset by the depletion of single stars through evaporation from the outer parts of the cluster. For each cluster, with a given mass and size, there is a critical binary fraction $f_{B,crit}$, forming a watershed between these two possibilities. For $f_B < f_{B,crit}$, primordial binaries will be burned up before the cluster as a whole dissolves in the tidal background field. In this case, the later stages of cluster evolution will show gravothermal oscillations (if $N$ is large enough). However, for $f_B > f_{B,crit}$, primordial binaries will remain present in the core until the end, and the overall binary fraction $f_B$ will in fact increase in the later stages of the cluster. Gravothermal oscillations will not occur during any stage of cluster evolution, in this case.

These two possibilities are illustrated in fig. 1, based on the simulations reported by McMillan and Hut (1994). The watershed value here is $f_{B,crit} \sim$ 15%. The precise value of $f_{B,crit}$ shows some dependence on the cluster parameters, especially on the size of the tidal radius $r_t$ (for $r_t < 0.8r_h$, the standard choice in our simulations, $f_{B,crit}$ will be smaller, since single star evaporation will be more rapid). In practice, $f_{B,crit}$ may well be lower than estimated here, if we take tidal shocking processes into account, which

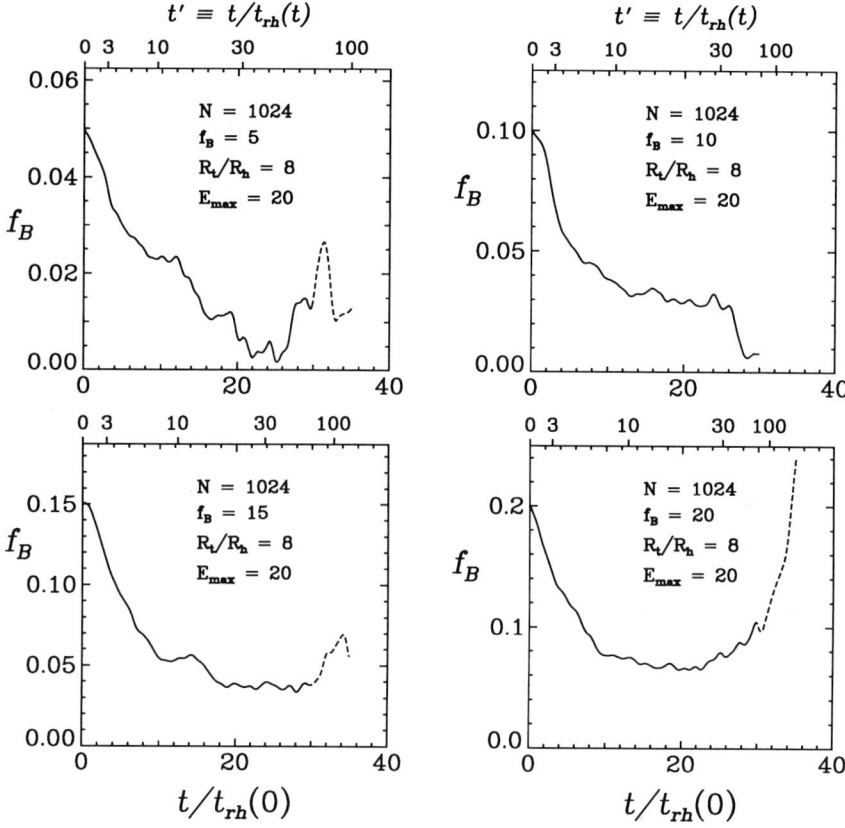

Figure 1. Binary fraction $f_B$ as a function of time $t$, in units of the original half-mass relaxation time. $N$ denotes the number of particles in the simulation, $R_t/R_h$ the ratio between tidal radius and half-mass radius, and $E_{max}$ the maximum value in the spectrum of initial binding energies of the primordial binaries. The dashed portion of the lines indicate where the total number of stars has dropped to such a low value that low-number statistics make the result unreliable.

recently have been shown to be more efficient than previously estimated (cf. Weinberg 1994; Kundić & Ostriker 1995).

In fig. 2, the core population of binaries is followed. Plotted is $f_{Bc}$, the fraction of objects in the core that are binaries, rather than single stars or multiple star systems containing more than two stars. Again, it is clear that for $f_B < 15\%$ the core binary population will eventually drop to very low values, whereas for $f_B > 15\%$ the core binary population will remain constant, within the numerical noise. The fraction of the cluster mass that is present in the core will increase towards the end of the lifetime of the cluster, as the outer regions are eaten away by the galactic tidal field (see McMillan and Hut 1994 for a detailed discussion). For values of $f_B$ above

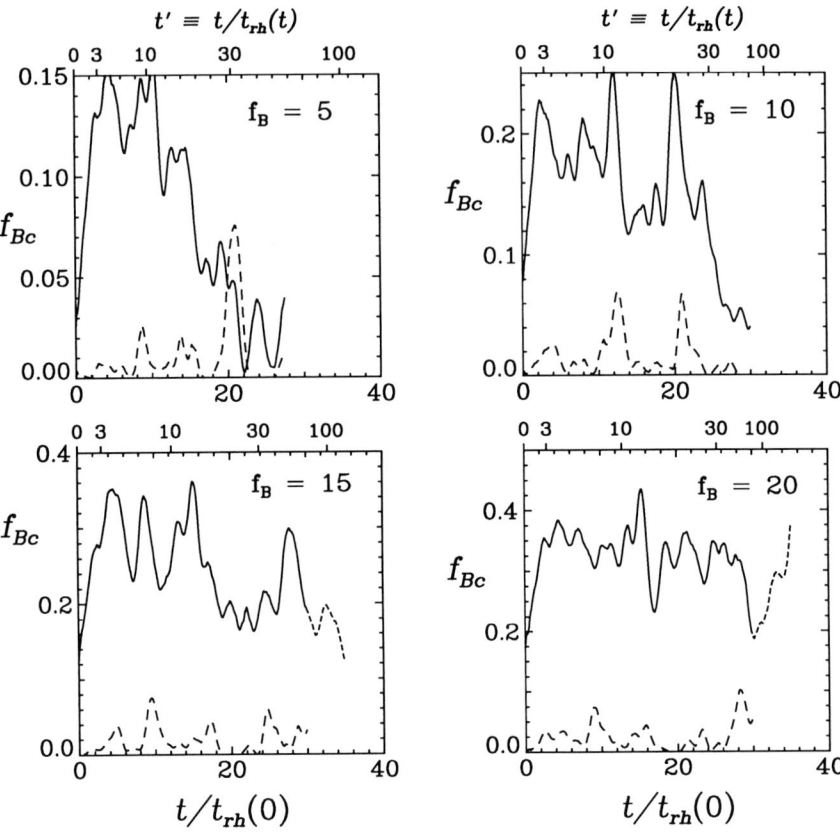

*Figure 2.* Binary fraction $f_{Bc}$ in the core, as a function of time. $f_B$ denotes the initial value of the overall binary fraction. The long-dashed line indicates the fraction of triple stars.

the watershed, this implies a growing core size, in units of the half mass radius, as is evident in Fig. 3. For smaller values of $f_B$, however, there is a near-cancellation of two tendencies, resulting in a near-constant value for $r_c/r_h$. An isolated cluster would show a shrinking core (fewer binaries lead to a higher density required for the same energy generation rate, fixed by the rate of energy loss at this outskirts), and an expanding half-mass radius. Both effects would lead to a decrease in $r_c/r_h$. A tidally truncated cluster, however, would show a decrease in $r_h$ as well as a decrease in $r_c$. In the case of the present simulations, these two effects turn out to be comparable in magnitude.

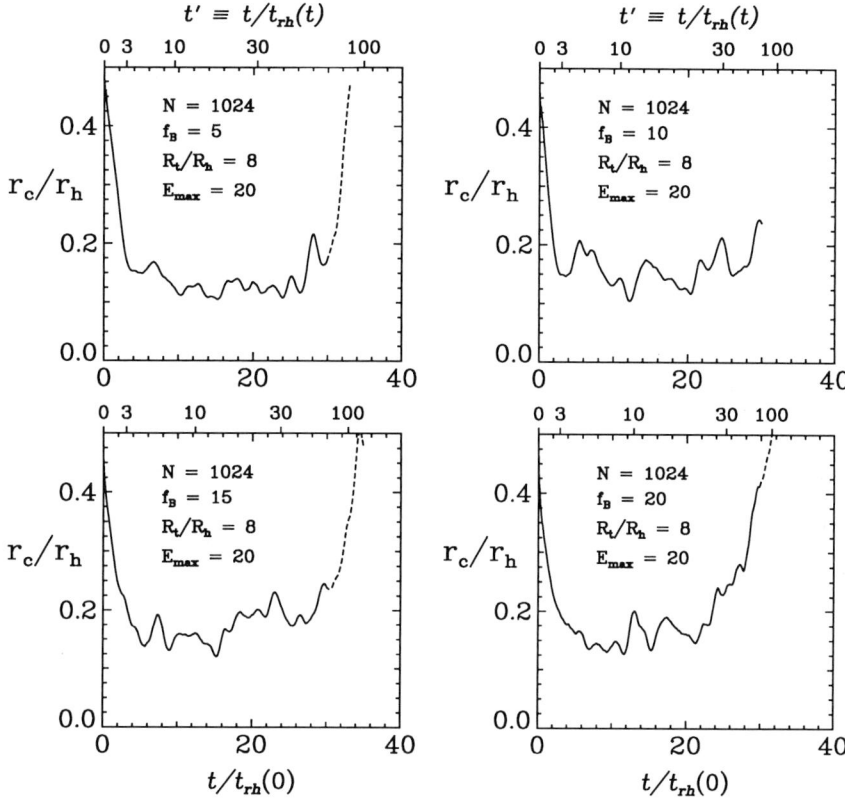

Figure 3. Core radius $r_c$ in units of the half mass radius $r_h$, as a function of time; labels as in fig. 1.

## 7. Summary and Discussion

It is quite possible that primordial binaries may provide the only type of fuel needed to power post-collapse evolution, for most globular clusters (McMillan & Hut 1994). In this case deep core collapse will not occur, and the core will contain a significant fraction of binaries at any time, even though the rest of the cluster may become more depleted in binaries through mass segregation.

For those clusters that do run out of primordial binaries, subsequent post-collapse evolution most likely will show chaotic core oscillations, with relatively large core radii almost all of the time, comparable to the case in which primordial binaries are still present. In both cases, $r_c/r_h \sim 10^{-2}$, within a factor of a few. The precise value of this factor is crucial for comparison with observations, but has not been determined so far from simulations. Fokker-Plank methods are inherently not well fit to deal with primordial binaries, and direct $N$-body calculations with primordial binaries have not yet been carried out for sufficiently large $N$ values. Fortunately, the GRAPE-4 computer will be up to the challenge, and is expected to provide us with the answer within a year or so. At that time, we will be able to judge whether the result reported by Guhathakurta (these proceedings), that in M15 the core radius $r_c < 0.1$pc, is compatible with either or both scenarios (primordial binary burning versus gravothermal oscillations).

## Acknowledgements

I thank Steve McMillan for his help in constructing the figures, based upon the runs reported in McMillan and Hut (1994). I thank Profs. Sugimoto and Makino, as well as the whole GRAPE group, for their hospitality during my visit to Tokyo University, when this paper was written. I am also grateful for the grant from JSPS which enabled me to make this visit.

## References

Bettwieser, E. & Sugimoto, D. 1984, MNRAS, 208, 439
Breeden, J.L., Packard, N.H. & Cohn, H.N. 1990, preprint
Breeden, J.L. & Cohn, H.N. 1995, ApJ, 448, 672
Cohn, H.N., Lugger, P.M., Grabhorn, R.P., Breeden, J.L., Packard, N.H., Murphy, B.W. & Hut, P. 1991, in The Formation and Evolution of Star Clusters, A.S.P Conference Series, 13, ed. K. Janes (ASP, San Francisco), p. 381
Gao, B., Goodman, J., Cohn, H. & Murphy, B. 1991, ApJ, 370, 567
Goodman, J. 1984, ApJ, 280, 298
Goodman, J. 1987, ApJ, 313, 576
Goodman, J. 1993, in Dynamics of Globular Clusters, eds. S. Djorgovski and G. Meylan (San Francisco: ASP), p. 87
Goodman, J. & Hut, P. (eds.) 1985, Dynamics of Star Clusters, IAU Symp. No. 113 (Dordrecht: Reidel)
Goodman, J. & Hut, P. 1989, Nature, 339, 40
Hut, P. & Inagaki, S., 1985, ApJ, 298, 502
Hut, P., McMillan, S. L. W., Goodman, J. G., Mateo, M., Phinney, E. S., Pryor, C., Richer, H. B., Verbunt, F., & Weinberg, M. 1992, PASP, 105, 981
Inagaki, S. 1986, PASJ, 38, 853
Kundić, T. & Ostriker, J. P. 1995, ApJ, 438, 702
Lynden-Bell, D. & Wood, R 1968, MNRAS, 138, 495
McMillan, S.L.W. 1986, ApJ, 307, 126
McMillan, S.L.W. 1989, in Dynamics of Dense Stellar Systems, ed. D. Merritt (Cambridge University Press), p. 207
McMillan, S.L.W., Hut, P. & Makino, J. 1990, ApJ, 362, 522
McMillan, S.L.W., Hut, P. & Makino, J. 1991, ApJ, 372, 111
McMillan, S. & Hut, P. 1994, ApJ, 427, 793
Murphy, B.W., Cohn, H.N. & Hut, P. 1990, MNRAS, 245, 335
Spitzer, L., 1987, Dynamical Evolution of Globular Clusters [Princeton Univ. Pr.]
Sugimoto, D. & Bettwieser, E. 1983, MNRAS, 204, 19p
Weinberg, M. D. 1994, AJ, 108, 1414

# STATISTICS OF SMALL-$N$ SIMULATIONS

DOUGLAS C. HEGGIE

*Department of Mathematics and Statistics, University of Edinburgh, King's Buildings, Edinburgh EH9 3JZ, UK*

**Abstract.** We first review the reasons for carrying out statistical analysis of results from large numbers of $N$-body simulations, including a summary of previous work. Then we describe some results about the behaviour of $N$-body systems which have been acquired by this technique. Finally we concentrate on the problems associated with the scaling of $N$-body data with $N$, with particular regard to the dissolution of systems in a tidal field when mass is lost by stellar evolution.

## 1. Introduction

One of the frequent complaints against data from $N$-body simulations is that they are "noisy". If one examines Lagrangian radii for the evolution of a 250-body system, for example, the inner radii show evidence of core collapse but it is not easy to measure the rate of collapse, or even the time at which collapse ends. One way in which the data can be improved is by combining the results from many simulations. Experience shows that the statistical noise is reduced as one would naively expect, i.e. in proportion to the square root of the number of simulations. Thus combining results from 50 simulations improves the signal-to-noise by a factor of approximately 7. Every observer knows how much more science he could do with such an improvement, but theorists have been slower in seeing the benefits.

This paper first reviews this and other reasons for taking a statistical approach to the study of $N$-body simulations. After summarising some previous investigations carried out in this spirit, we turn to some recent work which highlights some of the advantages and pitfalls of this kind of work.

## 2. Advantages of a Statistical Approach

### 2.1. THEORETICAL ARGUMENTS

It has been known for a long time that the detailed results of $N$-body calculations, i.e. the positions and velocities of the individual stars, are wrong. The early work of Miller (1964, 1971) showed that the growth of errors is extraordinarily rapid, with consequences which were illustrated by Lecar (1968) and Hayli (1970). Recent studies (Kandrup & Smith 1991, 1992a,b) have extended this work to larger $N$ and given it a theoretical foundation (Goodman et al. 1993).

In the face of this observation, the justification for the validity of $N$-body simulations is that the *statistical* results of the computations are nevertheless correct (Aarseth & Lecar 1975). In statistics, however, it is usual to take many measurements, in a compromise between cost and precision. Since the only valid results of $N$-body simulations are statistical, the same approach should be used there.

The assertion that statistical results of $N$-body simulations are valid is an article of faith among practitioners, but it has been given some support recently by "shadowing lemmas" (Quinlan & Tremaine 1992; Hayes 1995). Roughly speaking, these show that there exists an exact $N$-body model which lies close to a given computed model, despite the numerical errors in the latter.

### 2.2. PRACTICAL ARGUMENTS

The strategy of computing many realisations of a single system, differing only in the random numbers used to create initial conditions, is ideally suited to implementation on parallel computers or even networks of workstations. Communications are a negligible part of the effort, and so virtually 100% efficiency is achieved. The main practical difficulties, in fact, stem from storing and analysing the rather large quantities of data which emerge.

### 2.3. SCIENTIFIC ARGUMENTS

#### 2.3.1. *Improved signal-to-noise*

As already stated in the Introduction, the advantages of a statistical approach stem from the great reduction in statistical fluctuations. It is not the only way in which this can be achieved, however.

A first alternative is simply to run a single computation with much larger $N$. Suppose we compare a single calculation with $N$ particles with statistical results from $n$ simulations with $N/n$ particles each. Naively we may expect that the signal-to-noise of some measured quantity (such as the half-mass radius) will be comparable in the two sets of calculations. However, the single large calculation will have taken far longer, even if the $n$ small simulations were carried out sequentially, simply because the effort of a simulation increases more quickly than linearly with $N$. Thus the advantage of the statistical approach is efficiency (Smith 1977).

A second alternative is to resample a single simulation. The assumption here is that successive samples are approximately independent. Little has been done to test this assumption (Miller 1975), and in some cases it is clearly false. For example, core collapse takes place at different times in different simulations, and it follows that successive measures of the core radius in a single simulation will be biased by the time at which it collapses.

### 2.3.2. *Parameter dependence*
Very often the purpose of $N$-body simulations is not simply to provide an answer to a question about a single system, but to study the way in which the behaviour of the system changes as we change its initial or boundary conditions (e.g. the initial structure of the model). Nothing can be said about this, however, if one is restricted to a single simulation for each set of conditions (McMillan et al. 1990), for it is then possible that the changes observed are simply a result of statistical fluctuations and are not causally related to the changes in the imposed conditions.

### 2.3.3. *Incoherent Phenomena*
There are some important investigations for which a statistical approach is inappropriate. Incoherent phenomena (e.g. gravothermal oscillations) cannot be studied in this approach, at least not in the naive manner of combining data from different simulations at the same time. As another example, consider the implications of the fact, already mentioned, that different simulations end core collapse at different times. Though the averaged system will exhibit a collapse, it will not be as deep as that of any of the systems in the ensemble.

## 2.4. PREVIOUS STATISTICAL STUDIES OF $N$-BODY SYSTEMS

Numerous statistical studies of the three- and four-body problems have been reported in the literature. Such quantities as exchange cross sections could not, however, be obtained in any way other than by statistical study of large numbers of individual cases.

The earliest study in the spirit of this review was a set of twelve 5-body simulations carried out by Agekyan & Baranov (1969). Their aim was to measure the density distribution. This topic was one of several taken up subsequently by Smith (1977, 1979, 1985) who studied results from a small number of cases with $N$ in the range $8 \leq N \leq 64$. One aim of his investigations was to determine empirically whether integration errors would have a noticeable effect on the statistical results. Heggie (1991) continued this theme with somewhat larger numbers of simulations and stars. A more systematic attack with statistical methods was initiated by Casertano (1985) and McMillan et al. (1988), whose main aim was to study the $N$-dependence of relaxation. This was also one of the early aims in the rather extensive studies by Giersz & Heggie, which will be summarised in more detail below. Finally, mention may be made of the recent study by Kroupa (1995) of clusters initially consisting entirely of binaries. He studied between 3 and 20 systems with $N = 400$ each, with a view to investigating the distribution of binaries in the field, and related problems.

## 2.5. THE STUDY BY GIERSZ & HEGGIE

The final goal of this investigation was to compare the results of Fokker-Planck and $N$-body simulations. The former was used by Chernoff & Weinberg (1990) in a landmark study of the evolution of globular clusters under the action of (i) mass loss by stellar evolution, (ii) two-body relaxation, and (iii) a steady tide. Because of the approximations used in the Fokker-Planck treatment it seemed worth while to check its results with $N$-body models, but the approach adopted was a gradual one, various dynamical processes being added one by one.

The first set of results analysed in detail had equal masses with no tide, and was studied in two papers: one (Giersz & Heggie 1994a; hereafter Paper I) dealing with evolution up to core collapse, and a second (Giersz & Heggie 1994b, = Paper II) dealing with core bounce and post-collapse evolution. Next came a study of systems with a mass spectrum, taken to be a power law (Giersz & Heggie 1995a, = Paper III). Later papers (Giersz & Heggie 1995b,c, = Papers IV, V) will add the effects of a steady tide (from which a small sample of results can be found in Giersz & Heggie 1993) and then mass loss from stellar evolution, some results from which are discussed below (§3).

In our work values of $N$ in the range $250 \leq N \leq 2000$ were used, the number of runs in each series varying from 16 (at the largest $N$) to about 90 (at smaller $N$). Note that, from a naive statistical point of view, 16 simulations with $N = 2000$ yield results of comparable statistical quality to a single run with $N = 32000$. In addition to the runs with either equal masses or a power-law mass function, some sets of runs of two-component models were studied and have been partially published elsewhere (Spurzem & Takahashi 1995).

Among the topics discussed in the course of these investigations have been the following.

TABLE 1. Some Homological Solutions for Cluster Evolution

| Source | Phase | BC | Model |
| --- | --- | --- | --- |
| Hénon 1961 | pc | tidal | Fokker-Planck |
| Hénon 1965 | pc | isolated | Fokker-Planck |
| Lynden-Bell & Eggleton 1980 | cc | power-law | gas |
| Inagaki & Lynden-Bell 1983 | early pc | power-law | gas |
| Goodman 1984, 1987 | pc | isolated | gas |
| Heggie & Stevenson 1988 | cc, early pc | power-law | Fokker-Planck |
| Louis 1990 | cc | power-law | anisotropic gas |
| Takahashi & Inagaki 1992 and Takahashi 1993 | cc, early pc | power-law | Fokker-Planck |

Explanation: BC, boundary condition; cc, core collapse; pc, post collapse.

- *The Coulomb logarithm* in the expression for the relaxation time. Usually taken as $\ln(\gamma N)$ with $\gamma = 0.4$ (Spitzer 1987), we found that a lower value of $\gamma \simeq 0.11$ fitted the data for core collapse better. Somewhat different values are indicated for post-collapse evolution, and smaller values for unequal masses (Papers I–III).
- *The escape rate* increases during core collapse, partly because of the change of structure, and partly because of the evolving anisotropy. It has also been measured for post-collapse evolution and for models with unequal masses and in a tidal field (Papers I–V).
- *Statistics of binaries*, including the maximum energies which they reach before ejection (Papers I–V).
- *Energy generation by binaries*, whose effectiveness appears to increase with $N$ in the range studied (Paper II).
- *Mass segregation*, which appears virtually to stop by the close of core collapse (Paper III).
- *Homological evolution*. Though such solutions seem very special (Table 1), it is remarkable how much of the evolution of even quite a complicated model can be understood in these terms (Fig.1). Homologous models also provide a means of encapsulating the very complicated behaviour of $N$-body models in quite simple approximate formulae (Papers I–V).
- *Effect of different types of tidal model:* it turns out that it makes little difference to the lifetime of a model if the tide is modelled as a simple cutoff (Paper IV).
- *Anisotropy*: as expected, in the presence of a tide it is difficult to produce anisotropy from isotropic initial conditions (Papers IV, V).
- *Scaling to real clusters*, which will be discussed in the next section.

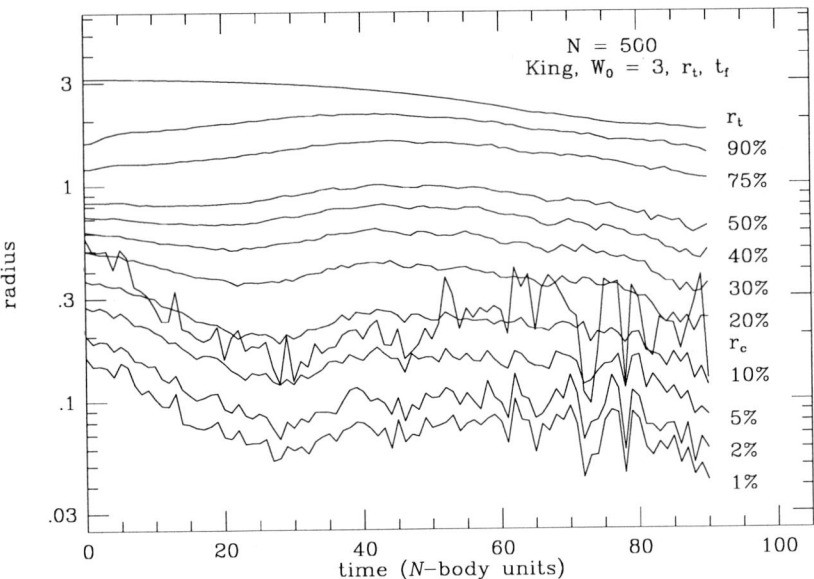

*Figure 1.* An $N$-body model exhibiting several types of nearly homologous behaviour (after Giersz & Heggie 1995b). Lagrangian radii corresponding to the stated mass fractions, and the core and tidal radii, are plotted against time in $N$-body units (Heggie & Mathieu 1986). The initial conditions are a 500-body King model with $W_0 = 3$, and a mass function with index 2.5 (Salpeter is 2.35) in a range such that $m_{max}/m_{min} = 37.5$ (cf. Chernoff & Weinberg 1990). The systems are tidally bound, but there is no mass loss from stellar evolution. Initially there were 56 cases, but the number of cases still running (determined by the condition $N > 25$) reduces rapidly between $t = 70$ and 90. This, together with the small number of very massive stars, accounts for the increasing noise in the innermost radii. Up to $t \simeq 30$ the core exhibits a nearly-homologous collapse. From then until $t \simeq 50$ there is almost self-similar post-collapse expansion as if the system were isolated. Thereafter the expansion is stopped by the contracting tidal radius, but the subsequent evolution is still nearly self-similar.

– *Validation of Fokker-Planck and gas models.* One of the significant applications has been to check and calibrate these simplified models (Papers I, II; Giersz & Spurzem 1994). Essentially this brings up to date the older comparison of Aarseth et al. (1974). One major implication which deserves to be highlighted is that these comparisons (e.g. between the results of $N$-body and Fokker-Planck models) lend strong support to the Chandrasekhar-Spitzer model of two-body relaxation, with the cutoff radius at a value comparable with the system size, despite claims to the contrary (e.g. Gurzadyan & Savvidy 1986, Smith 1992).

## 3. The Scaling Problem

One of the principal aims of $N$-body modelling is the understanding of globular clusters and their dynamical evolution. Even with modern special-purpose hardware, however, the values of $N$ for which reasonable numbers of simulations are feasible still fall short of the real numbers of stars in globular clusters. Therefore it is necessary to consider how the results scale with $N$. The essential problem is that evolution is driven by a variety of processes which scale in different ways with $N$ (McMillan 1993). In this section we illustrate this problem in connection with evolution driven only by two-body relaxation and mass-loss from stellar evolution in the presence of a galactic tide, with unequal masses.

This problem was considered by Chernoff & Weinberg (1990; hereafter CW) in their well known survey (see also Weinberg 1993). Consider as an example a cluster with initial mass $1.5 \times 10^5 M_\odot$, galactocentric distance 4kpc, initially a King model with scaled central potential $W_0 = 3$, and an initial mass function $dN/dm \propto m^{-2.5}$ in the range $0.4 M_\odot < m < 15 M_\odot$. We adopt the parameters for stellar mass loss and the galactic tide from CW.

The $N$-body results of Giersz & Heggie (1995c), with $N = 500$ and $1000$, showed several phases of evolution, which could be readily interpreted theoretically. Initially the cluster expanded, in response to the epoch of major mass loss by stellar evolution. This was followed by almost homologous contraction, as stars escaped across the tidal boundary. After 100 $N$-body units of time ($\simeq 1.7 \times 10^{10}$yr) there were the first signs of core collapse. Unfortunately, CW's Fokker-Planck result was quite different: they found that the cluster dissolved after $2.8 \times 10^8$yr!

The problem here is one of scaling, as Table 2 illustrates. The conversion between $N$-body time units and astrophysical units which was assumed above was guided by the correct scaling of the relaxation time, and is denoted $N$-body-A in the table. Because the crossing time scales differently with $N$, however, this means that the crossing time in the $N$-body model does not then scale to the correct value in years. As can be seen in the Table, this implies that mass-loss is impulsive in the $N$-body model, i.e. it takes place on a time scale short compared with the crossing time, whereas it is adiabatic in the "real" cluster. At first sight it is therefore puzzling that it is the $N$-body model which survives better: for an isolated cluster, simple virial estimates show that arbitrary amounts of mass may be lost adiabatically, but only 50% impulsively. It is possible that this distinction between the effects of adiabatic and impulsive mass loss is altered in the presence of a tide.

In the real cluster the three time scales are well separated, and it is

TABLE 2. Scaling of $N$-body models

| System | $t_{cr}$ | $t_{se}$ | $t_{rh}$ | $N$ |
|---|---|---|---|---|
| Cluster | $5 \times 10^6$ yr | $10^8$ yr | $3 \times 10^9$ yr | $10^5$ |
| $N$-body-A | 3 | 0.5 | 14 | 500 |
| $N$-body-B | 3 | 60 | $\gg 250$ | 16000 |

Explanation of symbols and units: $t_{cr}$, crossing time; $t_{se}$, time of mass loss by $4M_\odot$-star; $t_{rh}$, half-mass relaxation time; $N$, number of stars; for the $N$-body models the unit of time is standard (Heggie & Mathieu 1986).

clearly necessary to use an $N$-body simulation with the same ordering of time scales. This is difficult without increasing $N$ substantially. Nevertheless this is now quite feasible, thanks to the development of GRAPE special-purpose hardware (Ebisuzaki et al. 1993; papers by Makino and Taiji, these proceedings.) Using a low-precision GRAPE-3AF at Edinburgh it was possible to increase $N$ large enough so that the crossing- and stellar-evolution times scaled consistently, and the relaxation time was large enough not to affect the evolution. This model is denoted $N$-body-B in the table. (Note that, in the Fokker-Planck model, the evolution is over in about $0.1 t_{rh}$.) The result (Fukushige & Heggie 1995) is that the cluster dissipates after about $9 \times 10^8$ yr, i.e. longer than the lifetime obtained by CW, but by a factor of only about 3. The residual disagreement is thought to be due to the fact that the disruption takes place on a time scale which is not sufficiently large compared with the crossing time. Under these circumstances one of the assumptions underlying the use of the orbit-averaged Fokker-Planck equation, which were stated carefully and explicitly by CW, is not satisfied.

## 4. Conclusions and Reflections

The following conclusions are justified by the arguments of the foregoing sections.

First, there are several reasons for arguing that not just one but several $N$-body simulations should be carried out with the same parameters, differing only in the random numbers used in generating initial conditions. In particular, there are significant scientific benefits in combining results from large numbers of simulations.

The second set of conclusions concerns the problem of scaling results from simulations to real clusters, when the simulations are carried out with modest numbers of particles (i.e. small compared with the number of stars in a real cluster). In particular, we have shown that, while mass-loss by stellar evolution is important, it is necessary to scale the results so that the time scale for mass loss and the crossing time both scale consistently to the relevant astrophysical values. Thereafter, when mass loss becomes relatively unimportant, the scaling would have to alter so that the relaxation time scales correctly. Similar problems are presented by the modelling of disk shocking, whose inclusion in fully-fledged $N$-body simulations has not yet been attempted. In both cases, there may be sets of parameters for which no meaningful $N$-body simulations can be attempted with modest values of $N$ (i.e. thousands).

A third and rather minor conclusion is that some caution should be exercised in using the orbit-averaged Fokker-Planck equation for the modelling of very young clusters. More generally, it is well known that such models involve several simplifying assumptions and depend heavily on our theoretical understanding of the dynamical processes which drive the evolution of clusters. By contrast it is sometimes asserted that $N$-body models are essentially "assumption-free". It is worth pointing out, however, that $N$-body models also require considerable theoretical input, as we have seen, if it is desired to scale the results intelligently to real star clusters.

## Acknowledgements

The author thanks D. Sugimoto and his colleagues in the GRAPE group for their enthusiastic help in setting up GRAPE hardware in the UK. The author's own recent visits to Tokyo have been supported by travel grants under a cooperative scheme of the Royal Society, the Japanese Society for the Promotion of Science, and the British Council (Tokyo).

## References

Aarseth S.J., Hénon M., Wielen R., 1974, A&A, 37, 183
Aarseth S.J., Lecar M., 1975, ARAA, 13, 1
Agekyan T.A., Baranov A.S., 1969, Astrophys., 5, 144 (Astrofiz., 5, 305)
Casertano S., 1985, in Goodman J., Hut P., eds, Dynamics of Star Clusters, IAU Symp 113. Reidel, Dordrecht, p. 305
Chernoff D.F., Weinberg M.D., 1990, ApJ, 351, 121
Ebisuzaki T., Makino J., Fukushige T., Taiji M., Sugimoto D., Ito T., Okumura S.K., 1993, PASJ, 45, 269
Fukushige T., Heggie D.C., 1995, MNRAS, 276, 206
Giersz M., Heggie D.C., 1993, in Smith G.H., Brodie J.P., eds, The Globular Cluster-Galaxy Connection, ASP Conf. Ser. 48. ASP, San Francisco, p.713.
Giersz M., Heggie D.C., 1994a, MNRAS, 268, 257 (Paper I)
Giersz M., Heggie D.C., 1994b, MNRAS, 270, 298 (Paper II)
Giersz M., Heggie D.C., 1995a, MNRAS, submitted (astro-ph/9506143, Paper III)
Giersz M., Heggie D.C., 1995b,c, in preparation (Papers IV, V)
Giersz M., Spurzem R., 1994, MNRAS, 269, 241

Goodman J., 1984, ApJ, 280, 298
Goodman J., 1987, ApJ, 313, 576
Goodman J., Heggie D.C., Hut P., 1993, ApJ, 415, 715
Gurzadyan V.G., Savvidy G.K., 1986, A&A, 160, 203
Hayes W., 1995, MSc Thesis, Dept. of Computer Science, U. of Toronto
Hayli A., 1970, A&A, 7, 249
Heggie D.C., 1991, in Roy A.E., ed, Predictability, Stability and Chaos in $N$-body Dynamical Systems. Plenum, New York, p.47
Heggie D.C., Mathieu R.D., 1986, in Hut P., McMillan S.L.W., eds, The Use of Supercomputers in Stellar Dynamics. Springer-Verlag, Berlin, p.233
Heggie D.C., Stevenson D., 1988, MNRAS, 230, 223
Hénon M., 1961, Ann. d'Astrophys., 24, 369
Hénon M., 1965, Ann. d'Astrophys., 28, 62
Inagaki S., Lynden-Bell D., 1983, MNRAS, 205, 913
Kandrup H.E., Smith H., Jr., 1991, ApJ. 374, 255
Kandrup H.E., Smith H., Jr., 1992a, ApJ, 386, 635
Kandrup H.E., Smith H., Jr., 1992b, ApJ, 399, 627
Kroupa P., 1995, MNRAS, in press
Lecar M., 1968, Bull. Astron., 3, 91
Louis P.D., 1990, MNRAS, 244, 478
Lynden-Bell D., Eggleton P.P., 1980, MNRAS, 191, 483
McMillan S.L.W., 1993, in Djorgovski S.G., Meylan G., eds, Structure and Dynamics of Globular Clusters, ASP Conf. Ser. 50. ASP, San Francisco, p.171
McMillan S.L.W., Casertano S., Hut P., 1988, in Valtonen M.J., ed, The Few Body Problem, IAU Coll 96. Kluwer, Dordrecht, p.313
McMillan S., Hut P., Makino J., 1990, ApJ, 362, 522
Miller R.H., 1964, ApJ, 140, 250
Miller R.H., 1971, J. Comp. Phys., 8, 449
Miller R.H., 1975, in Hayli A., ed, Dynamics of Stellar Systems, IAU Symp 69. Reidel, Dordrecht, p.65
Quinlan G.D., Tremaine S., 1992, MNRAS, 259, 505
Smith H., Jr., 1977, A&A, 61, 305
Smith H., Jr., 1979, A&A, 76, 192
Smith H., Jr., 1985, ApJ, 288, 117
Smith H., Jr., 1992, ApJ, 398, 519
Spitzer L., Jr., 1987, Dynamical Evolution of Globular Clusters. Princeton UP, Princeton.
Spurzem R., Takahashi K., 1995, MNRAS, 272, 772
Takahashi K., 1993, PASJ, 45, 789
Takahashi K., Inagaki S., 1992, PASJ, 44, 623
Weinberg M.D., 1993, in Smith G.H., Brodie J.P., eds, The Globular Cluster-Galaxy Connection, ASP Conf. Ser. 48. ASP, San Francisco, p.689.

# GRAPE-4: A TERAFLOPS MACHINE FOR N-BODY SIMULATIONS

MAKOTO TAIJI, JUNICHIRO MAKINO[†],
TOSHIYUKI FUKUSHIGE, TOSHIKAZU EBISUZAKI AND
DAIICHIRO SUGIMOTO
*Department of Earth Science and Astronomy,*
[†]*Department of Graphics and Information Science,*
*College of Arts and Sciences, University of Tokyo*
*Komaba 3-8-1, Meguro, Tokyo 153, Japan*
*Internet : taiji@chianti.c.u-tokyo.ac.jp*

**Abstract.** We have developed a massively parallel special-purpose computer system for $N$-body simulations, GRAPE-4 (GRAvity-PipE 4). The GRAPE-4 system is designed for high-accuracy simulations of dense stellar systems. The GRAPE-4 calculates gravitational forces, their derivatives in time and potential energies. It has a hardware for prediction of positions and velocities, which is used for the individual timestep scheme. Using multi-chip module technology, we integrated 1692 chips of 640 megaflops performance. The peak speed of GRAPE-4 is 1.08 teraflops.

## 1. Introduction

Gravitational $N$-body simulation is one of the most important method for the study of star clusters. In $N$-body simulations, almost all computational time is spent in calculating gravitational forces, since the number of interactions between particles is proportional to the square of the number of particles. Thus, it requires huge computing resources beyond teraflops performance to simulate the post-collapse evolution of real clusters with $10^4 \sim 10^5$ particles. We have developed special purpose computer systems GRAPE (GRAvity PipE) to accelerate for $N$-body simulations of globular clusters, galaxies, cluster of galaxies, and the universe(Sugimoto *et al.* , 1990; Ebisuzaki *et al.* , 1993). In this paper we describe GRAPE-4, a teraflops massively-parallel special-purpose computer system for gravitational $N$-

body simulations (Taiji et al. , 1994a). GRAPE-4 is suitable for high-accuracy $N$-body simulations with the hierarchical timestep scheme.

GRAPE calculates only gravitational forces. The host workstation, which is connected to GRAPE, integrates orbits of particles. The GRAPE systems have very long pipelines specialized for the calculations of gravitational interactions. These pipelines perform a few tens of arithmetic operations per cycle. In the large $N$-body simulations, the force calculation, which costs $O(N^2)$ in time, dominates computational time. On the other hand, both the calculation in a host and the communication between a host and a GRAPE system cost $O(N)$. Therefore, commercial workstations can satisfy the requirements for the communication speed as well as for the calculation speed, although the calculation speed of GRAPE exceeds teraflops.

There are the tree algorithm (Barnes and Hut, 1986), the Particle-Particle Particle-Mesh (PPPM) method (Hockney and Eastwood, 1988), or the fast multipole method (FMM) (Greengard and Rokhlin, 1987), which reduce the cost of force calculations to $O(N \log N)$ or $O(N)$. However, these methods are very difficult to combine with individual or hierarchical timestep schemes, which are essential in simulations of dense stellar systems. Recently McMillan and Aarseth succeeded to implement a tree algorithm with a hierarchical timestep scheme (McMillan and Aarseth, 1993). However, their results indicate that the direct summation is more efficient for simulations at least with $N < 10^5$ on teraflops machines, if we take account of the overhead of parallelization. Thus, the direct summation is still the most efficient for simulations of dense stellar systems. In addition, we can use GRAPE to accelerate these clever algorithms. Makino implemented a tree algorithm (Makino, 1991b) and Brieu et al. implemented the PPPM algorithm on GRAPE-3 (Brieu et al. , 1995). Thus, GRAPE accelerates most of particle simulations of large-$N$ self-gravitating systems.

## 2. What GRAPE-4 calculates

GRAPE-4 calculates forces $\boldsymbol{f}_i$, force derivatives in time $\dot{\boldsymbol{f}}_i$, and potential energies $\phi_i$ of particle $i$ from the positions $\boldsymbol{x}_i(t_i), \boldsymbol{x}_j(t_j)$, the velocities $\boldsymbol{v}_i(t_i), \boldsymbol{v}_j(t_j)$, and the masses $m_j$ according to the following equation.

$$\begin{aligned}
\boldsymbol{f}_i &= \sum_j m_j \frac{\boldsymbol{r}_{ij}}{(r_{ij}^2 + \epsilon^2)^{3/2}}, \\
\dot{\boldsymbol{f}}_i &= \sum_j m_j \left[ \frac{\boldsymbol{v}_{ij}}{(r_{ij}^2 + \epsilon^2)^{3/2}} - \frac{3(\boldsymbol{v}_{ij} \cdot \boldsymbol{r}_{ij})\boldsymbol{r}_{ij}}{(r_{ij}^2 + \epsilon^2)^{5/2}} \right], \\
\phi_i &= \sum_j m_j \frac{1}{(r_{ij}^2 + \epsilon^2)^{1/2}},
\end{aligned} \quad (1)$$

where $r_{ij} = x_i(t_i) - x_j(t_i)$, $v_{ij} = v_i(t_i) - v_j(t_i)$, and $\epsilon$ is a softening parameter. Force derivatives in time are necessary in fourth-order Hermite scheme, which is more simple and accurate than schemes based on the Newton interpolation (Makino, 1991a; Makino and Aarseth, 1992; Aarseth, 1995). Here, the positions and the velocities at time $t_i$ ($x_j(t_i)$, $v_j(t_i)$) are calculated from those at time $t_j$ using third order predictors

$$\begin{aligned} x_j(t_i) &= x_j + v_j \Delta t + \frac{a_j \Delta t^2}{2} + \frac{\dot{a}_j \Delta t^3}{6}, \\ v_j(t_i) &= v_j + a_j \Delta t + \frac{\dot{a}_j \Delta t^2}{2}, \end{aligned} \quad (2)$$

where $\Delta t = t_i - t_j$. In the hierarchical timestep scheme, we have to evaluate predictors of *all* particles, while we only calculate forces and correctors of particles which share the block step. Therefore, predictor calculations becomes too expensive to be performed by the host computer as $N$ is increased. Since we have to send the predicted coordinates of *all* particles, the communication between the host and GRAPE-4 is also increased. To solve these problems, we added a hardwired pipeline for predictor calculations to GRAPE-4.

It also build the neighbor lists, which contains the indices of particles within a distance $h$ from particle $i$.

## 3. System Architecture

Figure 1 shows the block diagram of the GRAPE-4 system. It consists of a host computer, host interface boards (HIB), controller boards (CB), and processor boards (PB). This system has a two-level structure. It has four clusters, and each cluster has one CB and nine PBs. Each PB has 47 HARP (Hermite AcceleratoR Pipe) chips which calculates forces and their derivatives. The HARP chip has the performance of 640 megaflops. Therefore, the system has 1692 HARP chips and the peak performance of 1.08 teraflops. We use Digital Equipment Corporation (DEC) Alpha AXP 3000 series workstation with a TURBOchannel bus as a host computer. However, it is fairly easy to connect the system to another host computer. Since it consumes only 10 kW of power, no heavy cooling system is necessary. It cost about 1.2 million dollars to build the hardware of GRAPE-4. For comparison, a commercial massively-parallel-processors machine with 100–200 gigaflops of theoretical peak speed would cost about 30 million dollars in 1995. Figure 2 shows a photograph of the GRAPE-4 system.

The following two ideas are the keys to achieve such a high performance at a low cost. The first one is the very-long-pipeline architecture, which leads high-performance LSI chips. The second one is massive parallelization. We will explain these ideas in the following sections.

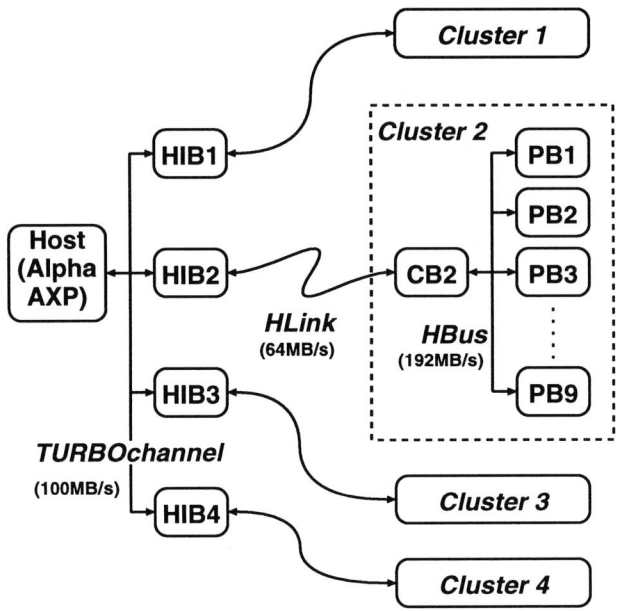

*Figure 1.* The block diagram of the GRAPE-4 system. HIB: host interface board, CB: controller board, PB: processor board.

*Figure 2.* Photograph of the GRAPE-4 system with one of the authors (JM).

## 4. Very Long Pipeline

Nowadays we can pack $> 10^7$ transistors into one silicon LSI. Since a floating point multiplier and adder need less than $\sim 10^{5.5}$ transistors even in

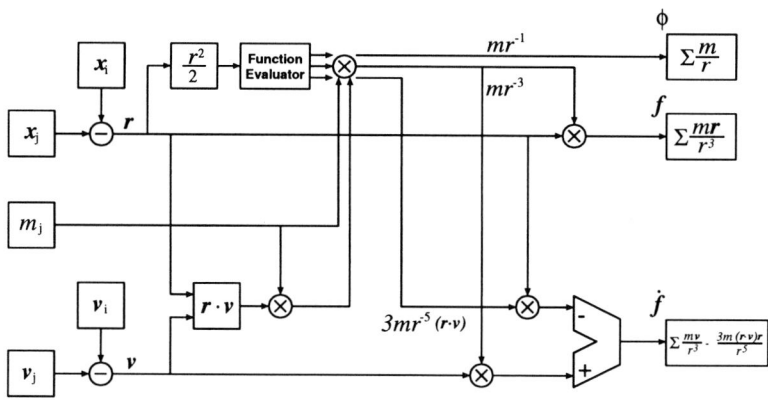

*Figure 3.* The block diagram of the HARP chip.

double precision, one LSI can have a few tens or hundreds of arithmetic units. It is very difficult to utilize so many arithmetic units efficiently. The very-long-pipeline architecture is one of answers to solve the problem. In this architecture, the whole force calculation are done in one or a few cycles using many arithmetic units. Figure 3 shows the block diagram of the HARP chip, which is developed for the GRAPE-4 system (Taiji et al., 1994b). The HARP chip calculates force $f_i$, force derivative in time $\dot{f}_i$, and potential energy $\phi_i$. Each arithmetic unit has the special role and they are connected in the specific sequence to calculate force. These units are fully pipelined so that all of them work simultaneously. The HARP chip calculate one interaction, which usually requires 50 ~ 60 operations, in three cycles. In other words, it performs about 20 operations per cycle. Thus its performance reaches 640 megaflops at 32 MHz. This is very high performance considering the fact that it is made by 1.0 $\mu$m technology, and operates at 32 MHz clock. We have developed another LSI for the evaluation of predictor, the PROMETHEUS chip, which also has the very-long-pipeline architecture.

This pipeline approach is used also in vector processors and RISC processors. Our approach to develop a special-purpose very-long-pipeline LSI has two major advantages over conventional vector processors or RISC pipelines. The first advantage of a special-purpose processor is that we can tune the precision and other features according to the application. The optimization of word lengths reduces the gate count significantly. The second advantage is that it requires rather low input/output (I/O) performance. In the case of general-purpose processors, it is difficult to use many arithmetic units efficiently without increasing the I/O bandwidth between the memories and the arithmetic units. On the other hand, in the case of a special-purpose processor, arithmetic units are connected in a specific topology following a data flow. Therefore, a LSI chip with many arithmetic units requires a moderate I/O bandwidth. In the case of processors specialized for the $N$-body simulations, we can use the "virtual multiple pipeline" architecture to further reduce the I/O bandwidth. These advantages make it possible to utilize almost all the transistors available on a chip for calculations. Thus, a special-purpose LSI can house hundreds times more arithmetic units than general-purpose processors.

## 5. Massive Parallelization

GRAPE-4 consists of 1692 HARP chips to achieve 1.08 teraflops. In this section we explain how we parallelize the task among many pipelines without losing efficiency. As we explained already, GRAPE-4 calculates only forces, their derivative, and potential energies. For example, the force $\boldsymbol{F}_i$ on particle $i$ is given by

$$\boldsymbol{F}_i = \sum_j \boldsymbol{f}_{ij}, \qquad (3)$$

where $\boldsymbol{f}_{ij}$ is the force exerted by particle $j$ on particle $i$. Thus, we have two ways of parallelization: parallelization over index $i$ ($i$-parallelization) and that over index $j$ ($j$-parallelization). GRAPE-4 uses both ways of parallelization.

### 5.1. $I$-PARALLELIZATION

The parallelization over index $i$ means parallel calculation of forces on many particles. This parallelization scheme is very important in the GRAPE architecture. In this method different pipelines calculate forces on different particles. The coordinates and masses of $j$-particles are supplied from the predictor (memory) unit to the pipelines. Since these pipelines calculate forces on different particles, we can use the same coordinates of $j$-particles for calculation of all these forces. Therefore, all the inputs of pipelines can share the common output from the predictor unit. Such a memory architecture cannot be used in general-purpose computers, since it can be applied only for some special applications like classical force calculations.

In the HARP chip we also used the virtual parallelization technique (Makino et al. , 1993; Taiji et al. , 1994b). The HARP chip accumulates forces, their derivatives in time and potentials of *two* different particles simultaneously. The pipeline accumulate these two forces in turn. Again, we can use the same coordinates and masses of $j$-particles for calculation of these two particles. The parallelization on an index $i$ can be applied not only for physical pipelines but also for a time-shared virtual pipeline.

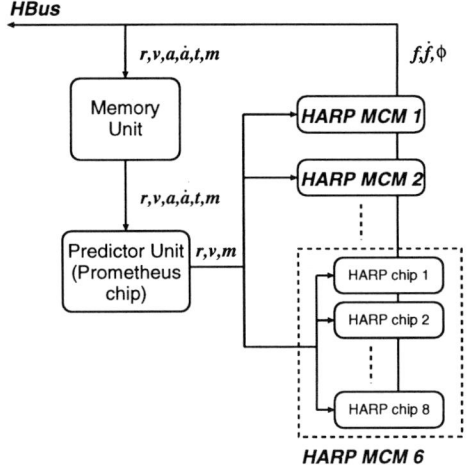

*Figure 4.* The block diagram of the GRAPE-4 processor board.

*Figure 5.* A photograph of a multi-chip module of the HARP chips. It has two cavities and each cavity can house 4 HARP chips.

Figure 4 shows the block diagram of the GRAPE-4 processor board (PB). It has a memory unit, a predictor unit (PROMETHEUS chip), and 47 HARP chips. We have developed a ceramic multi-chip module (MCM) of 8 HARP chips for efficient packaging. Figure 5 shows the photograph of the MCM. The total number of pins is only 240 since most of input/output pins are shared by all of the HARP chips. Thus, an efficient use of MCM is again the result of $i$-parallelization. The area of MCM (75 mm × 110 mm) is the half of 8 single-chip modules. A Multi-chip modules is very effective in reducing the complexity and size of the board. The processor board can mount 6 MCM (48 HARP chips). However, since we used a MCM with one defect for each board, the number of the HARP chips per PB is 47 in the current GRAPE-4 system.

The parallelization on $i$ is quite easy and effective to achieve high performance as explained, however, there are several problems which limit the number of $i$-parallel pipelines. In the hierarchical timestep scheme, the number of $i$-parallel pipelines must be smaller than the average number of particles at a block step, $N_{\text{step}}$. For simulations with $N < 10^5$, it becomes smaller than the number of (virtual) pipelines in GRAPE-4 (3384), since $N_{\text{step}}$ is roughly $10^3$ for $N = 10^5$ (Makino, 1991a). Therefore, we can not apply $i$-parallelization for all pipelines. In addition, it is physically difficult to connect thousands of pipeline chips to a single memory unit. Therefore, we use $i$-parallelization only to distribute tasks between pipelines on a processor board. The number of $i$-parallel pipelines is 94 in GRAPE-4.

## 5.2. J-PARALLELIZATION

To distribute tasks over processor boards, we use parallelization on index $j$. In the GRAPE-4 system with 36 processor board, The total force is divided into 36 partial forces. Each processor board calculates a different piece of the force. The total force is calculated as a sum of the partial forces

$$F_i = \sum_{j=1}^{n} f_{ij} + \sum_{j=n+1}^{2n} f_{ij} + \cdots + \sum_{j=35n+1}^{36n} f_{ij}. \quad (4)$$

This means that the communication between the host and GRAPE-4 is increased by a factor of 36. Since the summation of partial forces is done by the host, it also increases the amount of calculations in the host.

The controller board solves these problems. It sums up the partial forces calculated on processor boards and sends the result to the host. Thus, it can reduce the communication and the calculations which arise due to the $j$-parallelization by a factor of 9.

## 6. Software and Performance

In this section, we describe the procedure to use GRAPE-4, and discuss the behavior of the sustained performance. Since GRAPE-4 calculates only forces, their derivatives, and potential energies, it is easy to write programs for GRAPE-4. The following is the basic procedure to use GRAPE-4.

1. Select 94 particles in the current block step

2. Calculate predictors at the current time for these particles
3. Send $x$ and $v$ of these particles to GRAPE
4. Calculate forces by GRAPE
5. Receive $f, \dot{f}$, and $\phi$ from GRAPE
6. Calculate correctors and update $x, v$
7. Send $x, v, a$, and $\dot{a}$

More details are described in the paper by Aarseth (1995).

Next, we discuss the sustained performance of GRAPE-4. The computing time per particle per timestep is expressed as

$$T_{\text{step}} = T_{\text{host}} + T_{\text{comm}} + N t_{\text{force}} \qquad (5)$$

where $T_{\text{host}}$ is the time for calculations on the host computer, $T_{\text{comm}}$ is the time for communication between the host and GRAPE-4, and $N t_{\text{force}}$ is the time for force calculation on processor boards. Since $N t_{\text{force}}$ is proportional to the number of particles $N$ but $T_{\text{host}}$ and $T_{\text{comm}}$ are constant, the sustained performance strongly depends on $N$. Thus, the efficiency $\eta = T_{\text{force}}/T_{\text{step}}$ behaves as

$$\eta = \frac{N}{N + N_{\text{half}}}, \qquad (6)$$

where $N_{\text{half}}$ is the number of particle which is necessary to obtain the efficiency of 50%. For the full GRAPE-4 system, we obtained $N_{\text{half}} \sim 4 \times 10^5$. $N_{\text{half}}$ is scaled roughly to the peak performance of the GRAPE system. More details will be discussed in another paper (Taiji et al., 1996).

## 7. Summary and future prospects

We have developed GRAPE-4, a special-purpose computer system for astrophysical $N$-body simulations. It calculates gravitational forces and its time-derivative and has a hardware for predictor calculations. It is suitable for simulations of dense stellar systems using the hierarchical timestep scheme. The GRAPE-4 system is a massively parallelized system of 1692 force calculation pipeline chips. Its peak performance reached at 1.08 teraflops and it cost 1.2 million dollars.

The GRAPE architecture will become more important in future. General-purpose computers cannot utilize efficiently the increasing amount of transistors. On the other hand, GRAPE can use all of them to increase arithmetic units. The speed of GRAPE will increase by $10^3$ in ten years, though general-purpose computers become about 100 times faster at the same period. GRAPE-4 used the technology available at 1990. If we start to develop the next generation GRAPE (GRAPE-TNG) at 1998, we will be able to build a pipeline chips with 5 times faster clock and 50 times more transistors. Thus we get a speedup by a factor of 250. Therefore, we will be able to build a petaflops GRAPE in the end of the twentieth century for the price of 10 million dollars. GRAPE-TNG will open a new era of petaflops computing, and advance our understanding of dynamics of star clusters.

ACKNOWLEDGMENTS

This work was partially supported by the Grant-in-aid for Specially Promoted Research (04102002) of the Ministry of Education, Science, and Culture.

## References

* The viewgraphs of the presentation are available at
    "http://butterfly.c.u-tokyo.ac.jp:8080/pub/people/taiji/grape4".
Aarseth, S. J. 1995. *In this volume.*
Barnes, J., and Hut, P. (1986) *Nature*, **324**, 446.
Brieu, P. P., Summers, F. J., and Ostriker, J. P. 1995. *Astrophys. J.*, in press.
Ebisuzaki, T., Makino, J., Fukushige, T., Taiji, M., Sugimoto, D., Ito, T., and Okumura, S. K. (1993) *Publ. Astron. Soc. Japan*, **45**, 269, and references therein.
Greengard, L., and Rokhlin, V. (1987) *J. Comp. Phys.*, **73**, 325–348.
Hockney, R. W., and Eastwood, J. W. (1988) *Computer Simulation Using Particles.* Bristol: Adam Hilger.
Makino, J. (1991a) *Astrophys. J.*, **369**, 200.
Makino, J. (1991b) *Publ. Astron. Soc. Japan*, **43**, 621–638.
Makino, J., and Aarseth, S. J. (1992) *Publ. Astron. Soc. Japan*, **44**, 141.
Makino, J., Kokubo, E., and Taiji, M. (1993) *Publ. Astron. Soc. Japan*, **45**, 349.
McMillan, S. L. W., and Aarseth, S. J. (1993) *Astrophys. J.*, **414**, 200.
Sugimoto, D., Chikada, Y., Makino, J., Ito, T., Ebisuzaki, T., and Umemura, M. (1990) *Nature*, **345**, 33.
Taiji, M., Makino, J., Ebisuzaki, T., and Sugimoto, D. (1994a) *Pages 280–287 of: Proceedings of the 8th International Parallel Processing Symposium.* Los Alamitos: IEEE Computer Society Press.
Taiji, M., Makino, J., Kokubo, E., Ebisuzaki, T., and Sugimoto, D. (1994b) *Pages 302–311 of: Proceedings of the 27th Hawaii International Conference on System Sciences.* Los Alamitos: IEEE Computer Society Press.
Taiji, M., Makino, J., Ebisuzaki, T., and Sugimoto, D. 1996. *Publ. Astron. Soc. Japan*, to be submitted.

# GRAVOTHERMAL OSCILLATIONS

JUNICHIRO MAKINO
*Department of Graphics and Information Science,*
*College of Arts and Sciences, University of Tokyo*
*3-8-1 Komaba, Meguro-ku, Tokyo 153, Japan*

**Abstract.** We present the first clear evidence that the gravothermal oscillation takes place in $N$-body systems. We performed direct $N$-body simulations of systems of point-mass particles with particle numbers from 2,048 to 32,768. In the simulation with 32,768 particles, the central density shows an oscillation with an amplitude of $\sim 10^3$, which is similar to what was observed in more approximate models such as a conducting gas sphere and one-dimensional Fokker-Planck calculations. The amplitude is smaller for a smaller number of particles. The number of particles in the core at the maximum contraction is $\sim 10$ for all runs, while the number of particles at the maximum expansion is about $0.01N$. For 16,384- and 32,768-body runs, the temperature inversion during the expansion phase is clearly visible.

## 1. Introduction

Gravothermal oscillation was first found by Sugimoto and Bettwieser (1983), who modelled the post-collapse evolution of globular clusters using a conducting gas sphere with artificial energy production. In their model the energy production is expressed as $\epsilon = C\rho^k$, where $\epsilon$ is the energy production per unit mass per unit time and $\rho$ is the density. The power-law index they used is 1 or 2. They found similar oscillations in both cases, for a wide range of the value of coefficient $C$.

Before they found the oscillation, the standard picture of the evolution of a globular cluster had been the following two-stage one. The first stage is the gravothermal collapse, or core collapse, driven by the gravothermal instability. This collapse leads to a self-similar evolution, in which the core density reaches infinity and the core mass goes down to zero in a finite time

(Hachisu, et al. 1978, Lynden-Bell and Eggleton 1980).

In a real $N$-body system, contraction is halted by the energy production by 3-body binaries. It had been believed that the whole system expands homologously, driven by the energy generation from binaries after the contraction is halted (Goodman, 1984).

What Sugimoto and Bettwieser (1983) found is that this homologous expansion is unstable in a way similar to the way in which an isothermal sphere is unstable. A linearized stability analysis by Goodman (1987) showed that the expansion is unstable against thermal perturbation if the coefficient for the energy production is small. If translated back to the total number of particles, the system is unstable if $N > 7,000$. This result is confirmed also with Fokker-Planck calculations (Cohn et al., 1989).

To determine if such an oscillation actually takes place in real globular clusters, we have to perform a direct $N$-body simulation with a sufficiently large number of particles, since gas models and FP models have many simplifying assumptions that might make the evolution completely different. For example, both assume spherical symmetry, while in $N$-body simulation the core is known to wander around (Makino and Sugimoto 1987, Heggie et al. 1994). Both assume that the energy production by binaries is smooth, while in an $N$-body system binaries are formed stochastically.

An $N$-body simulation with sufficiently large number of particles has been impossible, simply because the requirement for computer power has been excessive. The largest simulation which has been tried on a general-purpose supercomputer is 10,000-body run by Spurzem and Aarseth (1996), which does not cover a long enough time after the first collapse.

In the present paper, we describe the result of $N$-body simulations with number of particles 2,048–32,768. All simulations were performed on GRAPE-4, a special-purpose computer for collisional $N$-body simulation. Our main result is the following. First, for large $N$, the core density and core mass exhibited an oscillation of large amplitude. The core mass at maximum expansion is almost independent of the number of particles, and is in good agreement with FP or gas model results (1-2% of the total mass). Second, we confirmed that the observed oscillation is driven by the gravothermal instability. There were several long expanding periods without any energy input. The temperature inversion is visible in such expansion phases. In addition, the behavior of the core density is strikingly similar to the result of FP calculations with a stochastic heat source (Takahashi and Inagaki 1991), suggesting that the mechanism is the same.

The structure of this paper is the following. In section 2, we describe the initial model, the numerical method and the computer used. In section 3 we present the results. Section 4 is for discussion.

## 2. Model and Numerical method

### 2.1. INITIAL MODELS AND THE SYSTEM OF UNITS

We followed the evolution of isolated systems of point-mass particles. For all calculations, we used random realization of the Plummer model as the initial condition. We used the standard system of units (Heggie and Mathieu 1986), in which $G = 1$, $M = 1$, and $E = -1/4$, where $G$ is the gravitational constant, $M$ and $E$ are the total mass and the total energy of the cluster. All particles have the same mass $m = 1/N$, where $N$ is the total number of particles. The half-mass crossing time $t_{hc}$ is $2\sqrt{2}$ in this unit.

### 2.2. NUMERICAL METHOD

For all calculations, we used NBODY4 (Aarseth 1995), modified for GRAPE-4 (Taiji et al. 1995). The numerical integration scheme adopted in NBODY4 is the 4th order Hermite scheme (Makino and Aarseth 1992). It implements the hierarchical (block) timestep algorithm (McMillan 1986, Makino 1991) to use the GRAPE hardware efficiently. Close two-body encounters and stable binaries are handled by KS regularization (Kustaanheimo and Stiefel 1965). Special treatment for compact few-body subsystems is also possible.

The outputs are taken at intervals of a fixed time, which is some fraction of the crossing time. We recorded the central density, core radius, number of particles in the core, the radii of Lagrangian shells, the velocity dispersion within Lagrangian shells, the binaries and their binding energies. The core parameters are calculated following Casertano and Hut (1985).

The accuracy of the time integration is adjusted so that the energy error between two outputs is smaller than a certain prescribed value. The value we used is $1 \times 10^{-5} \sim 1 \times 10^{-6}$ depending on $N$. We required higher accuracy for larger $N$, since the duration of the simulation is longer. When the energy error is very large, the program automatically reads the output at the previous checkpoint and restarts with a reduced accuracy parameter.

### 2.3. HARDWARE

For all calculations, we used GRAPE-4 (Taiji 1995). The GRAPE-4 is a special-purpose computer designed to accelerate the $N$-body simulation using the Hermite integrator and hierarchical timestep algorithm. The fully configured system has a theoretical peak speed of 1.08 Tflops. The simulations reported in the present paper were performed while the assembly and testing of the GRAPE-4 system were under way. Thus the number of processors varies during the calculation. For most of the 32k particle run, we used one quarter of the machine which has a peak speed of 270 Gflops.

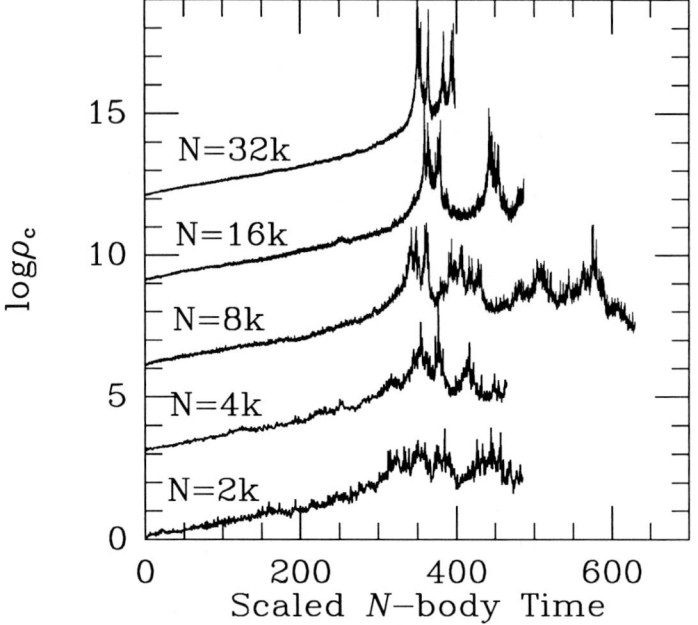

*Figure 1.* The logarithm of the central density plotted as a function of the scaled $N$-body time. Curves for different values of $N$ are vertically shifted by 3 units.

## 3. Result

### 3.1. CORE PARAMETERS

Figure 1 shows the time evolution of the central density for all runs. The time is scaled so that the thermal timescale is same for all runs. The scaling factor is $t_r(1000)/t_r(N) = 212.75\log(0.11N)/N$ (Giersz and Heggie 1994). The core density shows an oscillation with large amplitude in calculations with large $N$ (>16k). No matter what is the real nature of this oscillation, it is at least clear that the core density of the $N$-body system shows oscillation with the amplitude comparable to that observed in gas models or FP calculations. For runs with small $N$ (2k and 4k), there are some oscillation-like features but they are hardly distinguishable from fluctuations. For large values of $N$, however, the oscillation with large amplitude is clearly visible.

Note that in figure 1 there is no clear transition from stable expansion to oscillation or from regular oscillation to chaotic oscillation, which were observed in gas and FP models (e.g. Cohn *et al.* 1989, Heggie and Ramamani 1989). The reason is that binaries emit energy intermittently at the formation time and as a result of binary-single-body interactions.

Takahashi and Inagaki (1991) incorporated this stochastic nature of the

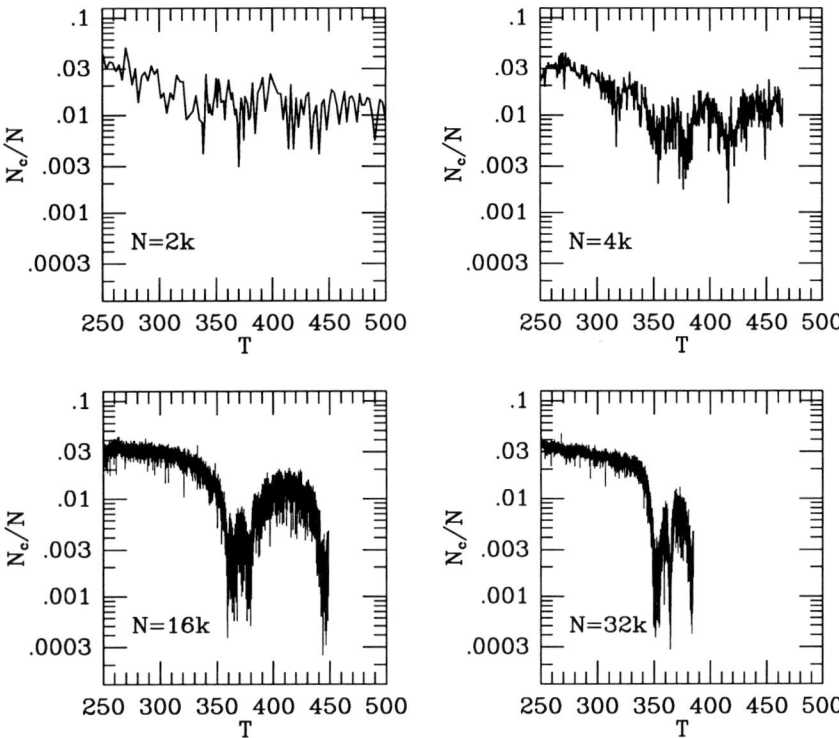

*Figure 2.* The number of particles in the core as the function of the scaled time, for simulations with 2k, 4k, 16k, and 32k particles

energy source into their FP model and found that the core density shows chaotic oscillatory behavior even if the energy production rate is larger than the critical value. They also found that the amplitude of the oscillation is smaller for larger energy input (smaller $N$), which is consistent with the present result. In fact, it would be difficult to distinguish between the result of $N$-body calculations and their stochastic FP result, except that their result is smoother while the central density is low.

Figure 2 shows the evolution of the number of particles in the core. For all runs, the number of particles at maximum contraction is of the order of 10, while that at maximum expansion is 1–2% of the total number of particles. This result is again in good agreement with gas models and FP calculations.

Figure 3 shows the fraction of time for which the number of particles in the core is smaller than the value $N_c$ as a function of $N_c/N$, for the post-collapse phase. For $N > 8192$, the median core mass is around 0.5–0.6%. This corresponds to $r_c/r_h \sim 0.01$. For $N = 32,768$, the core mass is

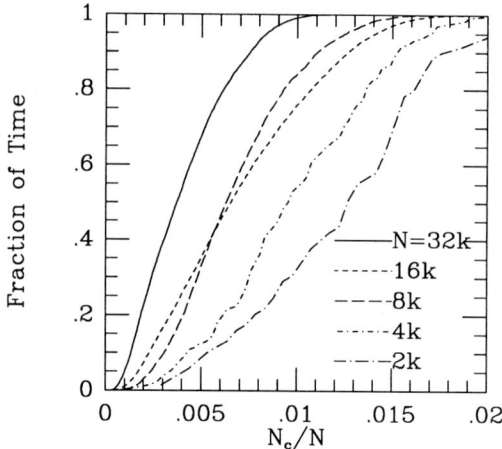

*Figure 3.* Fraction of time for which the number of particles in the core is smaller than $N_c$, as a function of $N_c/N$.

somewhat smaller than that for 16k or 8k runs. This difference is because the 32k run is not long enough.

Figure 3 indicates that the typical core mass for a post-collapse cluster would be less than 1%, if gravothermal oscillation occurs. One interesting question is whether it is possible to distinguish a core in gravothermal oscillation from a core dominated by primordial binaries. The theoretical prediction by Goodman and Hut (1989) gives $r_c/r_h \sim 0.02$, while $N$-body simulations by McMillan *et al.* (1990) give $r_c/r_h \sim 0.2$. McMillan *et al.* (1990) used 1136 particles. If their $N$-body results can be extrapolated to a larger $N$, cores with primordial binaries and cores in gravothermal oscillation would be clearly distinguishable. To obtain a definitive answer, we need to perform simulations of clusters with primordial binaries using a larger number of particles than employed by McMillan *et al.* (1990).

## 3.2. DETAILED VIEW OF THE 32K RUN

In the previous section, we gave an overview of the post-collapse evolution of $N$-body point-mass systems with 2k–32k particles. In this section, we take a closer look of the 32k-particle simulation to see whether we can find a direct signature of the gravothermal expansion. It is generally believed that a long expansion phase without significant energy input and a temperature inversion during the long expansion phase are the most direct signatures of gravothermal expansion (Bettwieser and Sugimoto 1984, McMillan and Engle 1995). In this section we investigate both of them.

Figure 4 shows an enlarged view of the time variation of the core radius

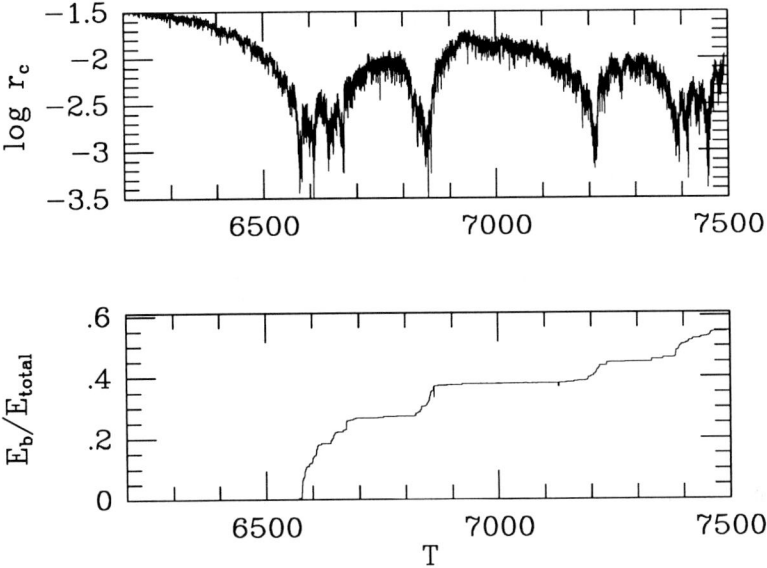

*Figure 4.* The core radius (top) and the binding energy of binaries (bottom) as a function of time for the 32k-particle run

as compared with the sum of the binding energies of all binaries. It is clear that most of the energy is generated when the core is very small. Five peaks account for most of the energy production.

Continued expansion without energy generation is considered to be one of the most direct signatures of gravothermal oscillation. In figure 4, we clearly see two such expansions, for $t = 6700 - 6740$ and $t = 6860 - 6920$. Both expansions continue for more than 10 half-mass crossing times, which is hundreds of the core relaxation time. These expansions cannot be driven simply by binary heating. If the expansion were driven only by binary heating, it could not continue without energy input for a timescale longer than the core relaxation time.

Figure 5 shows temperature profiles for the contracting and expanding phases. Near the end of the expanding phase a temperature inversion of the order of 5% is clearly visible. Since these profiles are time-averaged over 10 time units (80 snapshots), the actual inversion might be somewhat stronger. Note that the temperature inversion is visible only near the end of long expansion phases also in gas model and FP calculations (Bettwieser and Sugimoto 1984, Cohn et al. 1989).

Figure 6 shows the relation between the central density and the central velocity dispersion. The trajectory shows a striking resemblance to what is obtained by gas-model calculation (Goodman, 1987). The fact that the

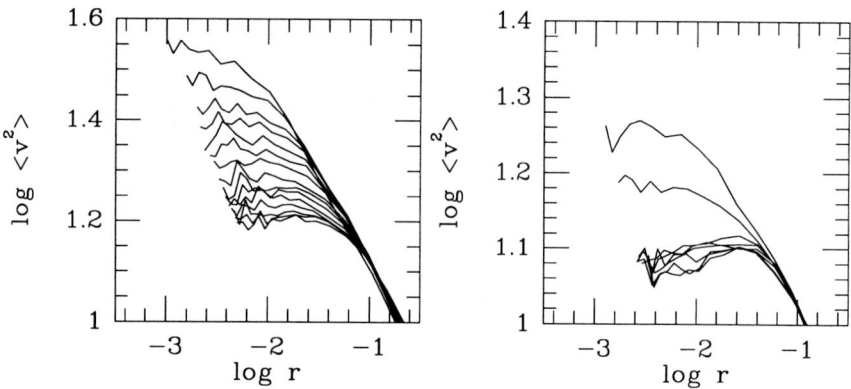

*Figure 5.* Velocity dispersion profiles for (a) contracting and (b) expanding phases. Each profile is obtained by time averaging over 80 snapshots (10 time units). The time interval between curves is 5 time units.

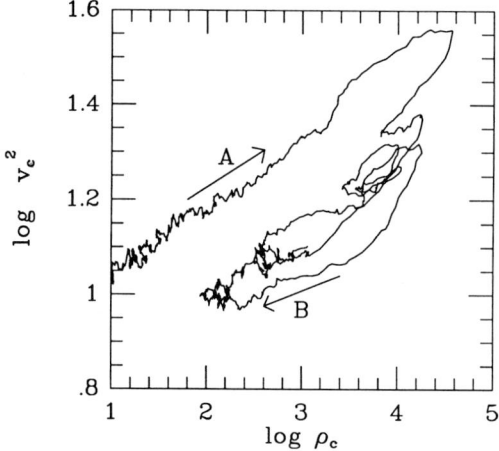

*Figure 6.* The change of the central density and the central velocity dispersion. Each data points is time-averaged value over 80 snapshots. Arrows indicate the direction of evolution.

trajectory shows clockwise rotations means that this is a refrigeration cycle, in which the central region absorbs the heat when the temperature is low, and release heat when the temperature is high (Bettwieser and Sugimoto 1984, Bettwieser 1985). In particular, the later phase of the large expansions (indicated by the arrow marked "B") is nearly isothermal. Therefore this phase is driven by the heat supply. Since the binding energy of binaries is unchanged during this phase, the heat is supplied from outside the core. In other words, the expansion is gravothermal.

## 4. Discussion

We have performed direct $N$-body simulation of the post-collapse evolution of globular clusters. We confirm that gravothermal oscillation actually takes place in a point-mass $N$-body system.

Whether real globular clusters undergo gravothermal oscillation or not is a question which requires further research. If clusters contain many primordial binaries, even after the core collapse they might still be burning the primordial binaries. In addition, the effect of two-body binaries on the evolution of the cluster is still unclear. In fact, the cross section for binary formation by tidal capture and for merging is not fully understood yet (Mardling 1995a, 1995b).

The most straightforward way to study the effect of primordial binaries or two-body capture is direct $N$-body simulation. In principle, we can put primordial binaries and their evolution into FP calculation by following the distribution of internal binding energy of binaries. However, the standard one-dimensional FP calculation, which assumes an isotropic distribution, is not appropriate for following the binary population, since most binaries are formed in the core and their orbits are nearly radial. Thus, we have to solve the FP equation at least in three dimensions ($E$, $J$ and the binary binding energy $E_b$). This would require prohibitively large computer power. One could also use the Monte-Carlo approach, but its result must be compared with $N$-body simulation anyway.

We have demonstrated that $N$-body simulation is now possible with a number of particles close to that in real globular clusters, thanks to the extremely powerful special-purpose computer and its full-time availability. We are now able to use direct $N$-body simulation to study various aspects of the evolution of globular clusters.

The present 32k-particle calculation took about three months of CPU time on 1/4 of GRAPE-4. If we tried to do a similar calculation on a Cray T90 vector supercomputer, it would have taken several years of CPU time. This is simply impossible to do on present-day supercomputers. If we want to finish the calculation in, say, one CPU month, we need a computer 50–100 times faster than a Cray T90, which will be available 10 years from now.

The CPU time of three months is still very long. However, for many simulations, we do not need as many as 32k particles. A 16k-particle calculation was finished in 2-3 weeks, on 1/8 of GRAPE-4. Thus to run many simulations of 16k-particle systems is now practical.

If we can continue the development of the special-purpose computer, we will have a system 100–1000 times faster than the present GRAPE-4 in the next five years. Such a system will make it possible to run 50–100k-particle simulation routinely, while 500k–1M-particle simulations will still take months.

## Acknowledgements

It's a pleasure to thank Daiichiro Sugimoto for his effort on GRAPE project, and also for many useful discussions and comments on the manuscript. I'm grateful to Sverre Aarseth for his NBODY4 code, Makoto Taiji, Toshiyuki Fukushige and Toshikazu Ebisuzaki for developping GRAPE-4 along with myself, Piet Hut and Steve McMillan for many useful discussion and Ivan King for comments on the manuscript. This work was supported by the Grant-in-aid for Specially Promoted Research (04102002) of the Ministry of Education, Science, and Culture.

## References

Aarseth, S. J. (1985) in *Multiple Time Scales*, eds. J. U. Brackhill and B. I. Cohen, Academic, New York, p. 377.
Bettwieser, E. (1985) in *Dynamics of Star Clusters, IAU Symposium No. 113*, eds. J. Goodman and P. Hut, Reidel, Dordrecht, p. 219.
Bettwieser, E. and Sugimoto, D. (1984) *MNRAS* **208**, 493.
Casertano, S. and Hut, P. (1985) *Ap. J.* **298**, 80.
Cohn, H., Hut, P., and Wise, M. (1989) *Ap. J.* **342**, 814.
Giersz, M. and Heggie, D. (1994) *MNRAS* **268**, 257.
Goodman, J. (1984) *Ap. J.* **280**, 298.
Goodman, J. (1987) *Ap. J.* **313**, 576.
Goodman, J. and Hut, P. (1989) *Nature* **339**, 40.
Hachisu, I., Nakada, Y., Nomoto, K., and Sugimoto, D. (1978) *Prog. Theor. Phys.* **60**, 393.
Heggie, D. (1986) in *The Use of Supercomputers in Stellar Dynamics*, eds. S. McMillan and P. Hut, Springer, New York, p. 233.
Heggie, D. and Ramamani, N. (1989) *MNRAS* **237**, 757.
Heggie, D., Inagaki, S., and McMillan, S.L.W. (1994) *MNRAS* **271**, 706.
Kustaanheimo, P., and Stiefel, E. (1965) *J. Reine. Angew. Math.* **218**, 204.
Lynden-Bell, D. and Eggleton, P.P. (1980) *MNRAS* **191**, 483.
Makino, J. (1986) in *The Use of Supercomputers in Stellar Dynamics*, eds. S. McMillan and P. Hut, Springer, New York, p. 151.
Makino, J.. (1991) *Publ. Astron. Soc. Japan* **43**, 841.
Makino, J. and Aarseth, S. J. (1992) *Publ. Astron. Soc. Japan* **44**, 141.
Makino, J. and Sugimoto, D. (1987) *Publ. Astron. Soc. Japan* **39**, 589.
Mardling, R. A. (1995a) *Ap. J.* **450**, 722.
Mardling, R. A. (1995b) *Ap. J.* **450**, 732.
McMillan, S. L. W. (1986) in *The Use of Supercomputers in Stellar Dynamics*, eds. S. McMillan and P. Hut, Springer, New York, p. 156.
McMillan, S. L. W. and Engle, E. A. (1995) in this volume.
McMillan, S. L. W., Hut, P. and Makino, J. (1990) *Ap. J.* **362**, 522.
Spurzem, R. and Aarseth, S. J. (1996) preprint.
Sugimoto, D. and Bettwieser, E. (1983) *MNRAS* **204**, 19p.
Takahashi, K. and Inagaki, S. (1991) *Publ. Astron. Soc. Japan* **43**, 589.
Taiji, M. (1995) in this volume.

# STAR CLUSTER SIMULATIONS ON HARP

SVERRE J. AARSETH
*Institute of Astronomy, University of Cambridge, UK*

**Abstract.** We describe some aspects of implementing star cluster simulations on HARP. The code NBODY4 employs the Hermite scheme with hierarchical block-steps for direct integration. The algorithms have been optimized for parallel processing with the eight pipeline HARP-2 delivering a peak performance of about 1.7 Gflops for $N = 10^4$ particles. Hard binaries are studied by KS regularization which also uses the Hermite scheme, whereas strong interactions between $3-5$ particles are treated by chain regularization. Astrophysical processes modelled include mass loss by stellar evolution, two-body tidal interaction, Roche lobe mass transfer, common envelope evolution, magnetic braking and gravitational radiation. Consistent values of stellar radii and evolution type are obtained by fast look-up. A new formulation of collision outcomes yields blue stragglers and other exotic objects. Some recent results for an open cluster model are presented.

## 1. Introduction

The advent of the HARP special-purpose computer represents a watershed in the history $N$-body simulations. In the present paper, we describe a new star cluster project undertaken with the eight pipe-line HARP-2 version which was installed at the Institute of Astronomy in August 1994. Until now, it has been feasible to perform such calculations with up to $N = 2500$ members on dedicated workstations, with just one model requiring several months of continuous work (Heggie and Aarseth 1992). Only by the most heroic effort on a supercomputer (i.e. more than 1000 CPU hours over two years) has it been possible to reach the magical limit $N = 10^4$ (Spurzem and Aarseth 1996). Using HARP-2, we can now investigate a range of different parameters for star cluster models with (say) $N = 10^4$. In order to reach this objective, it has been necessary to develop software for all the relevant processes occurring in star clusters and to adapt the code in a form suitable for HARP. The main processes which have been implemented for studying open clusters can be summarized as follows:

- Galactic tidal field
- Primordial binaries
- Mass loss from evolving stars
- Velocity kick during neutron star formation
- Tidal circularization
- Roche lobe mass transfer
- Common envelope evolution
- Magnetic braking
- Gravitational radiation
- Physical stellar collisions
- Formation of exotic objects.

In the subsequent sections we discuss briefly the astrophysical processes listed above. However, we begin by describing the main technical methods concerned with the actual integration since these aspects have required much new work for applications on HARP.

## 2. Hermite Integration Scheme

The basic integration scheme for HARP computers exploits the fast evaluation of the total force (per unit mass) and its first time derivative by employing a fourth-order formulation (Makino 1991). We write a Taylor series for each of these quantities as

$$\mathbf{F} = \mathbf{F}_0 + \dot{\mathbf{F}}_0 t + \frac{1}{2}\ddot{\mathbf{F}}_0 t^2 + \frac{1}{6}\dddot{\mathbf{F}}_0 t^3 \tag{1}$$

$$\dot{\mathbf{F}} = \dot{\mathbf{F}}_0 + \ddot{\mathbf{F}}_0 t + \frac{1}{2}\dddot{\mathbf{F}}_0 t^2. \tag{2}$$

These equations can be inverted to obtain the second and third time derivatives in terms of $\mathbf{F}$ and $\dot{\mathbf{F}}$ at the begining and end of a time-step, which can then be used to correct the coordinates and velocities.

The so-called Hermite integration scheme yields increased efficiency by the introduction of hierarchical time-steps $\Delta t_n = 1/2^{n-1}$, combined with a hardware chip for fast prediction of all coordinates and velocities to order $\dot{\mathbf{F}}$. Depending on the natural time-step obtained from some suitable convergence criterion, the new step may then be reduced by 2 (or even 4) or remains the same each time, but can only be increased by 2 every other step to maintain time commensurability. We note that this hierarchical treatment achieves a performance which is weakly dependent on the time-step range (i.e. depth of the potential well) for the same particle number. Moreover, a given fraction of the peak performance is reached at larger values of $N$ when the number of pipe-lines is increased to match the increased number of particles in the same block level.

Close binaries and hyperbolic fly-bys are integrated by Kustaanheimo-Stiefel (1964, hereafter KS) regularization. The equations of motion are

$$\mathbf{u}'' = \frac{1}{2}h\mathbf{u} + \frac{1}{2}R\mathcal{L}^T\mathbf{P} \quad (3)$$

$$h' = 2\mathbf{u}' \cdot \mathcal{L}^T\mathbf{P} \quad (4)$$

$$t' = \mathbf{u} \cdot \mathbf{u}. \quad (5)$$

Here $\mathbf{u}$ denotes the new coordinates which satisfy the relation $\mathbf{u} \cdot \mathbf{u} = R$, $h$ is the two-body binding energy per unit mass, $\mathcal{L}(\mathbf{u})$ is a $4 \times 3$ matrix and $\mathbf{P}$ is the physical tidal perturbation exerted by the other particles. Decision-making for existing KS solutions is based on the relative perturbation, $\gamma = PR^2/m_b$, with $R$ the separation and $m_b$ the combined mass of the components. Further discussions together with the relevant transformations can be found elsewhere (Aarseth 1985, 1994).

In order to achieve uniformity, we replace the standard difference scheme (Aarseth 1985) by a Hermite KS formulation. This entails differentiating Eqns. (3) and (4), making use of the linear matrix property $\mathcal{L}'(\mathbf{u}) = \mathcal{L}(\mathbf{u}')$, whereas the time transformation (5) yields $\mathbf{P}' = R\dot{\mathbf{P}}$. Although the new equations of motion contain more terms and also require velocity predictions of perturbers as well as a KS velocity transformation for evaluating $\dot{\mathbf{P}}$, the additional effort is to some extent compensated by a simpler treatment.

All the calculations relating to KS are carried out on the host computer and may therefore present an unbalanced load when integrating many perturbed binaries. This problem is alleviated by employing a slow-down scheme (Mikkola and Aarseth 1996) for weakly perturbed binaries (say $\gamma < 5 \times 10^{-5}$). The main idea is to exploit the adiabatic invariance such that one KS orbit with augmented perturbation may represent several physical orbits, while maintaining a correct treatment of the secular effects. As before, binaries with small relative perturbations ($\gamma < 10^{-6}$) are assumed to be unperturbed and advanced one or more periods without any step-wise integration. Moreover, hierarchical triples are treated as composite KS binaries while satisfying a criterion of dynamical stability (Eggleton and Kiseleva 1995). Finally, we remark that an energy stabilization term is included in Eqn. (3) which introduces a small systematic error in the eccentricity. However, this effect is reduced considerably when using the slow-down scheme.

## 3. Close Encounters

The treatment of close encounters on HARP requires many new considerations. Here we summarize some of the technical problems which must be overcome. Since the fictitious time defined by Eqn. (5) is not a linear function of time, each KS solution is advanced one or more steps within a new block-step before integrating the single particles and corresponding centre-of-mass (c.m.) particles.

We distinguish between active and inactive c.m. particles, where the latter are associated with unperturbed KS solutions; hence their new force can be evaluated as for single particles. To obtain a consistent force on an active c.m. particle, we include the contributions from distant members in one summation and add the mass-weighted force on the two components due to the perturbers. The case of several active c.m. particles in the same block can then be treated by obtaining the force on HARP for each one separately after defining a mask (i.e. by specifying zero masses) for the relevant perturbers. However, in order to utilize more than one pipe-line simultaneously, we first form a joint perturber list which acts as a mask for the different c.m. forces on HARP, and then add all these perturber contributions on the host. Although this algorithm requires a careful sifting to avoid multiplicity, the additional effort on the host is minimized by taking into account the spacing between different c.m. particles and their corresponding maximum perturber distances which can be estimated from the selection criteria based on binding energy and mass ratio.

The strategy for determining KS candidates exploits the property that such particles have small time-steps and occupy the deepest block-step levels. Hence it is only necessary to form a candidate list when the current level has few members, thereby reducing the effort. Once two particles have been selected for KS treatment, the initialization proceeds as before (Aarseth 1994). However, the commensurability requirement sometimes leads to unduly small values of the c.m. time-step with loss of efficiency in the case of short-lived interactions (i.e. wide binaries).

It is desirable to delay the termination of a KS treatment until the end of a block-step. This entails advancing the regularized solution some fraction of the current block-step, where the latter is likely to be quite small; say, the corresponding c.m. time-step in the case of a strong interaction. Now the loss of efficiency is less noticeable since the new steps for direct integration are usually small.

Several tasks can be carried out while HARP is busy with force evaluations, such as the prediction of the next set of particles (if any) as well as any KS coordinates and velocities, the construction of the joint perturber list and noting particles with small steps. All these tasks are performed during the first HARP call, whereas the corrector is included for the previous set of particles on any subsequent calls. The full optimization by so-called low-level functions enhances the performance from about 1.1 Gflops to 1.7 Gflops for $N = 10^4$, which in practice means twice the throughput of a Cray-YMP supercomputer using the vectorized NBODY5 neighbour scheme code (Spurzem and Aarseth 1996).

Perturbed chain regularization (Mikkola and Aarseth 1993) has also been implemented. Typically one large simulation employs the chain procedures about 100 times; however, most of these multiple encounters of three or four particles involve extremely energetic interactions. For instance, the chain code is used to study binary-binary encounters. Such systems may decay into a compact triple and an escaping body; alternatively a triple or quadrupole may increase the membership by an intruding particle.

Finally, the practical usefulness of obtaining neighbour lists on HARP should also be emphasized. Several strategies have been developed in order to deal with the different requirements, such as finding the nearest neighbour or constructing new perturber lists.

## 4. Cluster Model

We now describe the initial conditions for an open cluster model with $N = 10^4$ members. The coordinates and velocities are generated from an equilibrium King model with central concentration $W_0 = 7$ in an external tidal field (Heggie and Ramamani 1992). We choose a modern IMF (Kroupa, Tout and Gilmore 1993) with masses in the range $10 - 0.2$ $m_\odot$. The corresponding tidal radius in the solar neighbourhood is then $R_t = 22$ pc, with half-mass radius $R_h \simeq 3$ pc ($r_h = 0.82$ in scaled units) and rms velocity $V_0 \simeq 2$ km/s, giving a crossing time of $\simeq 4 \times 10^6$ yr.

The main aim of the present project is to study open clusters containing a significant proportion of primordial binaries. Although the dynamical effects of such binaries have been investigated before (McMillan, Hut and Makino 1990, 1991, Heggie and Aarseth 1992), the emphasis here is to include all relevant astrophysical processes which are outlined in the subsequent sections (but also see Aarseth 1996 for an earlier review). We adopt an initial binary fraction $f_0 = N_b/(N_b + N_s)$ of 5 % with $N_b = 500$ binaries and $N_s = 9500$ single stars, and a period distribution based on the solar neighbourhood (Duquennoy and Mayor 1991). Here the individual binary components are selected from the same IMF as the single stars but even so their combined masses do not tend to exceed the latter because of sampling effects. First we restrict the uncorrelated period to a minimum value of one day and, for practical reasons, impose an upper cutoff at a semi-major axis of $a_{max} = 48$ AU which is somewhat smaller than that of a hard binary.

According to recent ideas (Kroupa 1995), the closest binaries experience some changes in their two-body elements during the pre-main-sequence evolution. Consequently, we modify the semi-major axis ($a$) and eccentricity ($e$) according to tidal circularization theory (Mardling 1995), assuming a time interval of $10^5$ yr and an initial stellar radius of fifteen times the ZAMS value. This procedure yields 25 circularized binaries, and other binaries with some modifications of $a$ and $e$. Hence the eccentricity distribution already contains a small proportion of short-period circular orbits.

## 5. Synthetic Stellar Evolution

The modelling of several astrophysical effects requires knowledge of the stellar mass and radius which change with time. The stars are characterized by an index $K = 0, 1, ..., 10$, according to their temporal location in the HR-diagram; from low-mass main-sequence to neutron stars and even black holes or exotic objects. We adopt a continuous mass loss by stellar winds according to the modified Reimers expression (Tout 1990) beyond the main-sequence ($K > 1$), $\dot{m} = 4 \times 10^{-13} R_* L/m \; m_\odot/yr$, where the radius $R_*$, luminosity $L$ and mass $m$ are in solar units and a suitable time average of $R_* L$ is used. We implement mass loss when the accumulated value exceeds one percent. In order to maintain total energy conservation, we include correction procedures for the change in potential energy, together with a modification of the force on neighbouring members or a complete initialization evaluated on HARP, depending on the type of event.

Convenient look-up tables for the evolution of population I stars (Eggleton, Fitchett and Tout 1989) are employed to obtain the stellar radius, luminosity and classification type as a function of the *initial* mass and age. A solution for the core mass which is used for various procedures has also been introduced (Tout 1995). Each characteristic stage is subdivided into small intervals at which the stellar parameters are updated smoothly.

Neutron star formation is modelled by assuming an asymmetric ejection of the excess mass. The resulting velocity kick is derived from a probability distribution based on recent observations (Lyne and Lorimer 1994). For computational convenience, single stars are restricted to the typical escape velocity, whereas only binary components are assigned the full value in order to test the retention hypothesis. Thus the consistent updating of the stellar parameters permit a synthetic HR-diagram of the remaining cluster members to be constructed as a function of age.

With realistic cluster parameters, physical collisions are inevitable. A general scheme for stellar collision has recently been introduced. We construct a $10 \times 10$ collision matrix for the outcome which depends on stellar type and includes a consistent treatment of common envelope evolution.

## 6. Tidal Circularization

The growth of stellar radii in close binaries may induce significant tidal dissipation. According to a new treatment for tidal interaction (Mardling 1995), this process may also lead to a chaotic phase of irregular eccentricity behaviour before the final slow approach to a circular orbit. The complete theory has already been implemented in NBODY4 and replaces a previous formulation based on classical tidal capture (Aarseth 1992).

There are two ways in which a tidal interaction may be initiated: (i) increase of eccentricity induced by external perturbations, and (ii) expansion of the stellar radius with time. It has been noted that hierarchical triples may act as catalysts for the former process by producing systematic increase of eccentricity. If a binary is a candidate for tidal interaction (i.e $a(1-e) < 10 max(R_1, R_2)$) and the circularization time is below $1 \times 10^8$ yr, we evaluate the critical eccentricity $e_{crit}$ associated with the boundary between chaotic and regular behaviour. The relatively infrequent case $e > e_{crit}$ leads to a short-lived chaotic interaction between the orbit and the tides. This implementation is based on a non-linear dissipation time scale and assumes total angular momentum conservation (see Mardling and Aarseth 1996 for details). The KS variables are then modified to the new energy and pericentre distance and corresponding polynomials are initialized during epochs of significant perturbation. Once this process terminates, or if $e < e_{crit}$ initially, the binary begins a slow periodic phase of circularization with small tides.

The treatment of tidal circularization employs a semi-analytical relation for the eccentricity: $e = f(q, n, a, t)$, with mass ratio $q$ and polytropic index $n$ (Mardling 1996). Here giants are modelled with an effective polytropic index which depends on the envelope mass (Mardling and Aarseth 1996). Angular momentum conservation yields the new pericentre $r_p = r_p^0 (1 + e_0)/(1 + e)$ in terms of a previous pericentre distance and eccentricity. We adjust the associated KS variables to the new pericentre distance and velocity. If mass loss occurs or the motion is perturbed, the relevant parameters are re-initialized at pericentre. Except for the case of significant orbital expansion (or the formation of a compact object), the procedure terminates when $e = 0$. Although the resulting shrinkage of the semi-major axis shortens the period, the condition for Roche mass transfer is usually not satisfied during circularization since, by angular momentum conservation, $a(1-e)$ actually increases.

## 7. Mass Transfer and Orbital Decay

Once a synchronous orbit has been achieved, Roche mass transfer may be initiated after further growth of the radius. The modelling of this process is based on a sophisticated scheme which distinguishes between the three relevant time scales (dynamical, thermal and nuclear) for the expansion rate of the stellar radius. The donor star is assumed to undergo Reimers-type mass loss in addition to a mass transfer rate of $1.0 \times 10^4 ln(R_1/R_L)^3 \, m_\odot/yr$, where $R_L$ is the corresponding Roche radius (Tout 1990), and a certain fraction of this mass is accreted by the secondary. A rejuvenation procedure is carried out after significant mass change for main-sequence stars.

Common envelope evolution for giant or supergiant donor stars has also been included. This process may lead to a significant shrinkage of the semi-major axis, with physical collision in some cases. The subsequent evolution of short-period binaries ($a < 3R_\odot$) may be subject to magnetic braking (Regös and Tout 1995) on a time scale $\tau_m \propto a^5$. If at least one binary component is a degenerate object, gravitational radiation may also lead to coalescence ($\tau_g \propto a^4$) with an uncertain end result defined as a new type. We note that correction procedures are included for all dissipative processes in order to maintain an energy-conserving scheme.

## 8. Recent Results

The code development and testing of the various astrophysical processes is now essentially completed and permit a preliminary inspection of the results. Here we summarize some relevant features pertaining to the initial conditions described in Section 3. This model was studied until complete disruption, with a total life-time of $5.7 \times 10^9$ yr. Because of various time-consuming astrophysical processes (e.g. Roche mass transfer), the actual CPU time was about 320 hr. By half-life ($t_{1/2} \simeq 1.9 \times 10^9$ yr), there were $\simeq 1.1 \times 10^9$ individual integration steps and $\simeq 2.6 \times 10^8$ KS steps. With some $3.8 \times 10^7$ block-steps, this corresponds to an average of about 30 particles per block-step and indicates an efficient usage of the eight pipes.

A simple summary is best carried out at the half-life ($N = 5250$). At this stage $N_b = 290$, suggesting that the binaries are not preferentially depleted and therefore retain the capacity for preventing core collapse. The cluster half-mass radius expands from $r_h = 0.8$ to 1.3, during which some 1400 $m_\odot$ are lost due to stellar evolution compared to a total initial mass of $6857 m_\odot$. Core collapse is only marginal when stellar evolution is included; the minimum core radius and membership noted was $r_c \simeq 0.06$ and $N_c = 19$ at $t \simeq 2.2 \times 10^9$ yr. As the most massive single stars complete their evolution, the mass segregation of binaries becomes important. This is reflected in the enhanced abundance of binaries in the core, which increased to about 20 % by half-life. Hence it does not seem necessary to include a large initial binary fraction in order to halt core collapse and yield a significant observed abundance in the central regions.

The behaviour of the binary distribution is of particular interest. On the dynamical side we note that 269 original binaries remain at $t_{1/2}$, with 20 exchanges and one new binary. Of the remaining population, there were 174 escaped binaries (one being a new binary and one a hierarchical triple); hence 36 binaries were either destroyed in dynamical interactions or suffered physical collision. The relatively small formation rate of binaries is noteworthy and can be understood as follows. During the process of hardening, a binary typically experiences $\simeq 10 - 20$ significant energy changes. Hence if only one of these interactions involve another hard binary, this can be sufficient to destroy the emerging binary, whereas in systems with no primordial binaries the probability of such encounters is much smaller. Interactions with binaries also produce some energetic or runaway escapers. Thus in the present model there where 54 dynamical ejections with terminal velocity $V_f/V_t > 5$, six of which were binaries.

The astrophysics of close binary interaction in a dynamical environment presents many challenges. By the half-life stage some 24 binary orbits had circularized, of which five began as chaotic events. Following circularization, 13 binaries underwent Roche mass transfer with a high percentage reaching common envelope evolution. Of the 12 collisions, eight were defined as accretion-induced collapse (AIC), with one blue straggler and one star exceeding turn-off mass during Roche mass transfer. By the end of the calculation, there were 18 collisions, with 15 Roche cases, nine AIC, five blue stragglers and two Roche turn-off events. Finally, we note at least eight examples with both companions being white dwarfs, two of which had short periods ($\simeq 0.2$ and $0.6$ days).

## 9. Future Prospects

The attempt to synthesize stellar dynamics and stellar evolution represents a formidable challenge to the modern astronomer. Although the main ingredients are now in place, further refinements are required before observational comparison can be undertaken with confidence. In particular, we need a theory for Roche mass transfer for eccentric orbits and an improved treatment of giants for tidal circularization. Before making applications to globular clusters, we also need a synthetic stellar evolution scheme for population II stars and a convenient formulation for eccentric cluster orbits in a 3D galactic potential. Hopefully both of these features will be implemented in the near future. This will enable small ($N \simeq 25,000$) globular clusters to be modelled on the more powerful HARP-3 (with 88 pipes) which has now been installed at Cambridge. However, open star clusters are also unique laboratories for studying various exotic objects and should not be overlooked. Of special interest here is the possibility of analysing such processes in detail, particularly the formation of blue stragglers and binaries containing degenerate components. Another important predicted cluster feature for comparison is concerned with the period-eccentricity relation which is now well established observationally (Duquennoy and Mayor 1991). Hence, we may say that a realistic star cluster simulation offers the opportunity of direct comparison with observations and in addition gives the theoretician a powerful tool for examining the interplay of complex processes involving both dynamics and astrophysics.

## Acknowledgements

I thank my colleagues Drs. Peter Eggleton, Onno Pols and especially Chris Tout for advice on aspects of stellar evolution and Dr. Rosemary Mardling for implementing tidal circularization. Discussions with Dr. Jun Makino concerning the block-step scheme during several visits to Tokyo were also highly beneficial. This project has been supported by a cooperative scheme of the Royal Society, the Japanese Society for the Promotion of Science, and the British Council (Tokyo).

## References

Aarseth, S.J. 1985, in Multiple Time Scales, ed. Brackbill, J.U. and Cohen, B.I. (Academic Press, New York), 377
Aarseth, S.J. 1992, in Binaries as Tracers of Stellar Formation, ed. Duquennoy, A. and Mayor, M. (Cambridge University Press), 6
Aarseth, S.J. 1994, in Galactic Dynamics and N-Body Simulations, ed. Contopoulos, G., Spyrou, N.K. and Vlahos, L. (Springer-Verlag), 277
Aarseth, S.J. 1996, in Binaries in Clusters, ed. Milone, E.F. (ASP series), in press
Duquennoy, A. and Mayor, M. 1991, Astron. Astrophys. 248, 485
Eggleton, P.P. and Kiseleva, L.G. 1995, Astrophys. J. in press
Eggleton, P.P., Fitchett, M.J., and Tout, C.A. 1989, Astrophys. J. 347, 998
Heggie, D.C. and Aarseth, S.J. 1992, Mon. Not. R. astr. Soc. 257, 513
Heggie, D.C. and Ramamani, N. 1992, Mon. Not. R. astr. Soc. 272, 317
Kroupa, P. 1995, Mon. Not. R. astr. Soc. in press
Kroupa, P., Tout, C.A. and Gilmore, G. 1993, Mon. Not. R. astr. Soc. 262, 545
Kustaanheimo, P. and Stiefel, E. 1965, J. Reine Angew. Math. 218, 204
Lyne, A.G. and Lorimer, D.R. 1994, Nature, 369, 127
Makino, J. 1991, Astrophys. J. 369, 200
Mardling, R.A. 1995, Astrophys. J. 450, 722
Mardling, R.A. 1996, in preparation
Mardling, R.A. and Aarseth, S.J. 1996, in preparation
McMillan, S.L.W., Hut, P. and Makino, J. 1990, Astrophys. J. 362, 522
McMillan, S.L.W., Hut, P. and Makino, J. 1991, Astrophys. J. 372, 111
Mikkola, S. and Aarseth, S.J. 1993, Celest. Mech. Dyn. Astron. 57, 439
Mikkola, S. and Aarseth, S.J. 1996, to be published
Regös, E. and Tout, C.A. 1995, Mon. Not. R. astr. Soc. in press
Spurzem, R. and Aarseth, S.J. 1996, to be published
Tout, C.A. 1990, Ph.D. thesis, University of Cambridge
Tout, C.A. 1995, personal communication

# HIGH RESOLUTION STUDIES OF COMPACT BINARIES IN GLOBULAR CLUSTERS WITH HST AND ROSAT

JONATHAN E. GRINDLAY
*Harvard-Smithsonian Center for Astrophysics*
*60 Garden Street*
*Cambridge, MA 02138, USA*

**Abstract.** The studies of compact binaries containing an accreting white dwarf or neutron star in the dense cores of globular clusters have made considerable progress in the past few years as a result of the high resolution images obtained with HST and ROSAT. It is now clear that cluster cores contain a significant population of these systems which must constrain the similarly large populations of millisecond pulsars as well as dynamical histories of clusters. The population of dim x-ray sources appears to be dominated by cataclysmic variables (CVs) formed by tidal capture and not exchange collisions. Our recent HST/FOS spectra of the first CVs in a cluster core, summarized here in more detail, suggest that cluster cores may contain a significant population of magnetic CVs. The required magnetic WDs may arise in spun-up cores of blue stragglers.

## 1. Introduction

The cores of globular clusters have been long recognized as the site of stellar encounters leading to binary production (and destruction) and the dynamical evolution of the entire cluster (cf. Hut et al 1992 for a general review). Indeed, even the evolution of the globular cluster system in the Galaxy is driven to some extent by the stellar dynamics, and compact binary content, of the cluster core with core collapse apparently more likely to occur in disk clusters subject to tidal shocking. One of the most direct probes of compact binaries in cluster cores are the dim x-ray sources (Grindlay 1994 and references therein), which were discovered in the first imaging x-ray survey of globulars (Hertz and Grindlay 1983; HG) and have now been seen in much greater numbers (at least 27; cf. Johnston, Verbunt and Hasinger 1996) with more sensitive ROSAT observations. The other major tool is of course HST, for which the unprecedented spatial resolution has allowed both emission line object imaging searches and spectroscopic identification of both dim sources and compact binaries generally deep within the cores of core collapsed globulars.

In this paper we summarize first the increasingly rich studies of compact binaries in globulars being carried out with HST and then the ROSAT data and the possible dependences on cluster parameters of the integrated core x-ray luminosities for the dim sources. We provide more details for our recent HST spectra of the first CVs, or (more exactly) CV candidates, in the core of the core collapsed cluster NGC 6397 together with a first look at our very deep ROSAT HRI observation of the same cluster. At least 2-3 additional dim sources, for a total of 5-6, are now detected within the central $\sim 10''$ of the cluster center. Our HST/FOS optical spectra of the three brightest candidates suggest these CVs are magnetic, or of the DQ Her type. This would suggest that magnetic white dwarfs (WDs) are somehow enhanced in globular clusters over their relative abundance in the field. We outline a possible model for the production of magnetic WDs in globulars as the WD remnants of rapidly rotating cores of blue stragglers.

## 2. HST Studies of Compact Binaries in Globular Cluster Cores

2.1. WFPC1 AND WFPC2 OBSERVATIONS OF NGC 6397 AND NGC 6752

In order to search for CVs in globulars to test the original proposal that they account for the (vast) majority of dim x-ray sources in clusters (HG, Grindlay 1994), we have developed and conducted a photometric imaging search for H$\alpha$ emission line objects in cluster cores using the WFPC1 and WFPC2 cameras on HST. All classes of CVs show H$\alpha$ in emission (except for some dwarf novae during outburst maxima, when the line can go into absorption) so that emission lines rather than blue continuum colors (though this is also checked) or variability are our primary search criterion for selection of CV candidates. The implementation of this narrow-band photometry with HST using DAOPHOT as well as the calibrations and uncertainties are described in detail by Cool (1993).

Our WFPC1 results for the nearby core collapse cluster NGC 6397 are presented by Cool et al (1995; CG95). Three significant detections of H$\alpha$ bright objects were found in the error circles ($\sim 6''$ radius) of the three brightest dim x-ray sources in the cluster core discovered with ROSAT (Cool et al 1993; CG93). The three objects have apparent R magnitudes of 17.8, 18.8 and 19.5 corresponding to absolute magnitudes $M_R = 5.6, 6.6$ and 7.3. Thus all three are typical of nova-like CVs and not dwarf novae (in outburst) and yet are still somewhat brighter than the median absolute magnitude (~8) of disk CVs. All three were also uv bright, as verified from comparison with archival FOC images taken in the uv (cf. CG95). In addition there were three additional possible CV candidates which showed either possible H$\alpha$ emission or uv excess or both. One or more of these may be now identified with still fainter ROSAT sources as discussed below.

In NGC 6752, we obtained WFPC1 images in our first exploratory H$\alpha$ search with HST (in Cycle 1). The relatively shallow exposures (6 × 1300 s in H$\alpha$ ; 2 × 500 s in R) only reached equivalent absolute magnitudes $M_R \sim 5$ for the detection of H$\alpha$ bright objects (Cool 1993) and was thus not sensitive enough to see CVs except dwarf novae in outburst. However in deeper Cycle 5 observations with WFPC2 (cf. Bailyn et al 1996), we have found two H$\alpha$ objects in the central core (within the PC chip only). One of these CV candidates is in the error circle of one of the two (brightest) dim sources in the core (Grindlay 1993a, Grindlay and Cool 1996).

## 2.2. FOS SPECTRA OF CV CANDIDATES IN NGC 6397

We have recently reported (Grindlay et al 1995; GC95) the first spectra of CV candidates (or any stars below the ms turnoff) in the central core of a globular cluster, NGC 6397.

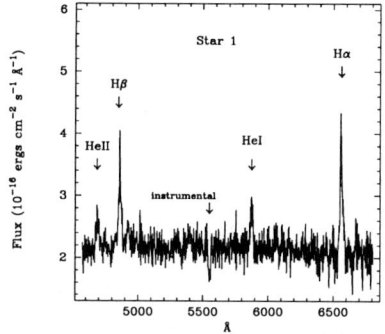 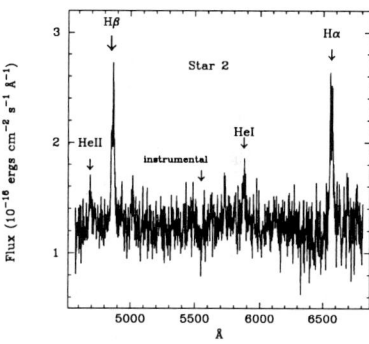

*Figure 1.* HST/FOS spectra of star 1 (left) and 2 (right) in NGC 6397.

We observed with the HST/FOS on UT 2 April 1995 the three brightest H$\alpha$ emission line candidates discovered with our WFPC1 photometry (CG95). Spectra were obtained not only to confirm the suspected H$\alpha$ emission but to further test whether these objects are indeed CVs or could be quiescent LMXBs (see below). The FOS spectra are all remarkably similar (cf. Figure 1) and show not only the bright Balmer lines but also He I and He II emission as is often detected in magnetic CVs of the DQ Her type (cf. Patterson 1994).

At least some of the Balmer emission lines and He II line are double-peaked (cf.Figure 2), suggesting these objects undoubtedly contain accretion disks with characteristic velocity half-widths of ∼600 km/s (cf. GC95), or Keplerian radii (for the emission line regions, and an assumed 1 $M_\odot$ compact object) of ∼ 3 × $10^{10}$ cm. This in turn suggests a minimum orbital period for the probable binary companions in these systems of ∼ 2.4 hours if their disks extend out to ∼0.5 their binary companion separations as is typical in compact binaries. These binaries could, therefore, be very compact though probably not as compact as AM Her type CVs which have an orbital period distribution peaked at $\lesssim$ 2 hours.

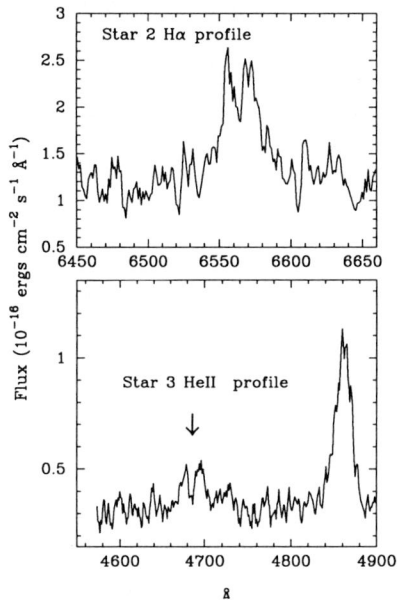

Fig. 2: Double-peaked emission profiles.

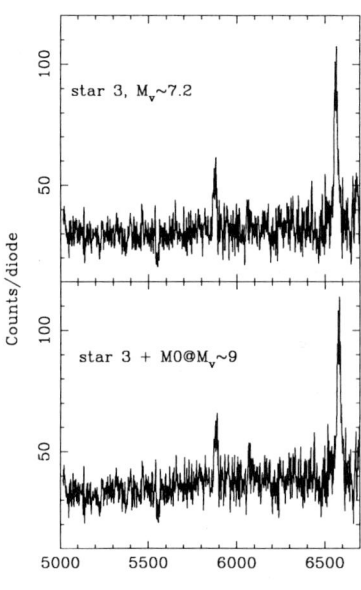

Fig. 3: Star 3 (top) vs. simulated spectrum including M0 star

A crude upper limit for the binary periods ($\lesssim$ 6 h) can be estimated from the fact that the secondary mass must be $\lesssim$ 0.45 $M_\odot$ due to the absence of any detectable absorption lines or red continuum in the spectra (GC95), as shown in Figure 3. The simulated spectrum (bottom) of star 3 with a M0 companion (with $M_V$ ∼ 9) shows a readily detectable red continuum not seen in the actual data (top), thus constraining the secondary mass for at least this faintest of our three CV candidates in NGC 6397.

## 2.3. DWARF NOVAE SEARCHES WITH HST

Searches for variable, and uv-bright, objects have now been conducted in several globulars to measure the number of dwarf novae. Paresce and De Marchi (1994) have found (just) one DN in the core of 47 Tuc, and indeed a repeat outburst of this same object was found (and no others) in more extensive coverage of the cluster core by Shara and collaborators (these proceedings). A search of a portion (~0.25) of the central core of NGC 6752 has also been reported by Shara et al (1995); no CV candidates (or DN) were found suggesting that CVs may be depleted relative to model expectations (and see Shara, these proceedings). Whereas these studies are indisputable in their sensitivity to DN outbursts, which could not be missed, they are less restrictive in their sensitivity to CVs in quiescence since their sensitivity to the (often) modest variability ($\lesssim 0.1$ mag) and their relatively non-uniform sampling have limited their sensitivity to CVs overall.

## 3. ROSAT Studies of Compact Binaries in Globular Cluster Cores

As with HST, the ability to measure high resolution x-ray images of clusters has proven crucial to the identification and study of compact x-ray binaries in clusters. However, the best x-ray imaging resolution thus far — the HRI on ROSAT, with $\sim 3''$ resolution is of course still worse than even ground-based optical images. This is largely (but not totally) compensated by the enormous difference in source spatial density: at the faintest soft x-ray luminosities thus far reached ($L_x \sim 2 \times 10^{31}$ erg/s), the highest density of discrete sources previously reported is the ~3-4 sources within a $10''$ radius of the center of NGC 6397 (CG93). However it is clear that even more sources exist in the cores of high density clusters like 47 Tuc, where Hasinger et al (1994) find at least 4-9 discrete sources (with at least 4 in the core) superimposed on diffuse emission in the cluster core that very likely is due to a much larger underlying source population. We reported (CG93) a similar effect in NGC 6397 and now find (see below) at least 2-3 additional dim sources, for a total of 5-6, in just the cluster core region so that the total population is very much confusion limited even for the ROSAT HRI. Thus a true census of the cluster dim source population must await the significantly higher spatial resolution ($\sim 0.5''$) of AXAF.

### 3.1. INTEGRATED X-RAY LUMINOSITIES OF GLOBULAR CORES

Despite the need for very high spatial resolution to study the entire dim source population, studies of globulars with the lower spatial resolution ($\sim 30''$) PSPC detector on ROSAT can measure the total integrated x-ray luminosity of compact binaries in cluster cores (e.g. Johnston et al 1994, 1996). A survey of 9 globulars with large core radii gave detections of integrated core luminosities (within $\sim 3r_c$ of the cluster centers; denoted here as "extended core") of $L_{x,core} \sim 1.5$ and $0.5 \times 10^{32}$ erg/s for NGC 6366 and NGC 6809 (M55), respectively and upper limits typically $\lesssim 1 \times 10^{32}$ erg/s for the remaining 7 clusters (Johnston et al 1996). A complete list of all currently measured values of $L_{x,core}$ (given in erg/s in the ROSAT band, 0.2 - 2 keV and from HG, CG93, Grindlay 1993a, Hasinger et al 1994, Rappaport et al 1994 and Johnston et al 1994, 1996) for all 14 globulars detected is given below in Table 1.

Table 1: **Globular Cluster Dim Source Luminosity vs. T2 or T3 prediction**

| Cluster | $[F_e/H]$ | $\log(L_{x,core})$ | Pryor/Meylan: | | Djorgovski: | |
| --- | --- | --- | --- | --- | --- | --- |
| | | | $\log(T2)$ | $\log(T3)$ | $\log(T2)$ | $\log(T3)$ |
| NGC104 | -0.71 | 33.6 | 7.73 | 5.18 | 7.64 | 5.24 |
| 1904 | -1.69 | 33.7 | 6.75 | 4.65 | 6.62 | 4.62 |
| 5139 | -1.59 | 32.6 | 7.51 | 5.76 | 6.91 | 5.31 |
| 6304 | -0.59 | 33.0 | — | — | 6.98 | 4.73 |
| 6341 | -2.24 | 32.5 | 6.86 | 4.66 | 6.56 | 4.96 |
| 6366 | -0.99 | 32.2 | 4.77 | 3.82 | 5.48 | 4.33 |
| 6397 | -1.91 | 32.5 | 6.95 | 4.30 | 6.28 | 3.43 |
| 6541 | -1.83 | 33.2 | 7.69 | 4.94 | 7.37 | 5.12 |
| 6626 | -1.44 | 32.8 | 7.49 | 5.04 | 7.33 | 4.74 |
| 6656 | -1.75 | 32.3 | 7.29 | 5.29 | 6.72 | 4.87 |
| 6752 | -1.54 | 32.6 | 6.99 | 4.39 | 7.14 | 4.69 |
| 6809 | -1.82 | 31.8 | 6.26 | 5.01 | 5.25 | 4.20 |
| 7099 | -2.13 | 32.7 | 7.39 | 4.44 | 6.72 | 4.17 |
| PAL 2 | -1.68 | 34.0 | — | — | 7.64 | 5.39 |

As frequently pointed out (e.g. Hut et al 1992, CG95), 2-body ("T2") tidal capture production of the dim sources (as for the luminous LMXBs) would predict a total *number* of sources scaling as $\rho_c M_c / v_c$, where the c subscript refers to core values and the $\rho$ and M are cluster core density and total mass, and v is the central velocity dispersion. Since $M_c \propto \rho_c$, this then gives a scaling as $\rho_c^2$. A flatter dependence on $\rho_c$ ($\propto \rho_c^{1/2} M_c$) is found for millisecond pulsars (cf. Johnston et al 1992) which has usually been interpreted that they can form in lower density clusters by 3-body ("T3") exchange collisions of NSs into primordial binaries which can have longer periods in low density clusters without being disrupted. Thus in Table 1, we also tabulate for each cluster the quantities $\log(T2) \propto \rho_c^2 r_c^3 /v_c$ and $\log(T3) \propto \rho_c^{1.5} r_c^3 /v_c$, where the cluster parameters for $\rho_c$ and $r_c$ have been taken from Djorgovski (1993) or Pryor and Meylan (1993) and the central velocity dispersions from Pryor and Meylan (1993). Total dim source numbers may be estimated from the extended core luminosities only if the dim source luminosity function is known.

The deepest observations (NGC 6397; see below) suggest the number of dim sources increases down to luminosities of at least as low as $L_* \sim 2 \times 10^{31}$ erg/s and with a power law index $\alpha \gtrsim 1$ for an integral number of sources $N_{dim}(\gtrsim L_x) = A \cdot (L_x/L_*)^{-\alpha}$, where A is the normalization factor for which the dependence $A(\rho_c, M_c, v_c)$ is desired. Thus the extended core luminosity may be obtained by integrating the above integral luminosity function over $L_x$ (equivalent to integrating $(dN/dL_x) \cdot L_x \, dL_x$) so that $L_{x,core} = C \cdot A \cdot (L_x/L_*)^{-\alpha+1}$, where C is the integration constant. Thus for $\alpha \sim 1$, and assuming all clusters have the same luminosity function and thus value of $\alpha$ (which may also depend on the same cluster properties as does A), the normalization constant $A(\rho_c, M_c, v_c) \propto L_{x,core}$. The values of $L_{x,core}$ given in Table 1 thus trace the dim source content.

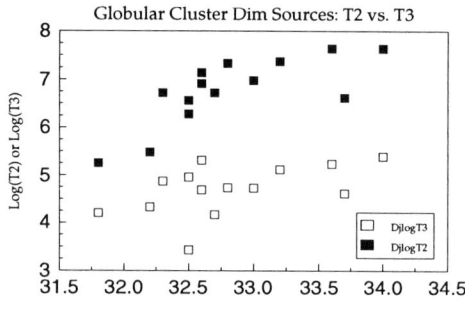

Figure 4. Dim source core x-ray luminosity vs. T2 or T3 scaling.

In Figure 4 we plot the values of log(T2) and log (T3) (using the Djorgovski 1993 values) vs. log $L_{x,core}$ for the currently known cluster dim source population. It is clear that the log(T3) scaling is much flatter (particularly for $L_{x,core} \lesssim 10^{33}$ erg/s) than the log(T2) values, or that the total (integrated) core luminosity does not depend on log(T3) but rather log(T2). Similar results, but with larger scatter, are found for the Pryor-Meylan values for T2 and T3. Thus the total number of cluster dim sources, and thus cluster CVs, appears to be better described by tidal capture formation than by exchange collisions between cluster WDs and primordial binaries.

We also list in Table 1 the cluster metallicities [Fe/H] (from Djorgovski 1993). These show that in contrast to the 12 clusters containing the luminous LMXBs, which are significantly metal rich (Grindlay 1993b) with <[Fe/H]> $\sim$ -1.17 (Bellazzini et al 1995), the dim source clusters have mean <[Fe/H]> $\sim$ -1.57, or approximately that of the cluster population at large. Thus, if the larger [Fe/H] for the LMXB globulars suggests they have flatter IMFs and an enhanced NS population, the (massive) WDs needed for the dim sources are a more constant fraction of cluster mass.

3.2. DEEP ROSAT OBSERVATION OF NGC 6397

Over the time period 17 March - 2 April, 1995, we obtained a deep (75 ksec) HRI exposure on NGC 6397. By good fortune the final 12 hours (on

April 2) overlapped with our HST/FOS exposures. Detailed analysis of this rich data set is in progress; space limitations do not permit presentation of images here. Qualitatively, the image looks similar to our 18 ksec image (CG93) recorded in two epochs (March 1991 and March 1992) although we detect additional emission from one or more sources in the vicinity of B and several in the C1-C3 complex. Source A is fainter and near the detection limit of $L_x \sim 1 \times 10^{31}$ erg/s, and a new source ("D") is detected $\sim 1'$ NE of the core. Of greatest interest is the additional structure now seen in the central 10-20" of the core, where fainter sources possibly consistent with the H$\alpha$ candidate stars 6 and 7 of CG95 are seen. The entire central source complex (5-6 sources) is clearly variable, with different sources dominating over the 2 week observation interval. Statistics limit short-time variability and comparison with the time-resolved FOS spectra. The new sources are all fainter than the original C1-C3, indicating the luminosity function continues to rise down to at least $L_x \sim 2 \times 10^{31}$ erg/s.

## 4. Nature of the CVs in Globulars: Magnetics ?

The presence of the He I and He II emission lines in the spectra of the first three CVs measured in the core of a globular cluster is striking. While occasionally dwarf novae or nova-likes show He emission lines, this is much more common in the magnetic CVs. The He lines are not strong enough (and the absolute magnitudes are in any case too bright) to be AM Her types, but the spectra are not dissimilar from the lower magnetic field IP class objects (e.g. Singh et al 1995), which have ratios of EW(He II)/EW(H$\beta$) $\sim 0.3$ - 0.5, or quite similar to the values for stars 1-3 (cf. GC95). However, given the direct evidence for accretion disks in the cluster objects, they most closely resemble DQ Her type CVs with still lower fields. This class is also favored since they generally obey the correlation between x-ray/optical flux continuum ratio vs. EW(H$\beta$) for CVs (cf. Patterson and Raymond 1985; PR) whereas the AM Her systems do not. As shown in the accompanying paper by Grindlay and Cool, the now-measured EW(H$\beta$) and visual flux (from spectral continuum) values put the three objects even closer to the PR correlation and distinct from the quiescent LMXB Cen X4.

The fraction of CVs in the field that are DQ Her type is not well known but is probably $\lesssim 0.25$ (cf. review by Patterson 1994). Thus finding (possibly) 3 for 3 in a globular would seem unlikely (binomial probability $\lesssim 0.02$). Why should there be more magnetic systems or at least magnetic white dwarfs (WDs) in globulars ? Prompted by the results for blue straggler formation presented by Rasio at this meeting (cf. these proceedings and Lombardi et al 1995), we propose the following heuristic model (or scenario) for enhanced magnetic CV formation. The basic idea is to use the

WD remnants from blue straggler stars (BSS) formed by the encounters of two stars each near the turnoff mass (or with one already a sub-giant). Such stars, with radiative cores and convective envelopes, will have rapidly rotating cores after the BSS formation. However, as Lombardi et al point out, their convective envelopes will be slowed by magnetic braking on timescales short enough ($\sim 10^5$ years) to account for the lack of rapid rotation typically observed in BSS. This combination of a rapidly rotating core and differentially (more slowly) rotating envelope must be an ideal dynamo generator: at least it contains the requisite ingredients of (rapid) differential rotation and convection. Therefore, we propose that the cores of BSS, which will evolve to WDs, are likely to have greater magnetic fields than main sequence stars of comparable total mass and thus preferentially produce magnetic WDs.

When captured by lower mass main sequence stars (with mass ratios $q \lesssim 0.8$ for stable mass transfer subsequently), these magnetic WDs will then give rise to magnetic CVs. Since these WDs are produced by stars of typically twice the current turnoff mass they will be more massive than "currently produced" WDs and thus favored for CVs. Furthermore, since Proctor, Bailyn and Demarque (1995) show that the colors of the BSS in NGC 6397 require that the Lombardi et al results, in which the merged stars are predicted to be unmixed, must be modified by either subsequent rapid mixing or the merger of 3 stars into a BSS (as from a star+binary merger), then if magnetic WDs are preferentially produced in spun-up cores of of BSS, they are again more likely to form observable CVs. Since in a collapsed core like NGC 6397 an appreciable fraction of the stars above the turnoff are BSS (some 9 of the $\sim 60$ stars above the turnoff within our (CG95) central $38 \times 35$ arcsec WFPC1 frame are BSS), the penalty for having to undergo two collision or capture processes (to first form a BSS and then have the massive WD captured into a CV) is not overwhelming. The observed BSS fraction, combined with the shorter BSS lifetimes, suggests that perhaps 25% of the post-main sequence stars in the central region of NGC 6397 could be processed through BSS so that the production of massive WDs at the current epoch is significantly enhanced over the baseline production from massive (e.g. $\gtrsim 2\ M_\odot$) progenitors if the mass function $dN/dm \propto m^{-(1+\eta)}$ has index $\eta \gtrsim 1$.

## 5. Conclusions

We conclude that the dim x-ray sources in globulars are primarily CVs formed by tidal capture. The CVs may contain a significantly enhanced magnetic fraction which could arise from production of magnetic, and more massive, WDs in the spun-up cores of BSS produced in dense cluster cores.

The lack of significant metallicity enhancement of the 14 clusters in which dim sources have thus far been detected suggests that the massive WDs likely incorporated into the cluster dim sources are distributed more uniformly (perhaps from BSS) than are the corresponding NSs in the LMXB clusters. Additional HST imaging and spectra, as well as higher sensitivity and resolution (AXAF) x-ray observations, can provide rigorous tests of this picture.

I thank Adrienne Cool for discussions. This work was partially supported by NASA grants NAGW-3280 and HST grant GO-5497.

## References

Bailyn, C. et al 1996, in preparation
Bellazzini, M. et al 1995, ApJ, 439, 687
Cool, A.C. 1993, *Ph.D. Thesis*, Harvard University
Cool, A. et al 1993, ApJ, 410, L103 (CG93)
Cool, A. et al 1995, ApJ, 439, 695 (CG95)
Djorgovski, S. 1993 in *Dynamics of Globular Clusters* (S. Djorgovski and G. Meylan, eds.), ASP Conf. Series, Vol. 50, p. 373
Grindlay, J.E. 1993a, in *Dynamics of Globular Clusters* (S. Djorgovski and G. Meylan, eds.), ASP Conf. Series, Vol. 50, p. 285
Grindlay, J.E. 1993b, in *The Globular Cluster–Galaxy Connection*, (G. Smith and J. Brodie, eds.), ASP Conf. Series, Vol. 48, p. 156
Grindlay, J.E. 1994, in *Evolution of X-Ray Binaries*, (S. Holt and C. Day, eds.), AIP Conf.Proc., 308, 339
Grindlay, J. and Cool, A. 1996, in preparation
Grindlay, J., Cool, A. et al 1995, ApJ, 455, L47 (GC95)
Hasinger, G.. Johnston, H. and Verbunt, F. 1994, A&A, 288, 466
Hertz, P., and Grindlay, J.E. 1983, ApJ, 278, 137 (HG)
Hut, P. et al 1992, PASP, 104, 981
Johnston, H. et al 1992, in *X-ray Binaries and Millisecond Pulsars* (E. Van den Heuvel and S. Rappaport, eds.), Kluwer, Dordrecht, p. 349
Johnston, H., Verbunt, F. and Hasinger, G. 1994, A&A, 289, 763
Johnston, H., Verbunt, F. and Hasinger, G. 1996, A&A, in press.
Lombardi, J. , Rasio, F. and Shapiro, S. 1995, ApJ, 445, 117
Paresce, F. and DeMarchi, G. 1994, ApJ, 427, L33
Patterson, J. 1994, PASP, 106, 209
Patterson, J. and Raymond, J. 1985, ApJ, 292, 535 (PR)
Proctor, A,, Bailyn, C. and Demarque, P. 1995, ApJ, submitted
Pryor, C. and Meylan, G. 1993 in *Dynamics of Globular Clusters* (S. Djorgovski and G. Meylan, eds.), ASP Conf. Series, Vol. 50, p. 357
Rappaport, S. et al 1994, ApJ, 423, 633
Shara, M., Drissen, L., Bergeron, L., and Paresce, F. 1995, ApJ, 441, 617
Singh, K. et al 1995, ApJ, 453, L95

# PULSARS IN GLOBULAR CLUSTERS

S. R. KULKARNI AND S. B. ANDERSON
*Division of Physics, Mathematics, and Astronomy,*
*California Institute of Technology, 150-24,*
*Pasadena, CA 91125*

**Abstract.**

Since the discovery of the first globular cluster pulsar in M28 (Lyne et al. 1987) a total of 33 pulsars have been found to reside within 13 seperate clusters. Many (but not all) of the cluster pulsars have properties similar to the millisecond pulsars in the disk: short period, binarity and low magnetic field strength. The common understanding is that these pulsars are primordial neutron stars (i.e. the remnants of massive stars in clusters) which have been spun up by accretion of matter from a companion. Therefore, in this framework, the cluster pulsars are descendents of Low Mass X-ray Binaries (LMXBs) (Alpar et al. 1982). This hypothesis is by no means accepted by all workers (e.g. Michel 1987, Ray & Kluzniak 1990, Romani 1990, Bailyn & Grindlay 1993). These workers have argued that at least some (if not all) cluster pulsars could be formed by accretion induced collapse of massive white dwarfs. In either case, it is clear from the sensitivity limits of current cluster searches, and the luminosity of field pulsars, that there are currently $\mathcal{O}(10^3)$ extant radio pulsars in the Galactic globular cluster system.

In this review, specifically targeted for astronomers working in the field of globular clusters, not pulsar astronomers, we argue that cluster pulsars have provided us with a new window into the population of long-dead massive stars and the physics of tidal capture. The precision with which pulsars can be timed has created new diagnostics: measurement of the mass distribution in the dense cores, measurement of orbital evolution on short timescales and precise determination of orbital characteristics. It is fair to say that all these diagnostics are unique, and not obtainable by other observations. Despite this, it is our assessment that the typical astronomer who works in the field of globular clusters is apparently unaware of these relevant contributions. Hopefully this review will bridge this gap. A complete copy of the review article may be found at http://astro.caltech.edu/~srk.

## 1. References

- Alpar, M. A., Cheng, A. F. Ruderman, M. A., & Shaham, J., Nature, 300, 728 (1982)
- Bailyn, C. D. & Grindlay, J. E., ApJ, 353, 159 (1990)
- Lyne, A. G., Brinklow, A., Middleditch, J., Kulkarni, S. R., Backer, D. C. , & Clifton, T. R., Nature, 328, 399 (1987)
- Michel, F. C., Nature, 329, 310 (1987)
  item Ray, A. & Kluzniak, W., Nature, 344, 415 (1990)
- Romani, R. W., ApJ, 357, 493 (1990)

# COMPARISON OF X-RAY SOURCES IN OLD OPEN AND IN GLOBULAR CLUSTERS

FRANK VERBUNT

*Astronomical Institute, Utrecht University*
*Postbox 80.000, 3508 TA Utrecht, the Netherlands*
*(verbunt@fys.ruu.nl)*

**Abstract.** Twelve bright ($L_x \gtrsim 10^{36}$ erg/s) X-ray sources have been detected in globular clusters: they are accreting neutron stars. Five of these are transients. Thirty less luminous sources have been detected in eighteen globular clusters. The luminosity function of these sources does not rise rapidly towards lower luminosities. The sources with $L_x \sim 10^{33}$ erg/s are probably transients in their low state; the least luminous sources currently detected may be cataclysmic variables. The old open cluster M 67 harbours chromospherically active binaries, with $L_x \sim 10^{30-31}$ erg/s. Globular clusters contain fewer of these binaries than estimated by scaling with mass from M 67.

## 1. Introduction

X-ray sources in globular clusters are amongst the brightest in the sky, and were discovered with the earliest X-ray instruments. The more recent highly sensitive X-ray satellites Einstein and, in particular, Rosat enable the study of less luminous sources in globular clusters. Such less luminous sources have also been discovered in old open clusters.

In this review I describe these sources, the bright X-ray sources in globular clusters in Section 2, and the less luminous sources in old open and globular clusters in Sections 3 and 4, respectively. In Section 5 the X-ray sources in old open and in globular clusters are compared.

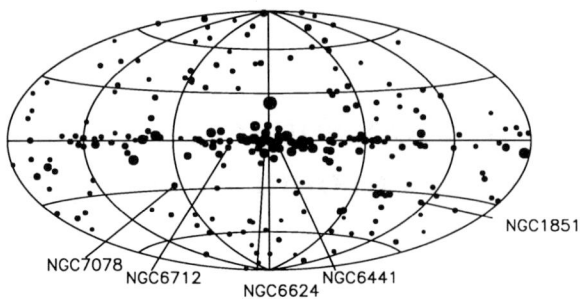

Ariel V 3rd Catalogue

*Figure 1.* Distribution over the sky of the X-ray sources contained in the third catalogue of the Ariel V satellite (McHardy et al. 1981). The size of the symbols indicates the flux of the X-ray sources. The central horizontal line is the Galactic Plane, with the Galactic Center in the middle. The sources are clearly concentrated towards the Galactic Center. Amongst the sources in our Galaxy, five can be identified with globular clusters.

## 2. Bright sources in globular clusters

Figure 1 shows the distribution in the sky of the brightest X-ray sources, as detected with the early X-ray satellite Ariel V. These sources are clearly concentrated towards the Galactic Plane, and in this plane, towards the Galactic Centre. Amongst the $\sim 200$ sources detected by Ariel V near the Galactic Plane, five can be identified with globular clusters.

The bright X-ray sources in globular clusters, with luminosities $L_x(2 - 10\,\text{keV}) \gtrsim 10^{36}\,\text{erg/s}$, are low-mass X-ray binaries, in which a neutron star accretes matter from a low-mass companion star and emits the liberated energy as X-rays. Pointed Einstein and Rosat observations have discovered a bright X-ray source in some thirty globular clusters of M 31 (Trinchieri and Fabbiano 1991; Magnier 1994). The high incidence in globular clusters of X-ray binaries is ascribed to the occurrence in their dense cores of formation processes of such binaries that do not operate in the Galactic Disk, viz. tidal capture of a neutron star by a passing main-sequence star (Fabian et al. 1975, see also Mardling these proceedings) or an encounter between a binary and a neutron star, in which the neutron star takes the place of one of the binary stars in an exchange encounter (Hills 1976).

Following the discovery with UHURU of six X-ray sources in globular clusters, six more have been found with later satellites; the most recent additions were detected in the Rosat All Sky Survey, in NGC 6652 and in Terzan 6 (Verbunt et al. 1995). The source in NGC 6652 was detected already with the HEAO-1 satellite, but the positional accuracy was insufficient for identification with the cluster (Hertz and Wood 1985). The source in Terzan 6 was not detected with HEAO-1, which indicates that it must have been much brighter in 1991 than a decade before. Comparison between the observations of the cluster sources obtained with various satellites show that at least five of them have been at luminosities below the detection limits of the early satellites, i.e. at $L_x(2-10\,{\rm keV}) \lesssim 10^{35.5}$ erg/s at one time or other. Such sources are called transients.

All well-studied bright X-ray sources in globular clusters have shown X-ray bursts, which proves that they are accreting neutron stars, rather than black holes. Low-mass X-ray binaries with neutron stars have spectral energy distributions characterized by blackbody (X-ray colour) temperatures $\lesssim 5\,{\rm keV}$. This is true also for the sources in globular clusters.

A source of a new class, which emits photons mainly at energies $\lesssim 0.5\,{\rm keV}$ has been detected in the Rosat All Sky Survey in NGC 5272. The nature of this source is not clear. For an assumed black-body temperature of 40 eV the bolometric luminosity during the Survey was $\sim 10^{35}$ erg/s (Verbunt et al. 1995). Repeated observations with the Rosat High Resolution Imager (HRI) show that this source is variable by at least a factor 30 (Hertz et al. 1993).

Rosat HRI positions show that all bright X-ray sources are in or close to the core of the globular cluster (Johnston et al. 1995a).

## 3. Dim sources in old open clusters

Rapidly rotating stars with convective envelopes have active chromospheres and emit X-rays. As they age, such stars lose their rapid rotation, and with it their chromospheric activity. As a consequence, relatively young clusters are usually selected for the study of stellar X-ray emission (e.g. Pye et al. 1994, Stern et al. 1993). If a star is tidally locked in a short-period binary, however, it may remain in rapid rotation even as it ages.

In a Rosat observation of M 67, an open cluster with an age of $\sim 5$ Gyr, a fair number of X-ray sources have been detected, and from their optical identifications it appears that most of these are indeed binaries (Figure 2). Because tidal forces on convective stars circularize a binary on a time scale given by $e/\dot{e} \propto (a/R)^8$, where $a$ is the semi-major axis, and $R$ the radius of the star (Zahn 1966), one expects short-period binaries with large, bright stars to be circularized and thus to be magnetically active X-ray sources. The brightest X-ray sources in M 67 are indicated in Figure 2 with nos. 4, 7, 8, 10 and 13 and have X-ray luminosities in the 0.5-2.5 keV bandpass of 5.5, 4, 8, 3.5 and 3 $\times 10^{30}$ erg/s, respectively. Of these, only X4, the blue straggler, is not known to be a binary. For X7, the radial velocity is known to vary, but the orbital period has not yet been determined. X8 was suggested by Belloni et al. (1993) to be a binary in the stage of mass-transfer. The eccentricity of its orbit appears to exclude this model, unless

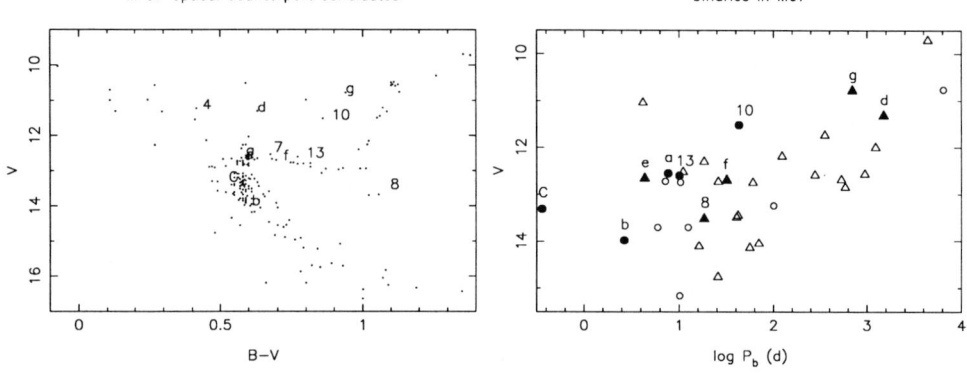

*Figure 2.* Left: Hertzsprung-Russell diagram of the old open cluster M 67. Optical counterparts of the X-ray sources are indicated with numbers for detections in a first observation (Belloni et al. 1993), and with letters for new detections in a longer, more recent observation (Belloni et al., in preparation); these include a contact binary (C). Right: Visual magnitude as a function of orbital period for known binaries in M 67 (data from Latham et al. 1992). Circles indicate binaries with (almost) zero eccentricity, triangles binaries with eccentric orbits. Filled symbols indicate binaries detected in X-rays.

the observed eccentricity is the consequence of distortion of the spectral lines by contribution of the gas stream.

It has been shown that eccentric orbits of short-period binaries and circular orbits of long-period binaries in open clusters can be understood if the dependence of the circularization time scale of the stellar radius $R$ is taken into account (Verbunt and Phinney 1995). Many of the X-ray-active binaries in M 67 have circular orbits, as expected, including the contact binary AH Cnc; a few of the least luminous sources, however, have eccentric long-period orbits (Figure 2). This is contrary to expectation, and we currently do not understand why these binaries are X-ray sources.

Two of the sources in M 67 have very soft X-ray spectra. One of these is a magnetic cataclysmic variable (Gilliland et al. 1991), for which the observation of M 67 was originally intended; the other one has been identified with a single hot white dwarf (Pasquini et al. 1994). In general, the X-ray luminosities of the chromospherically active binaries in the old open clusters M 67 and NGC 752 (with an age of $\sim$ 2 Gyr) appear to be in agreement with those of similar binaries in the Galactic Disk (Belloni and Verbunt 1995). The total X-ray luminosity of M 67 is about $3 \times 10^{31}$ erg/s.

## 4. Low-luminosity sources in globular clusters

Observations with the Rosat High Resolution Imager (HRI) have resolved the core of several globular clusters and shown that they contain multiple sources. Examples are NGC 6397 and NGC 6752 (Cool et al. 1993, Grindlay 1993). A total of some 30 dim sources, either single or multiple, have

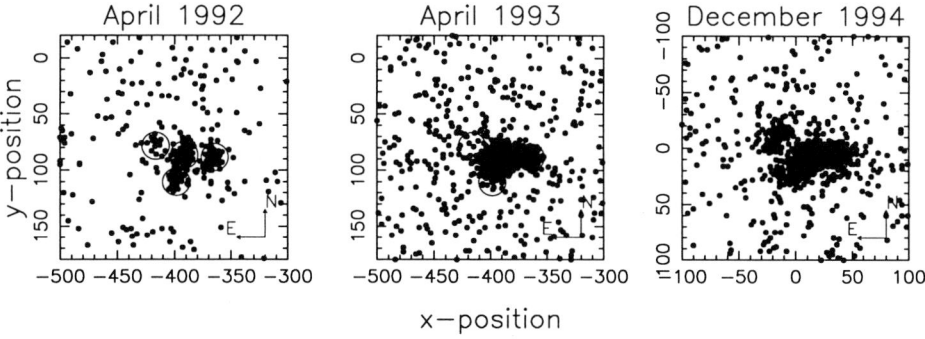

*Figure 3.* Photons in the inner 100 × 100 arcsec area of 47 Tuc detected in three ROSAT HRI observations showing the variability of the sources. To guide the eye, circles in the left two frames encircle photons of the four sources detected in the first observation.

now been detected in or near the the cores of 18 globular clusters, with luminosities in the range of $10^{31} - 10^{34}$ erg/s (Johnston and Verbunt 1996).

Repeated observations of the core of 47 Tuc show that the dim sources are highly variable: of the four sources detected in April 1992, only two are detected again in April 1993 together with a new source (Hasinger et al. 1994). One of the disappeared sources had reappeared when a third observation was obtained, in December 1994 (Figure 3). The absolute positional accuracy of the ROSAT HRI is about 5 arcsec; this precludes certain identification of any of the ROSAT sources with either the single X-ray source detected in 47 Tuc with the Einstein satellite (Hertz and Grindlay 1983) or with the ultraviolet variables detected with HST (Paresce et al. 1992, Paresce and DeMarchi 1994).

The accurate imaging of the ROSAT HRI allows subtraction from the image of the point sources. When this is done, a remaining extended emission is found in 47 Tuc, presumably due to unresolved, fainter point sources. The total luminosity of the extended emission is small, about $10^{33}$ erg/s, similar to the luminosities of individually detected sources, $L_x = 0.5 - 1.6 \times 10^{33}$ erg/s. ROSAT PSPC observations of ten clusters with large apparent cores find marginal evidence of extended emission in only two, viz. NGC 6254 and NGC 6352 (Johnston et al. 1996).

The ROSAT PSPC provides some information about the energy distribution of the X-ray photons. The sources in some clusters, e.g. $\omega$ Cen, have spectra that may be characterized with black body colour temperatures of $\gtrsim 0.5$ keV; those in others, e.g. 47 Tuc, have slightly softer spectra, of $\sim 0.2$ keV. For yet other sources only an upper limit to the colour temperature is obtained, and it cannot be excluded that these sources are as soft as the source in NGC 5272 (Johnston et al. 1994, Verbunt et al. 1995).

The X-ray luminosities of the dim sources in globular cluster cores pro-

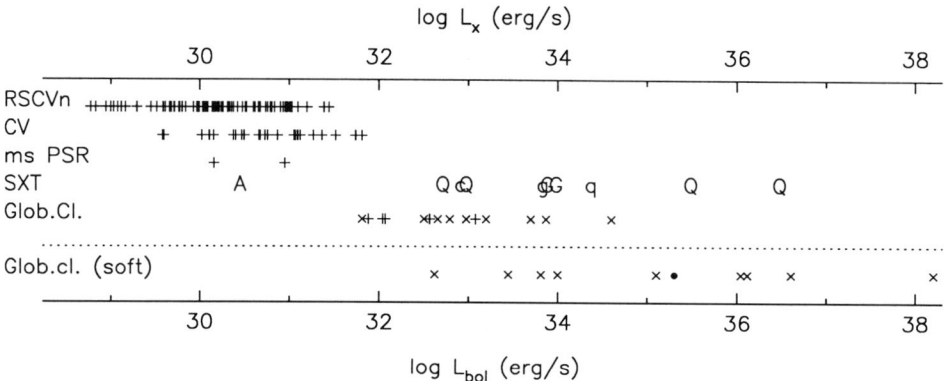

*Figure 4.* Luminosity distributions ($L_x$ between 0.5 - 2.5 keV) for chromospherically active binaries (RS CVn), cataclysmic variables, recycled radio pulsars (msec PSR), and soft X-ray transients (SXT) in the Galactic Disk, compared with the luminosity of the dim sources in globular clusters. Cluster sources marked with × only have upper limits to their X-ray colour temperature; their X-ray luminosity is shown for an assumed bremsstrahlung spectrum with $kT \sim 3$ keV, and their bolometric luminosity for an assumed blackbody with $kT = 40$ eV. (From Johnston et al. 1995b.)

vide a first clue to their identity. In Figure 4 we compare the luminosities of soft X-ray transients, recycled radio pulsars, cataclysmic variables, and chromospherically active close binaries (RS CVn systems), all detected in the Galactic Disk, and thus securely identified, with those of the dim sources in globular clusters. It is seen that only the soft X-ray transients cover the same luminosity range as the dim cluster sources.

A few words of caution are in place here. First, we know now that some sources previously detected as single sources are in fact multiple, and obviously the individual luminosities are smaller. This has been taken into account as far as possible in making Figure 4, but some sources may still be multiples. Second, the luminosity distributions of the cataclysmic variables and recycled pulsars in the Galactic Disk are incomplete. For many cataclysmic variables the distances are unknown; it is quite possible that some of them are faraway, luminous sources. The two recycled pulsars in the plot have relatively low rotational-energy loss-rates; it may well be that more rapidly rotating pulsars with larger period derivatives also have higher X-ray luminosities. It has, in fact, been plausibly suggested that the dim source in M 28 is the radio pulsar PSR 1821-24 in that cluster (Danner et al. 1994).

The luminosity distribution of cataclysmic variables in the Galactic Disk overlaps with lower end of the luminosity distribution of the dim X-ray sources in globular clusters. In NGC 6752, three dim X-ray sources have been optically identified with the Hubble Space Telescope, on the basis

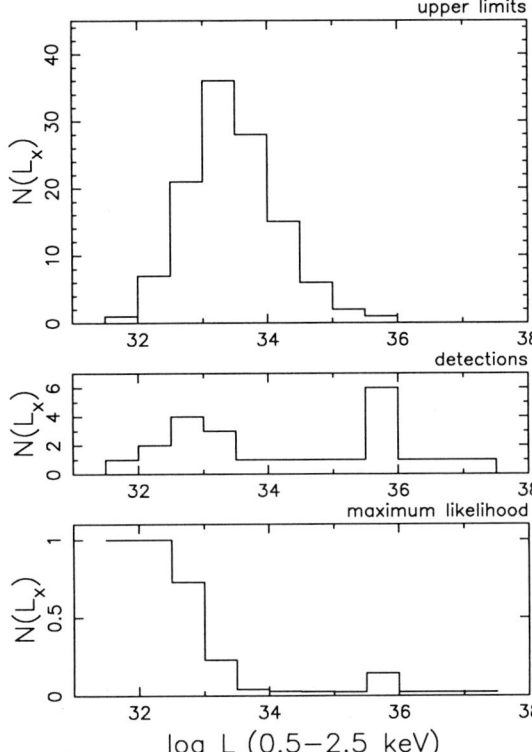

Figure 5. Luminosity distributions of X-ray sources in globular clusters obtained by combining Einstein and Rosat data. Top: upper limits. Middle: detected sources. Below: Maximum likelihood distribution derived from upper two graphs. After Verbunt et al. (1995).

of their H$\alpha$ emission (Cool et al. 1995). Their spectra, also obtained with HST, are very similar to one another, and to those of cataclysmic variables. They are also rather similar to optical spectra of soft X-ray transients in quiescence ... (Grindlay et al. 1995). For the moment, it appears reasonable to conclude that the dim X-ray sources in globular clusters with $L_x \gtrsim 10^{33}$ erg/s are probably soft X-ray transients in their low state. After all, we know that X-ray transients exist in globular clusters (as discussed in Section 2). The least luminous sources currently detected, such as those in NGC 6397 and NGC 6752, may well be cataclysmic variables.

With the growing number of dim sources detected in the cores of globular clusters, we can start drawing conclusions about their luminosity function. First, the ROSAT All Sky Survey has put stringent upper limits to the integrated X-ray luminosities of many globular cluster cores. More specifically, most globular cluster cores do not contain single or multiple sources with X-ray luminosities totalling more than $10^{33}$ erg/s. Combination of the upper limits with the known detections gives a maximum likelihood luminosity distribution

$$dN(L_x) = N_o L_x^{-\alpha} d\ln L_x \qquad (1)$$

with $\alpha \sim 0.7$ (see Figure 5). Second, from the observation that the extended emission due to unresolved point sources is less that that of individually detected sources, we derive that $\alpha < 1$.

A more sophisticated analysis may be made on the basis of the distribution of the detected luminosities with respect to the detection limits. In a flat luminosity distribution, i.e. equal numbers of sources in logarithmic luminosity intervals, we expect a homogeneous distribution of sources above the detection limit; in a skewed distribution in which most sources have low luminosities, most sources will be detected close to the detection limit. The analysis of the luminosity distribution of the dim X-ray sources in 11 globular clusters indicates $\alpha \simeq 0.5$. This value for $\alpha$ is compatible with the luminosity distributions in 47 Tuc and NGC 6752 (Johnston and Verbunt 1996).

The constant $N_o$ in Eq. 1 is proportional to the formation rate of dim X-ray sources in globular clusters. If such sources are formed by two-body encounters in the cluster core, their formation rate should be proportional to $\rho_c^2 r_c^3$, where $\rho_c$ and $r_c$ are the number density of stars in the cluster core, and the core radius, respectively (Verbunt and Hut 1986). More generally, one can write $N_o \propto \rho_c r_c^3 \rho_c^\gamma \propto M_c \rho_c^\gamma$ where $M_c$ is the mass of the core. Comparison of the collision numbers $M_c \rho_c^\gamma$ for various values of $\gamma$ with the distribution of the dim X-ray sources over the clusters indicates that $\gamma \simeq 0.5$ (Johnston and Verbunt 1996). A similar value of $\gamma$ has been found for the recycled pulsars in globular clusters (Johnston et al. 1992).

## 5. Old open clusters compared with globular clusters

Dim X-ray sources occur both in globular clusters and in old open clusters. The dim X-ray sources in old open clusters are binaries. The four brightest sources in M 67 are binaries with orbital periods that are sufficiently short that one expects such a binary to survive in a globular cluster (Figure 2). Naively, one may expect that the number of such chromospherically active close binaries in a globular cluster would scale with the mass of the cluster. As the mass of a typical globular cluster is more than a thousand times the mass of M 67, one would predict a number of chromospherically active binaries in the globular clusters more than a thousand times that in M 67, and thus an integrated X-ray luminosity in excess of $10^3 \times 3 \times 10^{31}$ erg/s. As shown in Figure 5, the ROSAT All Sky Survey puts an upper limit to the X-ray luminosity of the majority of globular clusters well below this predicted value.

As possible explanations for this surprising result, one may argue 1) that the binaries detected as X-ray sources in M 67 are *not* ordinary chromospheric binaries, or 2) that open clusters have lost a larger fraction of their initial mass than globular clusters, or 3) that binaries are less frequent in globular clusters than in open clusters. At the moment, we do not know which of these possibilities, if any, is the correct answer.

As an aside I would like to point out that there are probably rather many old open clusters in our Galaxy, quite possibly as many as globular clusters. After all, several of the oldest open clusters, like M 67 and NGC 188 are fairly close to the Sun, at 750 pc and 1500 pc, respectively, which suggests that many such clusters may exist at larger distances from the Sun. This relates to the question whether the formation of clusters has been continuous since the formation of our Galaxy, or alternatively that an initial period of formation of globular clusters ended well before the subsequent formation of open galactic clusters. If many old open clusters in our Galaxy exist, some of them may have masses rather larger than those of M 67 and NGC 188. If so, the suggestion by Mardling (these proceedings) that the low-mass X-ray binaries in the Galactic Disk were formed in open clusters gains in plausibility.

## 6. Summary and Conclusions

X-ray sources have been found both in globular clusters and in old open clusters. The bright sources in globular clusters are accreting neutron stars. A source with an extremely soft spectrum, highly variable, was detected in NGC 5272; the nature of this source is not known.

Most dim sources in open clusters are close, chromospherically active binaries, in which stars co-rotate with the orbit. Surprisingly, some very wide binaries, with eccentric orbits, have now also been detected as X-ray sources.

Dim X-ray sources in globular clusters are multiple and variable. After subtraction of the individually detected sources, little extended emission remains, and the total X-ray luminosity of a globular cluster is dominated by the brightest sources in it. The X-ray luminosity function of globular cluster sources may be written

$$dN(L_x) \propto M_c \rho_c^{0.5} L_x^{-0.5} d\ln L_x \qquad (2)$$

Some of the dim sources, especially the relatively luminous ones with $L_x \gtrsim 10^{33}$ erg/s, are probably soft X-ray transients in their low state, i.e. neutron stars accreting at low rates. Others, especially the fainter ones with $L_x \lesssim 10^{32}$ erg/s, may be cataclysmic variables.

Globular clusters in general have X-ray luminosities which are smaller than one would predict by multiplying the X-ray luminosity of M 67 with the ratio of the mass of a globular cluster to that of M 67. This indicates that chromospherically active binaries are relatively less frequent in globular clusters than in M 67.

## Acknowledgements

I thank Gene Magnier, Piet Hut and Rosemary Mardling for illuminating discussions about old open clusters, and Helen Johnston for comments on the manuscript. This research is supported by the Netherlands Organization for Scientific Research NWO under grant PGS 78-277.

## References

Belloni, T., Verbunt, F. 1995, A&A, in press
Belloni, T., Verbunt, F., Schmitt, J. H. M. M. 1993, A&A, 269, 175
Cool, A. M., Grindlay, J. E., Cohn, H. N., Lugger, P. M., Slavin, S. D. 1995, ApJ, 439, 695
Cool, A. M., Grindlay, J. E., Krockenberger, M., Bailyn, C. D. 1993, ApJ, 410, L103
Danner, R., Kulkarni, S. R., Thorsett, S. E. 1994, ApJ, 436, L153
Fabian, A., Pringle, J., Rees, M. 1975, MNRaS, 172, 15P
Gilliland, R. L., Brown, T. M., Duncan, D. K., Suntzeff, N. B., Lockwood, G. W., Thompson, D. T., Schild, R. E., Jeffrey, W. A., Penprase, B. E. 1991, AJ, 101, 541
Grindlay, J. E. 1993, in Structure and Dynamics of Globular Clusters, ASP Conf. Ser. 50, eds. S.G. Djorgovski and G. Meylan, p.285
Grindlay, J. E. 1995, submitted to ApJ
Hasinger, G., Johnston, H. M., Verbunt, F. 1994, A&A, 288, 466
Hertz, P., Grindlay, J. E. 1983, ApJ, 275, 105
Hertz, P., Grindlay, J. E., Bailyn, C. D. 1993, ApJ, 410, L87
Hertz, P., Wood,K. S. 1985, ApJ, 290, 171
Hills, J. 1976, MNRaS, 175, 1P
Johnston, H. M., Verbunt, F. 1996, A&A, in press
Johnston, H. M., Kulkarni, S. R., Phinney, E. S. 1992, in E. P. J. van den Heuvel, S. A. Rappaport (eds.), X-ray binaries and the formation of binary and millisecond pulsars, Kluwer, Dordrecht, p. 349
Johnston, H. M., Verbunt, F., Hasinger, G. R. 1994, A&A, 289, 763
Johnston, H. M., Verbunt, F., Hasinger, G. R. 1995a, A&A, 298, L21
Johnston, H. M., Verbunt, F., Hasinger, G. R., Bunk, W. 1995b, in IAU Symposium 165, Compact Stars in Binaries, ed. E. P. J. van den Heuvel, in press
Johnston, H. M., Verbunt, F., Hasinger, G. R. 1996, A&A, in press
Latham, D. W., Mathieu, R. D., Milone, A. A. E., Davis, R.J. 1992, in Binaries as tracers of stellar formation, eds. A. Duquennoy, M. Mayor, p. 132
Magnier, E. 1994, in S. S. Holt, C. Day (eds.), The Evolution of X-ray Binaries, AIP, New York, p. 640
McHardy, I. M., Lawrence, A., Pye, J. P., Pounds, K. A. 1981, *Mon. Not. R. astr. Soc.*, 197, 893
Paresce, F., DeMarchi, G. 1994, ApJL, 427, L33
Paresce, F., DeMarchi, G., Ferraro, F. 1992, Nat, 360, 46
Pasquini, L., Belloni, T., Abbott, T. M. C. 1994, A&A, 290, L17
Pye, J. P., Hodgkin, S. T., Stern, R. A., Stauffer, J. R. 1994, MNRaS, 266, 798
Stern, R. A., Schmitt, J. H. M, Rosso, C., Pye, J. P., Hodgkin, S. T., Stauffer, J. R. 1993, ApJL, 399, L159
Trinchieri, G., Fabbiano, G. 1991, ApJ, 382, 82
Verbunt, F., Hut, P. 1987, in D. J. Helfand, J.-H. Huang (eds.), The Origin and Evolution of Neutron Stars, IAU Symposium 125, Reidel, Dordrecht, p. 187
Verbunt, F., Phinney, E. S. 1995, A&A, 296, 709
Verbunt, F., Bunk, W., Hasinger, G., Johnston, H. M. 1995, A&A, 300, 732
Zahn, J.-P. 1966, Ann. d'Ap., 29, 313 & 489

# SEARCHES FOR BINARY STARS IN GLOBULAR CLUSTERS

C. PRYOR
*Rutgers, the State University of New Jersey*
*Dept. of Physics & Astronomy, P.O. Box 849, Piscataway, NJ 08855-0849 USA*

J.M. FLETCHER, J.E. HESSER, R.D MCCLURE, P.B. STETSON
*Dominion Astrophysical Observatory, HIA/NRC*

H.B. RICHER, G.G. FAHLMAN, R.A. IBATA, N.C. IVANANS, AND G. MANDUSHEV
*Univ. of British Columbia, Dept. of Geophysics & Astronomy*

R.A. BELL
*Univ. of Maryland, Dept. of Astronomy*

M. BOLTE
*Univ. of California, Lick Observatory*

H.E. BOND
*Space Telescope Science Institute*

W.E. HARRIS
*McMaster Univ., Dept. of Physics & Astronomy*

D.A. VANDENBERG
*Univ. of Victoria, Dept. of Physics & Astronomy*

AND

M.A. WOOD
*Florida Institute of Technology, Dept. of Physics & Space Science*

## 1. Introduction

Primordial binaries in globular clusters are important both because their properties are an integral part of the description of the stellar population and because they can strongly influence the dynamical evolution of the cluster (see Hut, this volume).

Binary stars have been identified in globular clusters using x-ray emission, eclipses, radial velocity variations, and their position in the color-magnitude diagram (cmd). The last few years have seen rapid progress in identifying eclipsing variables using large sets of CCD images. The deep cmds of the centers of clusters produced from *Hubble Space Telescope* (HST) images, many of them shown at this conference, are poised to revolutionize the identification of "binary sequences" in the cmd. Surveys for radial velocity variables are being extended to fainter stars and longer time baselines. Each search technique has its strengths and weaknesses. Surveys for eclipsing systems are currently a relatively unique window into the number of short-period binaries, binary sequences are sensitive to systems with a wide range of periods, and radial velocity surveys can, eventually, determine the distributions of orbital properties. The interesting task ahead for the next few years will be integrating these disparate data to build a more complete picture of the binary star populations in globular clusters.

What we want to know is the frequency of binaries. How does their number vary from cluster to cluster and how does it vary with position within a cluster? In conjunction with this, we also want to know the distribution of binaries in the "internal" properties of primary mass, mass ratio, period, and eccentricity. These are still relatively distant goals. This paper briefly reviews what we do know and the prospects for further progress.

## 2. Radial Velocity Variables and the Binary Frequency

Radial velocity surveys of luminous giants in a number of clusters (*e.g.*, Hut *et al.* 1992a, H92 hereafter; Fischer *et al.* (1993); Côté *et al.* (1994); Gebhardt *et al.* 1994; Mayor *et al.* 1995) find that a few percent of the stars show significant velocity variations when measured at 2–5 epochs over time baselines of 2–10 years. Converting this "discovery frequency" into a true binary frequency is a difficult problem because surveys can typically only detect a subset, often a small subset, of all binaries. The simplest and probably best procedure is to report true binary frequencies for just the ranges of orbital period and mass ratio to which the survey is sensitive.

Figure 1 shows the fraction of binaries discovered as a function of period and mass ratio for a slightly updated version of the Pryor *et al.* (1989) and H92 sample of clusters, which consists of 1329 radial velocities for 392 stars in the globular clusters 47 Tuc, M2, M3, M12, M13, and M71. These curves are the result of Monte Carlo simulations identical to those described in Olszewski *et al.* (1996) (see also H92).

Longer-period binaries have low discovery probabilities because of their slowly-varying velocities and small velocity amplitudes. The discovery efficiency declines at short period because of the decreasing fraction of these

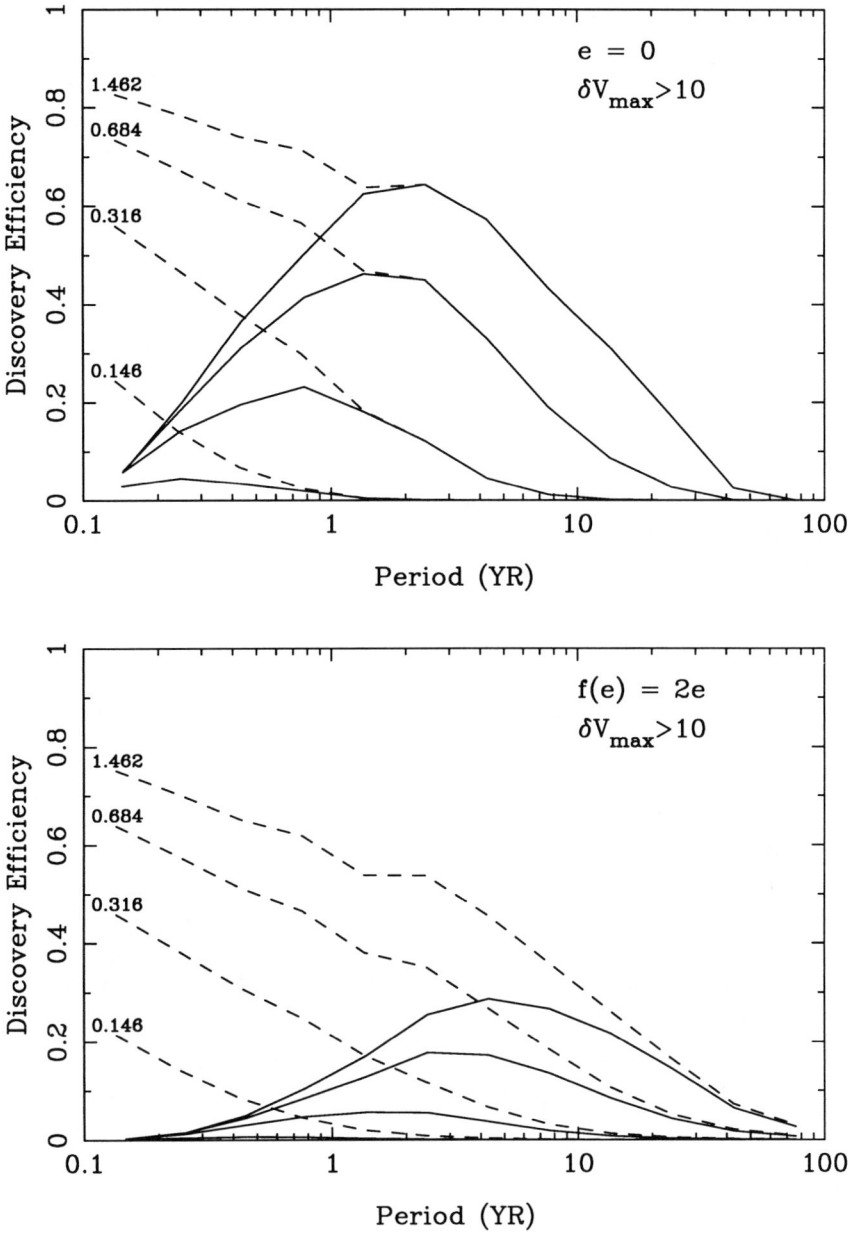

*Figure 1.* Binary discovery efficiency *vs.* orbital period for four bins in the mass ratio. The discovery criterion is a velocity range larger than 10 km s$^{-1}$. The dashed lines show the effect of ignoring the size of the giants (*i.e.*, these are the discovery efficiencies for main-sequence stars). Each of the four pairs of lines is for a different range of the binary mass ratio and is labeled with the mean value. The top panel assumes circular orbits, the bottom panel assumes a thermalized distribution of eccentricities.

luminous giants whose radii are small enough to fit in the small orbital separations implied by short periods. The four pairs of curves in the figure are the discovery efficiency for four equal bins in the logarithm of the mass ratio in the range $0.33 > \log(q) > -1.0$. Binaries with small mass ratios are difficult to detect because of the small velocity amplitude of the primary. The two panels of the figure show the significant uncertainty which the unknown eccentricity distribution implies for the discovery efficiency.

Figure 1 shows that this survey is most sensitive to binaries with orbital periods within a decade above and below the value of 2 yr. There are six stars with velocity ranges larger than 10 km s$^{-1}$, yielding a discovery fraction of 0.015 with the 95% confidence interval (0.0056, 0.033) (see H92). The average binary discovery efficiency between a period of 0.2 and 20 yr for the upper three mass-ratio bins of Fig. 1 is 0.29 for circular orbits and 0.10 for a thermalized distribution of eccentricities. These are slightly smaller than the values quoted in H92 because of small improvements in the simulations used to estimate the efficiencies. Correcting by these efficiencies yields binary frequencies of 0.027 per decade of period and 0.074 per decade for the two cases, respectively, for mass ratios larger than 0.22. In other words, 0.027 or 0.074 of the stars in these clusters are the primaries of binaries with periods in an arbitrary decade in the vicinity of 2.0 yr and with mass ratios larger than 0.22.

Mass ratios larger than 1.0 are rare in the solar neighborhood (Duquennoy & Mayor 1991, DM hereafter). In globular clusters, encounters between binaries and more massive single stars result in the heavier star being preferentially exchanged into the binary (Hills 1975; Sigurdsson & Phinney 1993). Even so, more realistic efficiencies may result from averaging the values for the middle two mass-ratio bins in Fig. 1. This yields efficiencies of 0.20 and 0.068 and binary frequencies of 0.037 per decade ($e = 0$) and 0.11 per decade ($f(e) = 2e$). The statistical uncertainties in all of these frequencies are about a factor of 2 at 95% confidence.

Nearby solar-type stars have a binary frequency of about 0.06 per decade in the vicinity of 2 yr periods (DM). Thus, the average binary frequency in the above sample of globular clusters is between a factor of two lower and a factor of two higher than that in this field population, depending on the mass ratio and eccentricity distribution of the globular cluster binaries.

The DM sample has a roughly Gaussian period distribution extending from 1 day to $10^7$ yrs, with the most likely period being about 170 yrs. In globular clusters, however, binaries can be disrupted by encounters with single stars and binaries (Heggie 1975; Hills 1975) and, the longer the period, the larger the cross-section for gravitational encounters. The shortest period for which the binary can be disrupted by an encounter with a typical single star (*i.e.*, binary binding energy equal to the typical kinetic energy of relative motion) is

$$P = (65 \text{ yr}) \left(\frac{5 \text{ km s}^{-1}}{\sigma}\right)^3. \tag{1}$$

Here $\sigma$ is the one-dimensional velocity dispersion of the cluster and all of the stars have been assumed to have a mass of 0.8 $\mathcal{M}_\odot$. This period is also the approximate "watershed" longer than which encounters make the binary less bound on average and shorter than which they make the binary more bound (Hut 1983, see also the discussion in H92).

Stars with periods equal to or longer than the limit given by eq. (1) are on average disrupted in $1.5 \times 10^{10}$ yr if $\sigma$ and the cluster number density of stars, $n$, satisfy

$$\left(\frac{n}{1 \text{ pc}^{-3}}\right) \left(\frac{5 \text{ km s}^{-1}}{\sigma}\right)^3 > 550. \tag{2}$$

This is based on the ionization rate of Hut & Bahcall (1983) and again assumes 0.8 $\mathcal{M}_\odot$ stars. This relation is satisfied at the centers of many globular clusters, though by widely varying margins. Equations (1) and (2) suggest that, even if globular clusters began with the same binary population as the DM sample, the overall binary frequency in globular clusters today would only be roughly half as large.

Binary stars with periods shorter than the boundary given by eq. (1) become more bound in encounters with single stars and this can eventually lead to their destruction through a stellar collision or to their ejection from the cluster by the recoil from an encounter. They can also be disrupted by encounters with other binaries. Particularly towards the centers of the denser clusters, binary stars of almost any period can have been destroyed by these processes. Mass segregation is expected to concentrate binaries towards the centers of clusters, so the observed numbers of binaries and their radial distribution will be the result of the interplay between segregation and destruction and this must be simulated in detail (Hut et al. 1992b; McMillan & Hut 1994; Sigurdsson & Phinney 1995).

Clearly it will be of interest to determine how the binary frequency varies between clusters with different densities, with radius in clusters, and with period. Evidence from radial velocity surveys on all of these points is still very preliminary. Pryor et al. (1991) report that the binary frequency seems somewhat higher in the low-density clusters NGC 5053 and NGC 5466. Yan (1995) is also surveying NGC 5053 and finding binaries, though no frequency has been reported yet. As summarized in H92, binary stars containing globular cluster giants have approximately the same radial distribution as the cluster light. Barden et al. (1995) report that velocity variables among stars near the base of the giant branch in M71 have the same radial distribution as the rest of the sample.

The Barden et al. (1995) sample and the Côté et al. (1995) sample of turnoff dwarfs in M4 are sensitive to shorter-period binaries than earlier studies (see also Yan 1995). They report preliminary binary frequencies of 0.05–0.08 per decade and about 0.07 per decade, respectively. These are for $q > 0.2$ and periods between a few days and a few years.

Studies with long time baselines can detect binaries with long periods. Côté et al. (1995) and Mayor et al. (1995) have recently reported studies for M22 and $\omega$ Cen, respectively, with time baselines of 10–20 yrs. Both studies find a binary frequency of 0.01–0.02 per decade for periods between a fraction of a year and 20–40 yr. These lower frequencies may be evidence for the destruction of longer-period binaries in globular clusters. However, note that the center of $\omega$ Cen does not satisfy eq. (2).

## 3. Searches for Eclipsing Variables

Mateo (1995) has recently reviewed this field and reports that there are now at least six globular clusters in which eclipsing variables are known and that this number is likely to increase rapidly. Accordingly, we only briefly discuss eclipsing binaries here (see also section 2.2 of H92; Shara, this volume; and Menzies, this volume).

Searches for eclipsing variables can only detect binary systems that are in contact or not far from it and thus have periods of at most a few days. The selection effects for detached systems favor the discovery of systems with separations of a few stellar radii and where both stars have similar luminosities. However, most of the systems discovered are contact main-sequence, or W UMa, systems. The strong distortion of the stellar shapes in these systems produce significant variations in the total luminosity even outside of the eclipses. Also, the thermal contact between the stars gives both stars equal surface brightnesses, making the detection of systems with large mass ratios quite feasible (see the discussion and references in Yan & Mateo 1994 and H92).

Unfortunately, converting the number of W UMa systems detected into the number of progenitors is difficult. Main-sequence systems slowly spiral together due to gravitational radiation and, more importantly, magnetized winds (see the discussion and references in Eggen & Iben 1989). Periods as long as 5 days might evolve into contact over the age of globular clusters. Even more uncertain is the lifetime of W UMa systems, with estimates ranging from $10^8$ to $10^{10}$ yrs (see Eggen & Iben 1989). The number of W UMa systems seen is the product of the rate at which orbit (and stellar) evolution produces contact systems and the system lifetimes. Thus, problems remain with producing binary frequencies that can be compared to those from other techniques.

Currently, the only survey of eclipsing systems in a globular cluster that has yielded a binary frequency is the study of M71 by Yan & Mateo (1994). They argue that all five eclipsing variables which they discovered are likely members of the cluster and calculate a binary frequency of about 0.013 for systems with any mass ratio and initial periods in the range 2.5–5 days, assuming a W UMa lifetime of $10^9$ yrs. The implied binary frequency per decade of period is 0.04. This is about a factor of 2 larger than the DM frequency in this period range, but studies of more clusters and a better understanding of the evolution of W UMa systems are needed before it will be clear whether this is significant.

## 4. Binary Sequences in the Color-Magnitude Diagram

Unresolved main-sequence binaries produce "stars" in the cmd that are brighter and redder than the main sequence. Even if there is no peak in the distribution of binary mass ratios at $q = 1$, there will be a peak in the number of binaries about $2.5 \log(2) \simeq 0.75$ mag above the main sequence (Romani & Weinberg 1991, section 2.3 of H92). The reason is that, at a fixed color, the binaries will consist of one star that is brighter and one star that is fainter than a main sequence star with that color. To first order, the combined magnitude of the two stars does not change as the difference in brightness between the two stars increases. Uncertainties in measured magnitudes and colors do tend to wash out this peak (see figures 9 and 12 in H92), but a plateau of stars above the main-sequence remains.

Because of the measurement uncertainties, particular cmds will maximize the visibility of the binaries consisting of two main-sequence stars. The steeper the main-sequence in the cmd, the larger will be the vertical width of the sequence due to the smearing by uncertainties in the color direction. A steeper sequence also means that the color uncertainty, which is always larger than the magnitude uncertainty, contributes more to the intrinsic width of the sequence. Because binaries are always at most about 0.75 mag brighter than the main-sequence, their visibility is greatest, for given measurement uncertainties, when the main-sequence has the shallowest slope in the cmd (e.g., Jaschek 1976). Thus, the redder color should always be used on the vertical axis of the cmd since this reduces the slope by a unit. Physically, this maximizes the contribution from the fainter and redder secondary.

These arguments also suggest that a long color baseline, such as $U - I$, would be good for detecting binaries in the cmd. The main-sequence colors from Edvardsson and Bell (1989) yield a slope in the $I$ vs. $U - I$ cmd of about 0.9 for the region from 1.0 to 3.0 magnitudes below the turnoff. The slopes for the $V$ vs. $B - V$ and $I$ vs. $V - I$ cmds are about 4.0. However, observations of globular cluster stars in the $U$ passband will tend to have larger uncertainties than those in redder bands, so quantitative estimates are necessary. The mass-magnitude relations also need to be taken into account since they determine the height above the main sequence of a binary with a specific mass ratio. Such quantitative comparisons of different cmds have not been carried out to our knowledge, but using $U$ magnitudes appears promising.

The largest problem with finding binaries in the cmd is the background of coalesced stellar images. This background can be evaluated using numerical experiments that add and recover artificial stars from the frames, but, if the crowding of images is significant, searching for binaries is very difficult (H92). The two globular clusters presenting the best cases for binary stars above the main sequence in ground-based data, E3 (McClure et al. 1985) and NGC 288 (Bolte 1992), have low central surface densities of stars, allowing data to be obtained close to the cluster centers. Thus, the advent of deep cmds for the centers of globular clusters based on high-angular-resolution HST images will have a large impact on this field. The analysis of binary sequences in these cmds is still in a preliminary stage and thus we focus on the expected strength of "binary sequences" in the cmd and a few preliminary results.

The orbital period of the binaries plays no role in detecting them in the cmd. Thus, in globular clusters, systems with periods in the approximately 4 decades between about 5 days and 70 yrs are expected to be present and could be detected. The range of mass ratios detectable depends on the mass-luminosity relation along the main sequence. The steep relation below masses of about 0.15 $\mathcal{M}_\odot$, visible for the first time in the deep cmd shown in Fig. 5 of Fahlman et al. (this volume), argues that binaries will be harder to detect fainter than about 5 mag below the turnoff. Brighter than that, however, the Alexander et al. (1995) $M_I$-mass relations suggest that stars differing by a factor of two in mass will have $I$ magnitudes that differ by about 2.0. For a factor of three in mass, the difference is about 3.3 mag. Thus, binaries with $q = 0.5$ and 0.33 will appear about 0.3 and 0.15 mag above the main-sequence in the $(V, I)$ cmd, respectively. The likely detection threshold in $q$ is likely to be between these two values.

About 0.4 of the DM field binaries have $q > 0.5$ and 0.6 have $q > 0.33$. So roughly half of the binaries in a population with the DM $q$ distribution would be detectable in the cmd. If the binary frequency of 0.05 per decade of period with $q > 0.2$ found from radial velocity studies extends over the whole 4 decade range of period and if the mass ratio distribution is similar to that from DM, then about 0.13 of the stars would be expected to appear in a "binary sequence" more than about 0.2 mag above the main sequence.

This estimate needs to be corrected downward slightly for systems with a white dwarf companion, which are detected by the radial velocity surveys and are not (at this location!) in the cmd. Ignoring mass transfer, assuming a power-law mass function with exponent $x$ ($x = 1.35$ being Salpeter), and again adopting the DM mass ratio distribution suggests that between 0.15 ($x = 2$) and 0.30 ($x = 1$) of the binaries will have a dark companion. Thus, the fraction of stars expected in a binary sequence is around 0.10.

Richer et al. (1995a) show the $U$ vs. $U - I$ cmd for stars at about 1 core radius ($1r_c$) in M4 based on photometry obtained with the HST. Richer et al. (1995b) report that approximately 4% of the stars in this diagram appear to be in a binary sequence. They also report two possible red dwarf – white dwarf binaries identified in the $U - V$ vs. $V - I$ diagram. Figure 5 in Fahlman et al. (this volume) shows the $V$ vs. $V - I$ cmd resulting from HST data for the $1r_c$ M4 field and another field at $4r_c$. The region of this cmd with $1.25 < (V - I)_0 < 1.9$ has a main sequence that is free of bulge contamination and relatively shallow. In the $4r_c$ field, the fraction of stars between 0.2 and 0.75 mag above the main sequence is 2.5%. Though the $V$ vs. $V - I$ cmd is less powerful than the $I$ vs. $V - I$ diagram, these two very preliminary results suggest that the number of stars in binary sequences in M4 is smaller than the 10% predicted in the previous paragraph.

Perhaps we are seeing the effect of the destruction of binaries with periods longer than the few years that are best detected by radial velocity surveys. Bolte (1992) found that 10% of the stars in a field about 2.0 core radii from the center of NGC 288 were in a binary sequence with $q > 0.7$. This cluster has a much lower central density than M4 and so it might be expected to have experienced less binary destruction. At least, it seems very likely that the distribution of binary periods in M4 is not as wide as in the solar neighborhood, in agreement with theoretical expectation. Much more work needs to be done to solidify these conclusions, however.

## 5. Conclusion

Binary stars clearly exist in globular clusters and there is tantalizing, but still very uncertain, evidence that the period distribution is different from that for binaries in the solar neighborhood. We are on the threshold of an exciting period where we will be able to use at least three different techniques to identify and study binaries: radial velocity variability, eclipses, and binary sequences in the cmd. These should be powerful tools for testing our theories for the dynamical evolution of globular clusters and their binary populations.

**Acknowledgments:** The research of CP, RAB, MB, and HEB on M4 is supported by NASA through grants from the Space Telescope Science Institute, which is operated by the Associated Universities for Research in Astronomy, Inc., under NASA conctract NAS5-26555. Support for the research of HBR, GGF, WEH, and DVB is provided by grants from the Natural Sciences and Engineering Research Council of Canada.

## References

Alexander, D. R., Brocato, E., Cassisi, S., Castellani, V., Ciacio, F., & Degl'Innocenti, S. (1996) submitted to *A&A*

Barden, S. C., Armandroff, T. E., & Pryor, C. (1995), in E. F. Milone (ed.), *The Origins, Evolution and Destinies of Binaries In Clusters*, in press

Bolte, M. (1992) *ApJS*, **82**, 145

Côté, P., Fischer, P., Pryor, C., & Welch, D. L. (1995), in E. F. Milone (ed.), *The Origins, Evolution and Destinies of Binaries In Clusters*, in press

Côté, P., Welch, D. L., Fischer, P., Da Costa, G. S., Tamblyn, P., Seitzer, P., & Irwin, M. J. (1994), *ApJS*, **90**, 83

Duquennoy, A., & Mayor, M. (1991), *A&A*, **248**, 485

Edvardsson, B., & Bell, R. A. (1989), *MNRAS*, **238**, 1121

Eggen, O. J., & Iben, I. (1989), *AJ*, **97**, 431

Fischer, P., Welch, D. L., Mateo, M., & Côté, P. (1993), *AJ*, **106**, 1508

Gebhardt, K., Pryor, C., Williams, T. B., & Hesser, J. E. (1994), *AJ*, **107**, 2067

Heggie, D. C. (1975), *MNRAS*, **173**, 729

Hills, J. G. (1975), *AJ*, **80**, 809

Hut, P. (1983), *ApJ*, **272**, L29

Hut, P. & Bahcall, J. (1983), *ApJ*, **268**, 319

Hut, P., McMillan, S., Goodman, J., Mateo, M., Phinney, E.S., Pryor, C., Richer, H.B., Verbunt, F., & Weinberg, M., (1992a), *PASP*, **104**, 981

Hut, P., McMillan, S., & Romani, R. (1992b), *ApJ*, **389**, 527

Jaschek, C. (1976), *A&A*, **50**, 185

Mateo, M. (1995), in E. F. Milone (ed.), *The Origins, Evolution and Destinies of Binaries In Clusters*, in press

Mateo, M., Harris, H. C., Nemec, J. M., & Olszewski, E. W. (1990), *AJ*, **100**, 469

Mayor, M., Duquennoy, A., Udry, S., Andersen, J., & Nordström, B. (1995), in E. F. Milone (ed.), *The Origins, Evolution and Destinies of Binaries In Clusters*, in press

McClure, R. D., Hesser, J. E., Stetson, P. B., & Stryker, L. L. (1985), *PASP*, **97**, 605

McMillan, S. & Hut, P. (1994), *ApJ*, **427**, 793

Olszewski, E. W., Pryor, C., & Armandroff, T. E. (1996), *AJ*, in press, Feb.

Pryor, C., McClure, R.D., Hesser, J.E., & Fletcher, J.M. (1989), in D. Merritt (ed.), *Dynamics of Dense Stellar Systems*, Cambridge University Press, Cambridge, p. 175

Pryor, C., Schommer, R. A., & Olszewski, E. W. (1991), in K. Janes (ed.), *The Formation and Evolution of Star Clusters*, ASP, San Francisco, p. 439

Richer, H.B. et al. (1995a), *ApJL*, **451**, L17

Richer, H. B. et al. (1995b), in E. F. Milone (ed.), *The Origins, Evolution and Destinies of Binaries In Clusters*, in press

Romani, R. W., & Weinberg, M. D. (1991), *ApJ*, **372**, 487

Sigurdsson, S., & Phinney, E. S. (1993), *ApJ*, **415**, 631

Sigurdsson, S., & Phinney, E. S. (1995), *ApJS*, **99**, 609

Yan, L., & Mateo, M. (1994), *AJ*, **108**, 1810

Yan, L. (1995), Caltech Ph.D. thesis

# DIAGNOSING STRUCTURE AND EVOLUTION OF CLUSTERS WITH NEUTRON STAR BINARIES

RALPH A.M.J. WIJERS
*Institute of Astronomy*
*Madingley Road, Cambridge CB3 0HA, United Kingdom*
R.Wijers@ast.cam.ac.uk

**Abstract.** A key problem in using binaries as a tool for diagnosing cluster evolution is that the tool itself is not very well understood. The theory of binary evolution, despite real successes that can be exploited, has serious problems in many areas relevant to cluster evolution. At least as important but often neglected are *connective* problems, which arise when theoretical model binaries need to be related to observed classes of object, which often requires poorly understood parts of their physics which can be quite irrelevant to their bulk properties. I shall discuss these issues in general briefly, and then illustrate them with the specific example of X-ray binaries and millisecond pulsars.

## 1. Introduction

Since the cluster dynamics meeting in 1984 in Princeton (Goodman and Hut 1985), binary stars have been found to exist in clusters in large enough numbers to influence the evolution of clusters (Pryor 1996), and for binary-single star and binary-binary encounters to compete with encounters between single stars in the formation of many interesting types of object (Davies 1996). Modelling a realistic cluster is beginning to come within the reach of our computing abilities (Aarseth 1996, Makino 1996) right at the time where new observing techniques are greatly expanding our knowledge of real clusters (Guhathakurta 1996, King 1996).

At the same time we face some (still) unresolved issues in both the theory of the evolution of clusters consisting of point masses and in the evolution of binary stars. Therefore, both those of us coming from the side of cluster dynamics and those coming from stellar evolution have our own skeletons to contribute to a still large joint cupboard. To arrive at meaningful conclusions about the workings of star clusters we will each have to find ways to isolate from the simulations those effects that are not so sensitive to the assumptions we make about as yet unresolved physics problems. Having thus gained insight into the role of the various physical inputs, we may search for types of observation that will guide us in solving some of the uncertainties in the input physics.

I discuss the binary stellar evolution end of such a programme of research. First a general outline is given of the problems from a stellar evolution point of view (sect. 2). Then I illustrate them with the evolution of X-ray binaries and millisecond pulsars, which have long been the focus of research on stellar evolution in clusters because they are so observationally conspicuous and overabundant in them. Our knowledge of their evolution contains examples of well-understood and reliably tested theory as well as a few fine skeletons (sect. 3). I then illustrate how the interaction between stellar evolution and cluster dynamics may manifest itself in the numbers and properties of observed millisecond pulsars and X-ray binaries (sect. 4).

## 2. Overview of the problem

We are to construct in our computer an artificial population of stars and evolve it to some time after its creation; then we compare our simulated population with the best available data and repeat the exercise until the real and simulated worlds look sufficiently alike. It sounds simple enough, but there are more pitfalls than nouns in this recipe. The pitfalls can be classified in three categories: theoretical, observational, and *connective*. By the latter I mean the type of problem that arises when a theoretical star or binary, defined by parameters such as masses, radii, temperatures and orbit is to be translated into an observational beast, usually defined by object class, X-ray variability, equivalent width of some emission line, types of outburst, etcetera. The theoretical problems are straightforward in the sense that the problem areas are usually easily identified by the disputes arising between theoreticians. Among the hardest problems in the theory of single and binary stars are those to do with hydrodynamics, such as convection and common-envelope evolution and encounters between stars, and with magnetic fields, stellar winds and angular-momentum loss. Observational problems are obviously related to trying to detect as many phenomena as possible and to characterising the sensitivities of observations well enough to get an accurate assessment of the selection effects involved in the construction of any sample of objects. The connective problems are often the most hazardous ones.

## 2.1. CONNECTIVE PROBLEMS

It is quite often non-trivial to endow the theoretically created population with the properties that are needed to properly compare them with observed samples. This is often simply because those properties are rather peripheral to the theoretical structure of the object and require expertise that is different from that needed to model the bulk properties of the object. Yet since observations in globular clusters are always done at the cutting edge of observational techniques and at the detection limits of present instrumentation, an accurate estimate of observable parameters of an object is needed. A classic example of this is the transformation from a theoretical Herzsprung-Russell diagram to an observed colour-magnitude diagram of a cluster. Good stellar atmosphere models are needed (and exquisitely calibrated data) in order to do the transformation, but one can perfectly well calculate the evolution of stars with very simple outer boundary conditions that leave the question of the eventual $V$ and $B-V$ of the star unanswered. However, stellar atmospheres can be calculated fairly accurately in many cases, so the problem can be solved.

The situation in compact-object binaries is worse: imagine my computer programme has created a close binary consisting of a white dwarf and a low-mass main-sequence star. It is tempting to call it a cataclysmic variable, but that will not do. To compare with observations, I will need to know whether this is a transient or a persistent source, a classical nova, a dwarf nova, an AM Her system, a novalike variable or what have you. And I need to know what its accretion rate will be, and understand the structure of the accretion disk, whose brightness dominates the total source brightness. Current theory is simply inadequate to make these predictions and the number of cataclysmic variables with well-known parameters is very small, so these matters have not been settled empirically either (Patterson 1984). Hence the debate at this meeting on whether the non-detection of certain types of CV in globular clusters implies a true paucity of main-sequence white dwarf binaries (Shara 1996).

Similar difficulties exist in the X-ray binaries. Two of the twelve bright ones have not been known very long because they are transients, and one may well ask how many more there are. (In the disk population transient sources outnumber persistent ones.) Many of the dimmer ones are transient or highly variable as well (Verbunt 1996). For the millisecond pulsars one encounters the usual pulsar problem that the radio luminosity is hardly predictable from the other parameters of the pulsar such as spin period and magnetic field: no theory exists for it, nor do we have a good empirical relation to use.

## 3. Formation of X-ray binaries and millisecond pulsars

The evolution of X-ray binaries and millisecond pulsars has been reviewed very well in the recent past (Bhattacharya and Van den Heuvel 1991, Phinney and Kulkarni 1994, Lewin, Van Paradijs, and Van den Heuvel 1995, Wijers, Davies, and Tout 1996). I summarise only the broad picture here and omit references to the original papers. To a create millisecond pulsar in the standard model is a two-step process. The first step is to make a binary with a normal neutron star like any young, high-field radio pulsar in orbit with a low-mass main-sequence star. In the second step, the binary is brought into contact and mass from the Roche-lobe filling star is transferred to the neutron star. This process causes a reduction of the dipole magnetic field of the neutron star to the low values observed for millisecond pulsars in a yet unknown way. It also causes the pulsar to spin much faster, as angular momentum is added to it with the accreted mass. This process of combined spin-up and field decrease is called *recycling*.

The first step can be accomplished in at least two ways. Beginning from a binary star that consists of an O or B star plus a low-mass main-sequence star, we obtain the desired neutron star binary as follows: the massive star evolves off the main sequence and at some point fills its Roche lobe. Since the mass transfer that ensues is to the lighter of the two stars it is dynamically unstable and the low-mass star plunges into the envelope of the massive star. If the binary is too close the two stars will coalesce and if it is too wide there will never be Roche contact. In a narrow intermediate separation range, a close binary can be formed consisting of the low-mass star and the helium core of the B star. If the B star was more massive initially than about $10\,M_\odot$, it will undergo a supernova explosion and leave a neutron star. A substantial fraction of the binaries may be disrupted at this time due to the sudden mass loss from the system and/or a natal velocity kick imparted to the neutron star. The surviving systems are the input for step two. In view of the limited range of initial binary separations allowed and the destructive effect on the binary of the supernova explosion, this formation channel is rare. This is actually as it should be because the fraction of neutron stars that ends up in a low-mass X-ray binary may well be as low as $10^{-5}$ in the Galactic disk. Another way of accomplishing step one, only applicable in dense stellar systems, is to create a neutron star separately first and let it encounter a binary or single star. Such encounters can result in a binary star of which the neutron star is one of the members (Davies 1996, Mardling 1996).

The proto-low-mass X-ray binary formed in step one may become a Roche-lobe filling system either because the two stars are brought closer together due to angular-momentum loss from the orbit or because the low-mass star expands to become a (sub)giant at the end of core hydrogen burning. Almost inevitably it will start with a substantial orbital eccentricity, but the time for the orbit to circularise is thought to be much shorter than the evolution time of the binary once either star comes close to its Roche lobe, so we can begin the next step with an orbit that has already been circularised. Let the neutron star have a mass $M_1$ and the companion (or donor) a mass $M_2$. For small values of $q \equiv M_2/M_1$ the Roche lobe radius is adequately approximated by $R_L/a = (q/(1+q))^{1/3}$, where $a$ is the distance between the stars. If we further assume that no mass is lost from the system during transfer and the donor radius is always close to the Roche lobe radius (a good approximation), we can write the mass transfer rate as

$$\left(\frac{-\dot{M_2}}{M_2}\right)\left(\frac{5}{6} - \frac{M_2}{M_1}\right) = \frac{1}{2}\left(\frac{\dot{R_2}}{R_2}\right) + \left(\frac{-\dot{J}}{J}\right). \quad (1)$$

The signs are chosen such that each term in parentheses is positive in the cases of interest here; $J$ is the orbital angular momentum of the binary. Among others, this requires that $M_2$ be less than $M_1$ by a margin that depends on the circumstances, or else mass transfer will be unstable and some form of common-envelope evolution results (as in step one above with the B star and the low-mass star). We can look at the two extreme cases in which the mass transfer is driven either by the first or by the second term on the right-hand side of eq. 1.

3.1. EVOLUTION DRIVEN BY GIANT EXPANSION

If we neglect angular-momentum loss and the donor is expanding because it is evolving off the main sequence, eq. 1 simplifies to

$$\left(\frac{-\dot{M_2}}{M_2}\right)\left(\frac{5}{6} - \frac{M_2}{M_1}\right) = \frac{1}{2}\left(\frac{\dot{R_2}}{R_2}\right). \quad (2)$$

Because the expansion due to nuclear evolution of a red giant is well understood quantitatively and no other degrees of freedom are left, the outcome is fully predictable: the giant will lose mass, spinning up the neutron star and reducing its field, until its envelope is exhausted and only a white dwarf core is left. The evolution is said to be *conservative*, as no mass or angular momentum leaves the system. Since the system must not come into contact until the (sub)giant phase, the orbital period at the time of first Roche contact must exceed about 10 hours in this approximation. Since orbital angular-momentum loss is uncertain (sect. 3.2), it is not clear how to translate that limit into one on the initial binary period. Probably an initial orbital period of a few days is required. And of course the donor star must be massive enough to reach the end of its main-sequence lifetime between its formation and now. In a globular cluster, that means it had to be at least the current turnoff mass, 0.7–0.8 $M_\odot$.

Because the radius of a giant at the time of envelope exhaustion only depends on the mass of the core, and the orbital separation is fixed by this radius and the fact that the giant must fill its Roche lobe, a unique relation is predicted between the orbital period and the mass of the white dwarf for the 15 or so known millisecond pulsars which have white dwarf companions in wide circular orbits. The measured values are indeed consistent with the prediction, as are the small eccentricities of the orbits which are caused by convective density fluctuations in the envelope of the giant (Phinney and Kulkarni 1994). It also appears that the observationally inferred birth rate of X-ray binaries with (sub)giant donors is similar to that of the wide ($P_{orb} \gtrsim 10$ d) millisecond pulsar binaries, which supports a one-to-one evolutionary connection between these two types of binary.

## 3.2. EVOLUTION DRIVEN BY ANGULAR-MOMENTUM LOSS

If the donor is on the main sequence when the system comes into contact, no radius expansion occurs spontaneously and mass transfer must be due to loss of orbital angular momentum. Then we have

$$\left(\frac{-\dot{M_2}}{M_2}\right)\left(\frac{5}{6} + \frac{n}{2} - \frac{M_2}{M_1}\right) = \left(\frac{-\dot{J}}{J}\right), \qquad (3)$$

where $n$ gives the relation between mass and radius of the donor ($R_2 \propto M_2^n$). Contact on the main sequence implies an orbital period less than about 10 hours for $M_2 < 1\,M_\odot$ (as required for stable mass transfer). The only well-understood mechanism for angular-momentum loss is gravitational radiation, but it can only drive mass transfer rates up to about $10^{-10}\,M_\odot\,\mathrm{yr}^{-1}$, whereas many low-mass X-ray binaries with short orbital periods have accretion rates up to a hundred times greater. Another mechanism for angular-momentum loss has been proposed by Verbunt and Zwaan (1981). It is based on the fact that low-mass stars have magnetic winds, which can carry away large amounts of angular momentum. Since the star is forced to co-rotate with the orbital period of the binary due to tidal friction any loss of angular momentum of the star will be taken out of the orbit eventually, and the orbit shrinks. The resulting mass loss can plausibly be enough to explain the accretion rates in low-mass X-ray binaries, but because the magnetic angular-momentum loss (called *magnetic braking*) depends on the parameters of the star and the orbit in rather uncertain ways, the theory makes no quantitative predictions that lend themselves to strong observational testing. The mass accretion seen in these X-ray binaries almost certainly leads to recycling of the pulsar, but although it is reasonably plausible that short-period binary (perhaps even single) millisecond pulsars can be the end product, the case for this has by no means been argued convincingly.

### 3.3. SPIN-UP OF THE MILLISECOND PULSAR

The accreted matter usually has more specific angular momentum than the neutron star, which will therefore be spun up. At some point, the torques exerted by the accreted matter and by parts of the magnetic field of the neutron star coupled to the accretion disk will balance, and the neutron star comes to its *equilibrium spin period* $P_{eq} = P_0 B_9^{6/7}$, where $B_9$ is the dipole field of the pulsar in units of $10^9$ G and $P_0$ is a constant that depends on the neutron stars properties and the accretion rate. For the maximum accretion rate, the Eddington rate, $P_0 \simeq 2\,\text{ms}$. We see immediately that the speed to which a pulsar can be spun up is limited by its magnetic field. Since the evolution of the latter cannot be predicted yet, neither can the final spin period. However, for each known pulsar with a measured field, we can calculate what its shortest possible spin period is likely to have been. This limit is known as the spin-up line. (It is clear from the known pulsars that they are not all spun up to the fastest possible spin.) The amount of material needed to spin up a pulsar to a given period also depends on the magnetic field, because the specific angular momentum of the accreted matter is set at the magnetospheric radius. For low-field pulsars, it will take 0.03–0.1 $M_\odot$ of material to spin it up to the spin-up line. One of the problems with the X-ray binary scenario for forming millisecond pulsars is the fact that the accretion should stop fairly suddenly because otherwise the pulsar will start spinning more slowly again to adjust its spin to a lower accretion rate. Especially with the angular-momentum loss driven evolution this requires extra ingredients to the scenario.

## 4. Some effects of encounters

The most obvious influence of the high stellar density in globular clusters is that stellar encounters add a new formation channel for X-ray binaries. They are in fact the archetypical encounter products, since it is the realisation of their overabundance in globular clusters 20 years ago (Gursky 1973, Katz 1975) and the tidal capture that was suggested as its explanation (Fabian, Pringle, and Rees 1975) that started this field of research. The problem with comparing numbers is that except in such a glaring case selection effects and problems in identifying observationally defined object types with corresponding model systems may prevent us from establishing abundance differences with any certainty. A case in point is the debate over whether there is a paucity of cataclysmic variables in globular clusters or not (Shara 1996). It may therefore be more profitable to look at the observed systems alone and see whether they are different from those in the disk in any way, in hopes of finding differences than cannot be attributed easily to selection effects.

### 4.1. DISK VERSUS GLOBULAR-CLUSTER RECYCLED PULSARS

The spin period distribution and binary fraction of recycled pulsars are noticeably different between the disk and clusters (fig. 1). Let us consider binarity first. Of the disk pulsars 8 out of 27 are single whereas the cluster pulsars have 21 singles in a total of 34, a rather larger fraction. The difference for slow pulsars ($P > 50\,\mathrm{ms}$) is even more striking : the only single disk pulsar could have come from a massive binary system, and the cluster membership of the only binary cluster pulsar (1718–19) is not very secure, although plausible. Hence it may be that among slower recycled pulsars, the disk ones are all binaries and the cluster ones are all single (see also sect. 4.2). The overall period distributions of disk and cluster pulsars do not, however, seem different. One might therefore imagine that on the whole they were formed in similar ways, but that in clusters there are more efficient ways of making them single. Simple disruption of binary stars is difficult because most binary millisecond pulsars in the disk would be hard binaries if put in a dense cluster. They could still lose their pulsar by exchange encounters, but since the pulsar is the most massive object it is more likely to stay in the binary, hence there would have to be more that are left in binaries than ejected ones.

The two globular clusters with many pulsar in them, M15 and 47 Tuc, are also shown separately in fig. 1. These two clusters have a rather different millisecond pulsar period distribution. Especially the absence of long-period pulsars in 47 Tuc is observationally significant: they would have been seen if they existed. One is left to speculate what causes the difference, and whether the post-collapse nature of M15 is responsible for the slow pulsars' presence.

### 4.2. THE SLOW GLOBULAR-CLUSTER PULSARS

The four slowest globular-cluster pulsars are listed in table 1, and are rather non-typical of globular clusters in that the measured fields are high and their lifetimes above the death line are therefore quite short compared to the age of the clusters. They were first pointed out as a class by Lyne. The next three slowest cluster pulsars are all in M15. One has a negative period derivative, hence its field cannot be measured; the other two have

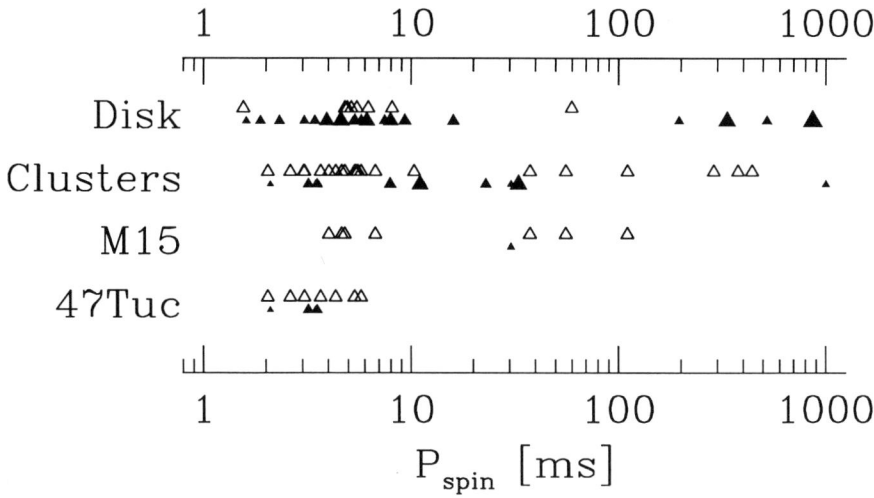

*Figure 1.* The spin period distribution of recycled pulsars in the disk of the Galaxy, in the total globular-cluster system, and in the two most pulsar-rich clusters, M15 and 47 Tuc. Open symbols are single pulsars, solid ones are binaries and the symbol size increases with orbital period. The disk data are from Lorimer (1994) and the cluster data are from Lyne (1994).

about $10^{10}$ G fields and ages of about $10^8$ yr. Only the cluster membership of 1718–19 is sometimes disputed, the others are secure associations. Their birth rate is reasonably high because of their short ages and because the high luminosities of the four known ones almost certainly mean that there are many more of them that have not been detectable. There must be a fairly efficient production mechanism for them. They are all in clusters which are centrally condensed and have high predicted tidal capture rates. One possibility could be that they are products of failed captures, in which either the initial approach was close enough to disrupt the extended star and leave the pulsar in a thick disk or wide enough that no bound system formed eventually (Mardling 1996) but some material may have been dumped on the pulsar anyway (see also Sigurdsson and Phinney 1995).

## 5. Discussion and conclusion

Since uncertainties in binary evolution are no less great than in cluster evolution, it is inherently risky to rely on exotic binary stellar evolution products to diagnose the evolution of globular clusters. A careful 'calibration' of binary evolution, to isolate trustworthy results from uncertain ones, is needed. Special attention must be paid to *connective* problems between theory and observation, which are the problems that arise when observable

TABLE 1. Properties of slow globular-cluster pulsars

| Name | cluster | $P$ (ms) | $\tau_c$ (Myr) | $B$ ($10^{12}$ G) | $L$ (mJy kpc$^2$) | $P_{\rm orb}$ (hr) |
|---|---|---|---|---|---|---|
| 1718–19 | 6342 | 1004 | 10 | 1.5 | 400 | 6.2 |
| 1744–24B | Ter 5 | 443 | – | – | 200 | single |
| 1745–20 | 6440 | 289 | 12 | 0.4 | 300 | single |
| 1820–30B | 6624 | 379 | 200 | 0.11 | 120 | single |

properties (often very superficial) are to be derived for theoretical systems.

The properties of millisecond pulsars show clear differences between the disk and clusters, and from cluster to cluster. This begs explanation in terms of processes that are unique to clusters, i.e. stellar encounters. The fact that different types of pulsars appear to come in different typs of clusters will help in telling which are the dominant production mechanisms for each of them.

## References

Aarseth, S. J.: 1996, this volume
Bhattacharya, D. and van den Heuvel, E. P. J.: 1991, *Phys. Rep.* **203**, 1–124
Davies, M. B.: 1996, this volume
Fabian, A. C., Pringle, J. E., and Rees, M. J.: 1975, *Mon. Not. R. Astron. Soc.* **172**, 15P–18P
Goodman, J. and Hut, P. (eds.): 1985, Reidel, Dordrecht
Guhathakurta, P.: 1996, this volume
Gursky, H.: 1973, Oral presentation at the 1973 Cambridge conference on 'Physics and Astrophysics of Compact Objects'
Katz, J. I.: 1975, *Nature* **253**, 698–699
King, I.: 1996, this volume
Lewin, W. H. G., van Paradijs, J., and van den Heuvel, E. P. J. (eds.): 1995, Vol. 26 of *Cambridge Astrophysics Series*, Cambridge University Press, Cambridge
Lorimer, D. R.: 1994, Ph.D. thesis, University of Manchester
Lyne, A.: 1995, in A. S. Fruchter, M. Tavani, and D. C. Backer (eds.), *Millisecond Pulsars: A Decade of Surprise*, Vol. 72 of *ASP Conference Series*, pp 35–45, Astronomical Society of the Pacific, San Francisco
Makino, J.: 1996, this volume
Mardling, R. A.: 1996, this volume
Patterson, J.: 1984, *Astrophys. J., Suppl. Ser.* **54**, 443–493
Phinney, E. S. and Kulkarni, S. R.: 1994, *Annu. Rev. Astron. Astrophys.* **32**, 591–639
Pryor, C.: 1996, this volume
Shara, M.: 1996, this volume
Sigurdsson, S. and Phinney, E. S.: 1995, *Astrophys. J., Suppl. Ser.* **99**, 609–635
Verbunt, F.: 1996, this volume
Verbunt, F. and Zwaan, C.: 1981, *Astron. Astrophys.* **100**, L7–L9
Wijers, R. A. M. J., Davies, M. B., and Tout, C. A. (eds.): 1996, *Evolutionary Processes in Binary Stars*, NATO ASI Series C, Dordrecht, Kluwer Academic

# COMBINING STELLAR EVOLUTION AND STELLAR DYNAMICS

PETER P. EGGLETON
*Institute of Astronomy*
*Cambridge, UK*

## 1. Introduction

There seem to me to be four approaches to the problem of computing the evolution of star clusters. Firstly, one might assume that our knowledge of the evolution of stars can be condensed into a subroutine that can be added to an N-body code. This subroutine would mainly have to give the radius and the time-dependent mass of a star as a function of its initial mass and its age. Secondly, standing this on its head, one might assume that our knowledge of N-body evolution can be condensed into a subroutine that can be added to a stellar evolution code. This subroutine would determine, probably in a Monte-Carlo fashion, whether the star had picked up, or lost, a binary companion, or whether the orbit of its companion was significantly changed; the probabilities would be determined by simple analytic approximations to the time-dependent distribution functions of stars (and binaries) of different masses and ages, and by interaction cross-sections as functions of density and 'temperature'. Thirdly, if the computing power is available, one might more simply unite an N-body code with a Stellar Evolution (SE) code, and follow both the dynamics and the internal evolution simultaneously. Fourthly, we might hope at some stage to put together simple analytic approximations both from N-body *and* from SE studies, to develop a unified simple model. I venture to say that it is only the last stage, if it is attainable, that would entitle us to say that we 'understand' the evolution of stellar clusters. 'Understanding', I think, means that we can extract some essential wisdom from large numerical simulations, and apply it on the back of the proverbial envelope.

All of these approaches have problems. One that is common to all four is that the SE of *interactive* binaries is nothing like as well-determined as the SE of single stars, or of stars that are 'effectvely single', *i.e.* in binaries with periods $\gtrsim 10^4$d. I shall return to this later; but, putting it aside for the moment in the hope that interactive binary SE can somehow be fudged, then the first approach is probably the easiest to implement. Aarseth (1995) has incorporated into his N-body code a subroutine based on the simple interpolation formulae of Eggleton, Fitchett & Tout (1989) for the evolution from ZAMS to neutron star or white dwarf of Pop I stars. We (Drs C. A. Tout, Zh. Han and myself) are currently trying to upgrade these formulae to incorporate more recent data on opacities (Rogers & Iglesias 1992, Alexander & Ferguson 1994) and the equation of state and other physical input (Pols *et al.* 1995). We also intend to generalise our formulae for a range of metallicities from Pop I to a fairly extreme Pop II.

The second approach does not yet appear to be feasible. One can hope that as N-body simulations approach $\sim 10^5$ bodies some systematic results will emerge regarding the way in which particles in different mass bins distribute themselves in phase space, as a function of time. However, since core collapse phases, and subsequent reexpansion phases due to the formation of central binaries, maybe be somewhat chaotic if the number of particles is large enough, it is not clear that simple interpolation formulae will ever describe such processes reliably. But I should emphasise that SE may also have chaotic aspects. It is not clear that mass loss is such a straightforward process that the mass of a star can be reliably predicted at later times from its mass at age zero; and certain aspects of interactive binary SE, especially those which depend on mass loss (ML) or angular momentum loss (AML), may also show chaotic behaviour.

The third approach, involving brute force, may well be feasible in a decade or two. It would need an SE code that is rather more reliable than any that exists now. My own code, developed over the last 25 years, is now (Han, Podsiadlowski & Eggleton 1994; Pols *et al.* 1995) almost reliable enough to follow a star, without manual intervention, from the ZAMS to carbon burning, except that the helium flash in low-mass stars is still an obstacle. It is not impossible, though, that we can fudge our way through that, and so make it fully automatic. On a modern work-station it takes about 2 hours to compute the entire evolution of a star, provided that for stars which have a helium flash ($\lesssim 2.3\,M_\odot$ for Pop I) we ignore the time it takes at present to restart the star manually on the horizontal branch. My SE code can also take care of certain kinds of Roche Lobe overflow (RLOF) in interacting binaries, but it cannot yet deal automatically with contact binary evolution – nor can anyone else's.

In the fourth approach, which I hope will take place in the fullness of time, we might get away with something like the 'gas-dynamical' models of Hachisu *et al.* (1978), Lynden-Bell & Eggleton (1980) and Bettweiser & Sugimoto (1984), in which the whole cluster is treated like a gas sphere with a 'temperature' and 'pressure'. The analogue of 'composition' in an SE code would be the abundance fractions, as functions of position, for different mass bins, and for binaries and triples. These different species would interact through the analogue of nuclear reactions, for which the rates would have to be estimated. If, as a last step, one is able to extract from such calculations a few simple insights, expressible as analytic approximations, one might finally feel that cluster evolution has been 'understood' – although I dare say not everyone would agree.

There are at least two ways in which one might transfer information on stellar evolution to an N-body code. One might try to express the SE results in the form of interpolation formulae (Eggleton, Fitchett & Tout 1989): as simple formulae as possible, though one should probably aim for an accuracy better than $\sim 10\%$. Alternatively, one can provide the information in tables (*e.g.* Schaller *et al.* 1992). Perhaps 30 different masses between 0.1 and 100 $M_\odot$, each tabulated at about 50 well-chosen evolutionary stages, might be adequate for an accuracy of between 1% and 10%. There is not much point in aiming for higher accuracy, since for some masses (particularly high masses) and some evolutionary stages there is much bigger disagreement between theoretical and observed models. Even where disagreement is not obvious, one should not suppose that the theoretical models are good. Although several main-sequence stars, *i.e.* those in eclipsing double-lined spectroscopic binaries, have masses, radii and luminosities measured to $\sim 1\%$ (Popper 1980, Andersen 1991), very few *evolved* stars, such as red giants and supergiants, Wolf-Rayet stars, hot subdwarfs and white dwarfs, have masses, radii and luminosities which are known to 10%. Even in rare cases where something like this accuracy is achieved for such stars, it is normally uncertain what the *initial* masses were, and so it is not clear that there is agreement between observation and theory.

Some major sources of uncertainty in SE calculations are:-

(a) Mass loss (ML). In low-mass ($\lesssim 2.3\ M_\odot$) and intermediate-mass ($\sim 2.3 - 8\ M_\odot$) stars there are probably two main types of ML. These are:

(a1) A weak, fairly steady wind which increases during the first giant-branch (FGB) and asymptotic giant-branch (AGB) stages. This wind may be driven by magnetohydrodynamic energy, as in the solar wind, although at high FGB and AGB luminosities it may be that magnetic energy is less important, and turbulent hydrodynamic energy in the surface convection zone may be the prime cause of the wind. Judge & Stencel (1991) found an empirical relation of the form $-\dot{m} \sim 10^{-13.6}/g^{1.43}\ M_\odot/\text{yr}$, with gravity $g$ in solar units. There is quite considerable scatter, however, and it is by no means clear that this can all be attributed to measurement uncertainties. Even near the top of the AGB, however, where $g \sim 10^{-5}\ g_\odot$, this ML rate is barely comparable to the nuclear timescale.

(a2) A 'superwind', which removes the remaining envelope at the top of the AGB rather rapidly, forming a planetary nebula with a hot subdwarf (or Wolf-Rayet-like) nucleus. This superwind may be ultimately due to the fact that thanks to progressive hydrogen recombination in the outer layers the total binding energy of the AGB envelope changes from positive to negative (Paczyński & Ziółkowski 1968, Han et al. 1994). The latter authors, by computing the binding energy of the envelope of a star as it climbs the AGB, and assuming that the envelope is lost at the stage when this energy passes through zero, found a relation between the initial mass (in the range $0.8 - 7\,M_\odot$) and the final (i.e. WD) mass of a star. This relation, which I would emphasise has no free parameters in it to be 'tuned', was in fairly good agreement with the semi-empirical relation of Weidemann & Koester (1983) between the initial and final masses of stars of stars in galactic clusters. But we should note that the relation of Weidemann & Koester (1983) has considerable scatter in it, some of which might be real; it is by no means established that the mass of a WD remnant is tightly correleted with initial mass for a given metllicity, and there may well be a chaotic aspect to the ML process which could lead to a real spread.

For massive stars ($\gtrsim 30\,M_\odot$) there may be four stages of ML:

(a3) A weak but growing wind, to some extent as in (a1) but with a much higher velocity ($\sim 1500$ km/s as against $30 - 300$ km/s). Radiation pressure is probably what accelerates the wind to the high speed observed, but the wind may be energised in the first place by MHD processes and by rapid rotation. Conti (1982) plotted ML rate against luminosity, and this relation can be fitted by $-\dot{m} \sim 10^{-16.7} L^{1.9}\,M_\odot$/yr, with luminosity $L$ in solar units. Once again there is considerable scatter. Though relatively stronger than (a1) at early evolutionary stages, this wind probably also does not affect a star strongly on nuclear timescales.

(a4) A P Cyg wind, which is very copious as well as fast, and which presumably can remove 20-80% of the star's mass on substantially less than a nuclear timescale. This requires $-\dot{m} \gtrsim 10^{-4}\,M_\odot$/yr. The P Cyg wind may start when the star reaches a gently sloping line in the HRD – the Humphreys-Davidson (1979) limit, Figure 1 – and may be due to some inherent instability perhaps related to the fact that the interiors of such stars are very near the Eddington limit. It appears to prevent the star from becoming a red supergiant, although not from becoming a yellow supergiant at least at the lower masses ($\sim 30 - 40\,M_\odot$).

(a5) A Wolf-Rayet wind, which may be a continuation of the P Cyg wind at slightly lower intensity ($-\dot{m} \sim 10^{-4.4}\,M_\odot$/yr; Willis 1985) once the star has been stripped nearly or entirely to its small, very hot helium core. This ML rate may be roughly comparable to the (helium burning) nuclear timescale of the star.

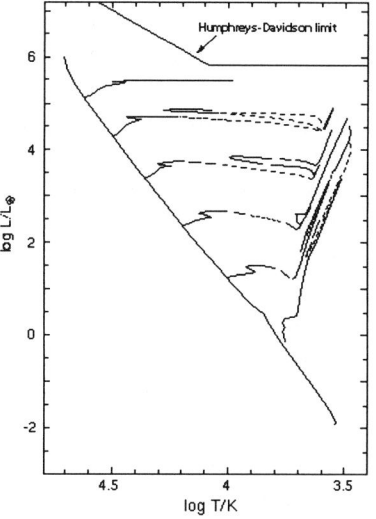 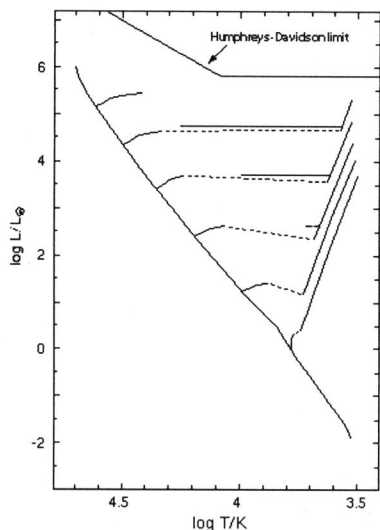

*Figure 1.* Comparison, in a theoretical Hertzsprung-Russell diagram, of some evolutionary tracks computed (left) directly with a stellar evolution code, and (right) with the simple interpolation formulae given by Eggleton, Fitchett & Tout (1989). Above the Humphreys-Davidson (1979) limit, virtually no stars are found in the Galaxy, presumably because of strong winds.

(a6) An SN ejection of the remaining envelope, virtually instantaneously. Stars in the mass range $\sim 8-30\ M_\odot$ probably do not go through the P Cyg phase (a4) or the WR phase (a5). Perhaps they start with (a3) and finish with (a6), passing through a combination of (a2) and (a4), but at a less intense level than in lower and higher mass stars respectively, on the way. It is far from clear how much mass they lose in this intermediate phase.

(b) Semiconvective mixing. Stars of virtually all masses require semiconvective mixing at some point in their evolution. Semiconvection is a difficult physical process, well discussed by Spruit (1992). There is no clear numerical model for this, although in my opinion a model which assumes that mixing always occurs at a rate that keeps the temperature gradient very close to the marginal adiabatic rate (Schwarzschild & Härm 1958), except in the surface layers, is the most likely to be realistic. It is quite probable that real semiconvection is a somewhat chaotic process. 2-D hydrodynamic, or better still 3-D models, may cast some light on this, but such modeling is in its infancy (Merryfield 1995).

(c) 'Convective overshooting'. To improve agreement between theory and observation of some of the most evolved stars within the MS band (Andersen 1991), it is helpful to assume that convective cores in the MS band are larger than is indicated by the standard Schwarzschild criterion. This is also the conclusion of a study of red giants and supergiants in $\zeta$ Aur binaries, *i.e.* those (rather few) binaries from which the most accurate masses, radii and luminosities of highly evolved stars are obtainable (Schröder 1995, private communication). It is also suggested by attempts to fit the HRDs of young clusters (Meynet, Mermilliod & Maeder 1993; Pols 1995, private communication). The problem is not one of semiconvection, since semiconvection does not normally occur to a significant extent in the portion of the MS band at issue ($\sim 2-8\ M_\odot$). Larger cores can be obtained by *ad hoc* variants on the Schwarzschild criterion, although there is no good theoretical basis for those that have been proposed so far (Eggleton 1983a). Recent improvements in the opacity (Rogers & Iglesias 1992) have decreased the problem, but not resolved it. Although larger convective cores appear to be necessary above $\sim 2\ M_\odot$, lower mass stars ($\sim 1.25\ M_\odot$) as in M67 (Morgan & Eggleton 1978; Pols 1995, private communication) appear *not* to need larger convective cores. It is by no means obvious what physical process could achieve this.

(d) Convective heat transport. For most of a stellar interior it is not necessary to have a model of convection more definitive than the mixing-length theory (Böhm-Vitense 1958). But in the outer layers of red supergiants a better theory is necessary. The radii of such stars is therefore quite uncertain. In fact, such stars do not have a 'radius' in a very meaningful sense. Their outer layers probably consist of a small number of plumes rising and falling in a fairly chaotic way. However the uncertainty in their radii is probably conflated with (and perhaps related to) the uncertainty in the ML rate in (a), (a2) above.

(e) Bolometric correction. To convert from a theoretical HRD to an observed one, or *vice versa* requires rather detailed and reliable models of stellar *atmospheres*. This does not seem to have been achieved yet for cool stars, in particular those with atmospheric opacity dominated by molecular bands (and which coincidentally also have a temperature structure dominated by the uncertain process of convection). Now that it is possible with HST to measure even the *lower* MS of some globular clusters, the uncertainty in the bolometric correction in such late dwarfs is probably the main source of disagreement with theoretical models.

There are many other sources of uncertainty, but I believe the above five are the most relevant to single-star evolution. Given these uncertainties, I believe that the approximate (and very simple) formulae of Eggleton *et al.* (1989) are adequate for Pop I; but they will be improved in the near future. A comparison in the HRD of the results of SE computations and these approximate formulae is shown in Figure 1.

Even if single-star evolution were well understood, binary evolution is much more uncertain, and not just because there is a larger space of initial parameters to explore. All of the following kinds of binary are likely to be important at some stage in the evolution of a globular (or Galactic) cluster: contact binaries (WUMas), RS CVns, Algols, symbiotic stars, Ba or CH stars, cataclysmic variables (CVs), high and low-mass X-ray binaries (HMXBs, LMXBs), millisecond binary pulsars (MSBPs) and double degenerates (DDs). All have problems over and above the single-star problems described in the previous Section. Some of these problems are the following:

(i) ML is probably enhanced by the presence of a close companion, to a largely unknown extent (Eggleton & Tout 1989).

(ii) AML, which can take place even with negligible ML, is difficult to quantify. Although GR gives a definite rate, which could be important in some CVs and DDs, magnetic braking with tidal friction (Verbunt & Zwaan 1981) is quite uncertain, and may be dominant in most of the above types of binaries (WUMas, RS CVns, Algols, CVs, LMXBs, MSPBs) which have short periods and cool companions.

(iii) Common-envelope evolution (Paczyński 1976) is a very important process, but its outcome can only be modeled very crudely.

The above three processes, and several others which are uncertain but perhaps less so, can affect the evolution of binaries *qualitatively*, and not just quantitatively. The same initial binary might, with slightly different assumptions, become (Eggleton 1983b) either a CV (with period usually $\lesssim 0.5d$) or a Ba star (with period usually $\gtrsim 400d$). For the present, we have to be content with crude recipes for many stages of binary evolution.

Let us suppose, for the sake of argument, that the evolution of single stars, and also of interacting binaries with circular orbits, can be modeled to a sufficient accuracy for inclusion in an N-body code. It will be necessary to modify the standard equations of motion

$$m_i \ddot{\mathbf{r}}_i = \mathbf{F}_i^{(Gr)} \equiv -\sum_{j \neq i} \frac{G m_i m_j \mathbf{r}_{ij}}{r_{ij}^3} \ , \quad \mathbf{r}_{ij} \equiv \mathbf{r}_i - \mathbf{r}_j \ , \quad (1)$$

for the following reasons. Firstly, we need a term for tidal friction (TF), which will ensure, in those binaries where at least one star grows large enough to interact with its companion, that the orbit is gradually circularised. Neither primordial binaries, nor those that are formed later by exchange reactions, by three-body interaction, or by tidal capture, will be circular to start with, but they will cicularise on a timescale that, empirically, appears to be fairly short compared with the nuclear timescale, provided that at least one component has a radius comparable to the separation of the components at periastron. Secondly, we need a term which operates during the mass-transfer phase (RLOF) that can be expected to occur not long after the circularisation of the orbit by TF. This term must conserve angular momentum in a binary which is undergoing mass *transfer* with no net mass *loss*. The second correction is necessary because equation (1), as applied just to a binary, but to a binary whose components possibly vary in mass, implies that the angular momentum *per unit reduced mass* is a constant; whereas in a binary undergoing conservative RLOF it is the *total* angular momentum that is constant. In other words, equation (1) with varying mass supposes that mass which leaves one star carries away the same specific momentum as the star it leaves, and if it arrives on the other star then it adds a *different* amount of momentum, to wit the specific momentum of the gainer rather than the donor.

I believe the following dynamical equation is probably the simplest that encompasses both requirements:

$$m_i(t)\ddot{\mathbf{r}}_i + \dot{m}_{iT}(\dot{\mathbf{r}}_i - \mathbf{V}) = \mathbf{F}_i^{(Gr)} + \mathbf{F}_i^{(TF)}, \quad i = 1, 2 \quad (2)$$

where

$$\mathbf{V} \equiv \frac{m_1 \dot{\mathbf{r}}_1 + m_2 \dot{\mathbf{r}}_2}{m_1 + m_2}, \quad (3)$$

$$\dot{m}_1 = \dot{m}_{1W} + \dot{m}_{1T}, \quad \dot{m}_2 = \dot{m}_{2W} + \dot{m}_{2T}, \quad \dot{m}_{2T} = -\dot{m}_{1T}, \quad (4)$$

suffix $W$ referring to wind which leaves the star and goes to infinity, and suffix $T$ referring to the mass which is transferred from one star to the other. Finally,

$$\mathbf{F}_i^{(TF)} \equiv -\sum_{i \neq j} \frac{\lambda_i R_i^8 + \lambda_j R_j^8}{r_{ij}^8} \frac{\mathbf{r}_{ij} \mathbf{r}_{ij} \cdot \dot{\mathbf{r}}_{ij}}{r_{ij}^2}, \quad (5)$$

where $R_i$ is the radius of the $i$-th star, and the $\lambda$'s are TF coefficients. The $\lambda$'s are of course dependent on the internal physics of the stars, but can probably be reasonably approximated on dimensional grounds as stellar mass divided by a timescale which is short compared with the thermal timescale of the star. This TF term has been incorporated into a 3-body code by Dr L. G. Kiseleva, and is found to give interesting and useful results in the context of triple stars as well as binaries. It is the simplest term we can think of that
(i) is radial, and antisymmetric in $i, j$, so that it conserves linear and angular momentum;
(ii) opposes the *radial* component of velocity, and so goes to zero as the orbit becomes circular;
(iii) crudely depends on the ratio of stellar to orbital radius in the way determined by Zahn (1977) from a detailed physical analysis of TF.
Note that the above TF term is not intended as an accurate representation of the very complicated physics that occurs during a tidal capture process

(Mardling, this Symposium), where possibly a great amount of kinetic energy is dissipated chaotically on a short timescale. It is only intended as a prescription for a gradual circularisation.

The mass transfer term in equation (2) has the property that if $\dot{m}_{1W} = 0 = \dot{m}_{2W}$, so that all the mass lost by one star is gained by the other, then *total* angular momentum is conserved, and we get $P^{-1} \propto m_1^3 m_2^3$, $a^{-1} \propto m_1^2 m_2^2$, where $P$ and $a$ are the period and semimajor axis, the latter being the constant radius of a circular orbit by virtue of the TF term in the previous evolution. These are the normal results of conservative RLOF. If, on the other hand, $\dot{m}_{1T} = 0 = \dot{m}_{2T}$, so that there is only wind from one or both stars and no RLOF, then as usual the angular momentum per unit reduced mass is constant and so we get $P^{-1} \propto (m_1 + m_2)^2$, $a^{-1} \propto m_1 + m_2$, $e =$ const., as expected; though we should note, for wide orbits, that these equations for the expansion of an orbit in response to ML assume that the ML does not vary significantly during an orbit. If there is instaneous mass ejection as in an SN explosion, then equation (2) will still apply and will give the right results for that limit. Of course, it is implicit in such modeling of ML that the mass is lost isotropically (in the frame of the mass-losing star) at sufficiently high speed that it does not accumulate significantly within even the cluster, let alone the binary. It is not obvious to me how equation (2) might be further generalised to take into account the possibility of AML by magnetic braking, for example, although such a generalisation would seem to be necessary.

Note that in equation (3) the velocity **V** is obtained by summing only over the two stars that are transferring mass, and that because the two masses are variable **V** is not, as one might have supposed, the velocity of the CG of the binary. We can of course generalise so that we can write these equations down, with a rather more elaborate notation than I care to use here, for all $i$ rather than just $i = 1, 2$. However since the TF force is a short-range one it should only be necessary to include the nearest neighbour, and perhaps in the case of some hierarchical triples the next nearest neighbour. Note also that in equation (2) the rate of mass transfer $\dot{m}_T$ can be considered as a fairly simple function of the ratio of stellar radius $R_1$ (if *1 is the loser) to the Roche lobe radius of *1, the latter being itself a fairly simple function of separation $a$ and mass ratio $m_1/m_2$. The rate is usually taken to be zero if the radius ratio is less than unity, and a rapidly increasing function if the ratio is greater than unity (Paczyński & Sienkiewicz 1972). However, we cannot just use the stellar radius obtained from our simple SE subroutine, because the mass transfer itself affects the radius if it is comparable to or faster than the thermal timescale of the star. It is possible, however, that this can be allowed for with a very simple extra differential equation for the degree of Roche-lobe overfill (Whyte &

Eggleton 1985).

To sum up, we see that (i) back-of-the-envelope formulae exist for bulk stellar (Pop I) properties as functions of $m, t$. They are being upgraded for state-of-the-art internal physics, and generalised to a range of metallicities; (ii) evolution of *interacting* binaries is still quite problematic; and (iii) to accommodate mass loss by winds and mass transfer by RLOF, the N-body dynamical equations should be modified slightly. This includes adding a tidal friction term in order to circularise the orbits of really close binaries.

I am indebted to Drs S. J. Aarseth, Zh. Han, L. G. Kiseleva, O. R. Pols, K.-P. Schröder, and C. A. Tout for many helpful conversations.

## References

Aarseth, S. J. (1995) in *Binaries in Clusters*, ed. Milone, E. F., ASP series, in press
Alexander, D. R. & Ferguson, J. W. (1994) ApJ, 437, 879
Andersen, J. (1991) A&A Rev., 3, 91
Bettweiser, E. & Sugimoto, D. (1984) MN, 208, 493
Böhm-Vitense, E. (1958) ZsAp, 46, 108
Conti, P. S. (1982) in *Mass Loss from Astronomical Objects*, ed Gondhalekar, P. M., RAL 82-075, p45
Eggleton, P. P., Fitchett M. J. & Tout C. A. (1989) ApJ, 347, 998
Eggleton, P. P. (1983a) MN, 204, 449
Eggleton, P. P. (1983b) in *The Origin and Evolution of Cataclysmic Binaries*, eds Livio, M. & Shaviv, G., p239
Eggleton, P. P. & Tout, C. A. (1989) Sp. Sc. Rev., 50, 165
Hachisu, I., Nakada, Y., Nomoto, K. & Sugimoto, D. (1978) Prog. Theor. Phys., 60, 393
Han, Zh., Podsiadlowski, P. & Eggleton, P. P. (1994) MN, 270, 121
Humphreys, R. M. & Davidson, K. (1979) ApJ, 232, 409
Judge, P. G. & Stencel, R. E. (1991) ApJ, 371, 357
Lynden-Bell, D. & Eggleton, P. P. (1980) MN, 191, 483
Merryfield, W. J. (1995) ApJ, 444, 318
Meynet, G., Mermilliod, J.-C. & Maeder, A. (1993) A&AS, 98, 477
Morgan, J. G. & Eggleton, P. P. (1978) MN, 182, 219
Paczyński, B. (1976) in IAU Symp. 73, *Structure and Evolution of Close Binary Systems*, eds Eggleton, P., Mitton, S. & Whelan, J., p75
Paczyński, B. & Sienkiewicz, R. (1972) AA, 22, 73
Paczyński, B. & Ziółkowski, J. (1968) Acta Astr., 18, 255
Pols, O. R., Tout, C. A., Eggleton, P. P. & Han, Zh. (1995) MN, 274, 964
Popper, D. M. (1980) ARAA, 18, 115
Rogers, F. J. & Iglesias, C. A. (1992) ApJS, 79, 507
Schaller, G., Schaerer, D., Meynet, G. & Maeder, A. (1992) A&AS, 96, 269
Schwarzschild, M. & Härm, R. (1958) ApJ, 128, 348
Spruit, H, (1992) A&A, 253, 131
Verbunt, F. & Zwaan, C. (1981) A&A, 100, L7
Weidemann, V., & Koester, D. (1983) A&A, 121, 77
Whyte, C. A. & Eggleton, P. P. (1985) MN, 214, 357
Willis, A. J. (1985) in *Interacting Binaries*, eds Eggleton, P. P. & Pringle, J. E. p103
Zahn, J.-P. (1977) A&A, 57, 383

# GRAVITATIONAL SCATTERING EXPERIMENTS

STEPHEN L. W. MCMILLAN
*Department of Physics, Drexel University,
Philadelphia, PA 19104*

**Abstract.** We describe a fully automated gravitational scattering package capable of determining cross sections and reaction rates for binary–single-star scattering, and present some applications to systems of astrophysical interest.

## 1. Introduction

Interactions between binaries and other stars play a central role in determining both the dynamical evolution and the appearance of a globular star cluster. Estimates of the primordial fraction of "hard" binaries (systems with binding energy exceeding the mean stellar kinetic energy, $\frac{3}{2}kT$) in globular clusters now range from ~3 to ~30 percent, with 10–20 percent regarded as fairly typical (see Hut et al. 1992). The mean time between significant interactions for a $1\,kT$ binary is comparable to the local two-body relaxation time, which may be only a few million years in the densest cluster cores. Since mass segregation causes heavier-than-average objects to drift into the high-density central regions on time scales ~1 Gyr, we expect binaries in globular clusters to interact frequently with their environment.

Encounters between hard binaries and other cluster members tend to release energy, leaving the binaries more tightly bound and thus heating the cluster (Heggie 1975). Numerical simulations of clusters containing primordial binaries have shown that, so long as any binaries remain in the dynamically "active" energy range (1–50 $kT$, say, corresponding to orbital semi-major axes of 0.2–10 A.U. for a cluster 3-D velocity dispersion of 10 km s$^{-1}$), binaries control the cluster dynamics, and continue to do so until they are all destroyed by interactions with other binaries, or recoil out of the cluster after a triple or four-body encounter (McMillan et al. 1990,

1991; Gao et al. 1991). For many clusters, this binary depletion time scale may well exceed the age of universe (McMillan & Hut 1994). In addition to their dynamical importance, binary interactions also greatly increase the probability of stellar collisions, particularly in low-density systems (Hut & Inagaki 1985; Sigurdsson & Phinney 1993). A population of 10% binaries with separations < 1 A.U. can increase the stellar collision rate by up to two orders of magnitude (Verbunt & Hut 1987).

While large-scale N-body or Monte-Carlo simulations are probably necessary to study in detail the interplay between the many different physical processes occurring in star clusters, the very comprehensiveness of these approaches makes it difficult to disentangle competing physical effects. Scattering experiments provide an alternative, and much more controlled, means of investigating binary interactions, allowing the investigator to isolate and study specific processes in a systematic manner. The most common way of reporting the outcome of scattering experiments is to quote cross sections for processes of interest. These can then be translated into reaction rates for use in a variety of applications, ranging from heating and interaction rates for Fokker-Planck and Monte-Carlo simulations, to estimates of collision rates and branching probabilities for the production of specific classes of object. The first systematic direct determination of cross sections and reaction rates for binary–single-star encounters was made by Hut & Bahcall (1983). Recent surveys have been carried out by Hills (1992), Sigurdsson & Phinney (1993), and Davies (1995).

In the simple case of point-mass (and often identical) stars, it is possible to publish "atlases" of encounter outcomes covering most of the parameter space of interest (see, for example, Hut 1984). However, where there is a distribution of stellar masses, or when stellar evolution and physical interactions between stars must be taken into account, there are simply too many parameter combinations for such atlases to be feasible. Instead, specific questions generally require individual calculations. Rather than having a standard format for reporting cross sections, it is much more useful to have a standard, reliable means of obtaining new results as needed. We describe here a software system for performing these calculations in a robust and efficient manner, automating the many error-prone steps involved in performing a series of scattering calculations.

## 2. Computation of Cross Sections

A scattering experiment entails integrating a large number of individual three-body encounters, holding some parameters fixed and choosing others randomly from specified distributions. In a typical case, the binary semi-major axis $a$ and component masses $m_1$ and $m_2$ are held fixed, the initial mean anomaly is chosen uniformly on $[0, 2\pi)$, while the eccentricity $e$ is either held fixed or chosen randomly from a thermal distribution: $f(e) = 2e$. For the outer orbit, the incomer mass $m_3$ and relative velocity at infinity $v_\infty$ are specified, the orbital orientation is chosen randomly, and the impact parameter $\rho$ is chosen uniformly in $\rho^2$ between $\rho = 0$ and $\rho = \rho_{max}$ (a parameter whose value will be discussed in more detail in a moment). Finally, the initial separation $R_3$ is chosen to keep the initial tidal perturbation below some tolerance $\gamma \ll 1$: $R_3 = a\gamma^{-1/3}[1 + m_3/(m_1 + m_2)]^{1/3}$.

If $v_\infty$ is less than the critical velocity $v_c$, defined by

$$v_c^2(m_1, m_2, m_3, a) \equiv \frac{Gm_1m_2(m_1 + m_2 + m_3)}{(m_1 + m_2)m_3 a}, \qquad (1)$$

then the total energy is negative and the scattering will eventually result in a binary and an unbound single star. In that case, the calculation is terminated when the perturbation on the final binary again falls below $\gamma$. If $v_\infty \geq v_c$, disruption of the initial binary into three unbound stars is also energetically possible. In that case, the calculation stops when all three stars are receding from one another, are mutually unbound, and none is a significant perturber of the relative motion of the other two.

Encounters are classified as resonances or non-resonances, depending on whether or not the quantity $|\mathbf{x}_1 - \mathbf{x}_2|^2 + |\mathbf{x}_2 - \mathbf{x}_3|^2 + |\mathbf{x}_3 - \mathbf{x}_1|^2$ has more than one minimum (see Hut 1993). The final state of the system is classified as a preservation, if the initial incomer escapes, an exchange, if one of the binary components escapes, or an ionization, if the binary is destroyed. When nonzero stellar radii are included, physical collisions between stars become possible, and additional final states must be defined.

For uniform sampling with $N$ trials between impact parameters $\rho = 0$ and $\rho = \rho_{max}$, the cross section for events of type X is simply

$$\sigma_X = \pi\rho_{max}^2 \frac{N_X}{N}, \qquad (2)$$

where $N_X$ is the number of times outcome X occurs. Differential cross sections are determined in an analogous way. The standard error in $\sigma_X$ is

$$\delta\sigma_X = \frac{\sigma_X}{\max(1, \sqrt{N_X})}. \qquad (3)$$

In practice, we have found it advantageous to incorporate the determination of $\rho_{max}$ into the actual calculation of cross sections (rather than relying on inspection of pilot calculations), as follows. We start by performing $n$ scatterings (where the "trial density" $n$ is a parameter controlling the overall accuracy of the final results), uniformly distributed in impact parameter over the range $0 \leq \rho < \rho_0 = 2a[1 + G(m_1 + m_2 + m_3)/av_\infty^2]^{1/2}$. The value of $\rho_0$ simply corresponds to a periastron separation of $2a$. We then systematically expand the impact-parameter range, covering successive annuli of outer radii $\rho_i = 2^{i/2}\rho_0$ with $n$ trials each, until no "interesting" interactions take place in the outermost zone ($i = i_{max}$) sampled. Typically, an interesting interaction is one in which the binary is significantly perturbed in energy or eccentricity, although the precise definition is easily tailored to the particular application at hand.

This procedure produces rapid convergence toward accurate cross sections, with a minimum of wasted effort. With this non-uniform sampling, expressions (2) and (3) become:

$$\sigma_X = \frac{\pi}{n} \sum_{i=0}^{i_{max}} (\rho_i^2 - \rho_{i-1}^2) N_{Xi}, \tag{4}$$

$$(\delta\sigma_X)^2 = \frac{\pi^2}{n^2} \sum_{i=0}^{i_{max}} (\rho_i^2 - \rho_{i-1}^2)^2 N_{Xi}, \tag{5}$$

where $N_{Xi}$ is the number of times outcome X occurs in zone $i$, and $\rho_{-1} = 0$.

## 3. The STARLAB Software Environment

The scattering programs described here operate within a programming environment known as STARLAB, a collection of modular software tools for simulating the evolution of dense stellar systems—star clusters and galactic nuclei—and analyzing the resultant data. It consists of a library of loosely coupled programs, sharing a common data structure, which can be combined in arbitrarily complex ways. Individual modules may be linked in the "traditional" way, as function calls to C++ (the language in which most of the package is written), C, or FORTRAN routines, or at a much higher level—as individual programs connected by UNIX pipes. The former linkage is more efficient, and allows finer control of the package's capabilities; however, the latter provides a quick and compact way of running test simulations. The combination affords great flexibility to STARLAB, allowing it to be used by both the novice and the expert programmer with equal ease. Its structure is described in more detail by McMillan (1996) and McMillan & Hut (1996).

The automated scattering software in STARLAB is constructed in several layers, each largely independent of the others. The system is designed to function with only high-level user input, but each layer remains accessible if necessary, and diagnostic data at each level can be obtained as desired, without the need for rewriting and recompiling existing sections of code.

The lowest level of the scattering subsystem consists of an orbit integration engine based on a fourth-order, variable-time-step Hermite integrator

(Makino 1992). The STARLAB scattering integrators augment this algorithm with a novel technique developed by Hut et al. (1995), which guarantees time symmetry in the integration and hence enforces exact energy and momentum conservation in periodic orbits. This results in spectacular improvements in the long-term stability of the integration scheme in all circumstances, even in the case of long-lived resonances.

The next software level consists of several scattering-specific layers. These include: (1) routines to create an initial scattering state, holding some parameters fixed while choosing others randomly, as discussed in §2; (2) checks to determine whether a given scattering experiment has reached its final outgoing state; (3) optimization features, such as analytical integration of inner and outer orbits of hierarchical triple systems in which the outer orbital period greatly exceeds the inner orbital period, and unperturbed two-body motion near the pericenter of a close encounter; (4) diagnostic functions to store information describing the build-up of energy errors; (5) bookkeeping functions to chart the overall character of the orbits (e.g. "democratic" versus "hierarchical" resonances); (6) checks for stellar collisions, and routines to implement mergers as needed.

Above these layers lies the three-body scattering manager, which oversees the initialization, integration, and classification of a single three-body interaction. The masses and radii of the stars can be specified, as can the orbital parameters of the binary, the impact parameter of the encounter, and the incomer's relative velocity at infinity. The initial distance from which the integration starts is determined automatically, with a default $\gamma$ of $10^{-6}$. Our production runs usually achieve median relative energy errors on the order of $10^{-6}$; however, these could easily be kept as small as $10^{-10}$ if desired.

The next layer contains all the management software needed to conduct a series of scatterings, including the automatic feedback system described earlier to ensure a near-optimal choice of maximum impact parameter. The basic output at this level is a table of total cross sections and errors for all possible outcomes. However, the structure of the package is such that a more ambitious user can use the software as a reliable means of providing a correctly sampled environment in which more complex calculations may be performed, with proper statistical weights attached to each scattering event.

The present highest-level layer is a Maxwellian rate estimator, which uses repeated cross-section calculations to perform a Maxwellian average over velocity. With the complexity of orbit integration and scattering management hidden in the lower-level modules, it is relatively simple to implement new levels, and additional layers can be added with little additional investment in time.

## 4. Sample Applications of the Scattering Software

### 4.1. DETERMINATION OF COLLISION RATES

The most basic application of the STARLAB scattering package is the generation of total cross sections for specific processes of astrophysical interest. Here we present cross sections for physical collisions and non-colliding exchanges and resonances during encounters between main-sequence binaries and incoming stars, for parameters typical of a globular cluster core.

The binary components are taken to have masses $m_1 = 0.8 M_\odot$ and $m_2 = 0.4 M_\odot$, and radii $R_1 = 0.8 R_\odot$ and $R_2 = 0.4 R_\odot$, respectively. The initial binary orbital eccentricity is randomly chosen from a thermal distribution, with the proviso that the separation at periastron is at least twice the sum of the stellar radii, so that immediate collisions are avoided. The incomer is taken to be an intermediate-mass main-sequence star, of mass $m_3 = 0.6 M_\odot$, radius $R_3 = 0.6 R_\odot$, and velocity at infinity $10 \, \mathrm{km \, s^{-1}}$. The results reported here are intended mainly to illustrate the capabilities of the software; they may be compared with similar calculations presented elsewhere in the literature, most recently by Davies (1995).

The outcome of a collision between two main-sequence stars is fairly well known: if the two stars approach within roughly the sum of their radii, they merge to form a single object of approximately double the original radius, with negligible mass loss (see, e.g., Benz & Hills 1987; Lombardi et al. 1995). In order to determine whether or not a second merger occurs (if the first occurs with the third "spectator" star bound to the center of mass of the colliding pair), we assume the simple mass–radius relation $R \propto m$.

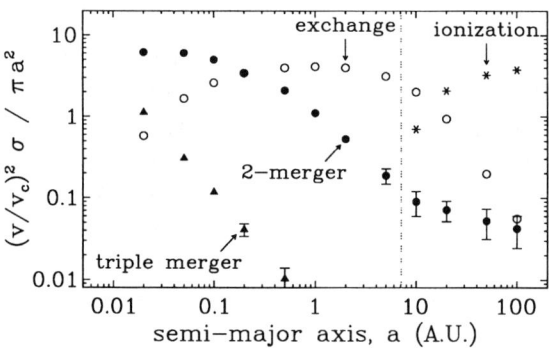

**Fig. 1:** Scaled cross sections for "clean" exchange (open circles), ionization (stars), two- and three-star mergers (filled circles, filled triangles), for the particular interaction described in the text.

The possible results of an encounter then are: (1) a non-colliding (or "clean") exchange, preservation, or ionization; (2) a two-body merger, leaving a 1.0, 1.2, or $1.4 M_\odot$ blue straggler, which may itself be part of a stable binary system; (3) a triple merger, forming a $1.8 M_\odot$ blue straggler. Figure 1 shows the cross sections for these processes (excluding preservations, whose cross section is obviously infinite), for binary semi-major axes ranging from 0.02 to 100 A.U. Ionization occurs only for semi-major axes such that $v_c(0.8M_\odot, 0.4M_\odot, 0.6M_\odot, a) < 10\,\mathrm{km\,s^{-1}}$, or $a > 7.1$ A.U., as indicated by the dashed line. In this figure, each set of normalized cross sections, along with error estimates, is generated by a single invocation of the cross section calculator described earlier. Here and in Figure 2 below, error bars are shown only where they exceed the size of the symbols used.

The suppression of clean exchanges (whose scaled cross section should be roughly constant in the point-mass limit) by collisions during close encounters is clearly evident in the figure. For the adopted set of initial parameters, collisions dominate over "clean" encounters for $a \lesssim 0.2\,\mathrm{A.U.}$, corresponding to initial binary periods of $\lesssim 30$ days. When the factor of $a$ implicit in the $a^2 v_c^2$ scaling is taken into account, the 2-body collision cross section is found to be roughly independent of $a$, at $\sim 20$ A.U.$^2$ for $0.2$ A.U. $\lesssim a \lesssim 10$ A.U. Within this range, the merger ("blue straggler formation") rate per binary in a cluster core of density $10^4\, n_4$ stars pc$^{-3}$ is $\sim 0.05\, n_4$ Gyr$^{-1}$. The constancy of the 2-body merger cross section is easily understood as the combination of two factors: the overall binary interaction cross section scales as $a$ because of gravitational focusing, while the probability of a collision during the course of an interaction scales roughly as $\langle R_* \rangle / a$.

The break in the slope of the "2-merger" cross section at $a \sim 10\,\mathrm{A.U.}$ is the result of our particular choice of initial binary eccentricities, which always permitted "almost colliding" systems, regardless of $a$. Collisions with $a \gtrsim 10$ A.U. are mainly "induced mergers," in which the components of a very eccentric binary are perturbed onto a collision course by the passage of the third star. In part because of these induced mergers, collisions between the original binary components tend to dominate; other merger events occur at significant rates only in democratic resonances, and for small $a$.

Figure 2 shows the fraction of two-body mergers resulting in an unbound final system (an isolated blue straggler). Only induced mergers are likely to be unbound; resonant mergers almost always lead to a bound final system. Triple collisions account for only a negligible fraction ($\lesssim 5\%$) of the total, except for $a = 0.02$ A.U., where they represent about 15% of all mergers.

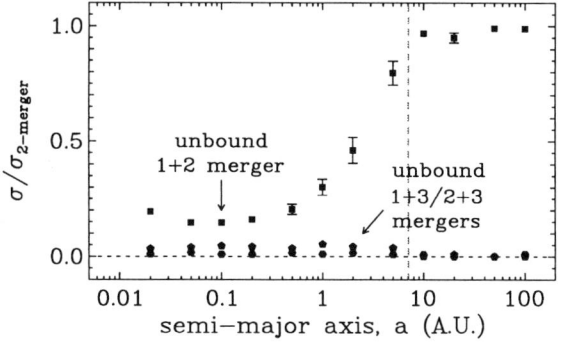

Fig. 2: Branching ratios within the class of two-body mergers (see Fig. 1) for the formation of unbound merger products. Only "1+2" mergers contribute significantly to the total.

## 4.2. FORMATION OF THE TRIPLE SYSTEM B1620-26

A rather different application of the STARLAB scattering package is described by Rasio, McMillan, & Hut (1995), who have used the Maxwellian rate estimator to study the "inverse scattering problem" of determining what initial conditions, if any, could have led to the formation of the binary millisecond pulsar (BMP) system B1620-26 in the globular cluster M4. B1620-26 has a low-mass companion (most likely a white dwarf, of mass $m_2 \approx 0.3\,M_\odot$ for a pulsar mass $m_1 = 1.35\,M_\odot$) in a nearly circular orbit of period $P_1 = 0.524$ yr (Lyne et al. 1988; McKenna & Lyne 1988), corresponding to a separation $a_1 \approx 0.8$ AU. The pulsar timing data indicate the presence of a second, more distant orbital companion (see Backer et al. 1993; Thorsett et al. 1993), with mass in the range $m_3 \sim 10^{-3}$–$1\,M_\odot$, with a corresponding semi-major axis $a_2 \sim 10$–$100$ AU for the outer orbit (Michel 1994).

We propose that the present triple system could have formed by exchange of an existing BMP into a wide primordial binary. The motivation for invoking such an explanation is as follows. If the second pulsar companion is indeed a main-sequence star of mass $\sim 0.5\,M_\odot$, then the eccentricity of the outer orbit must be large ($e_2 > 0.5$) and its semi-major axis cannot be much smaller than $\sim 50$–$100$ A.U., to be consistent with the pulsar timing data (Rasio 1994). Such a wide orbit is easily disrupted at its location near the core of M4: we estimate its lifetime to be $\sim 10^7$–$10^8$ yr, much shorter than the age of the BMP, which is most likely $\gtrsim 10^9$ yr (see Thorsett et al. 1993). Thus the pulsar could not have formed inside the triple system. Instead, the BMP must have become a member of the triple by exchange after its formation.

This scenario is supported by the tentative identification by Bailyn et al. (1994) of an optical counterpart to the triple companion, with mass $\sim 0.5\,M_\odot$. For the most likely parameters of the outer orbit, the perturbation on the inner binary due to the triple distant companion can also account for the binary's unexpectedly large eccentricity of 0.025.

We thus wish to determine which primordial binary parameters could have led to an exchange interaction that neither disrupted the wide primordial binary, nor perturbed the BMP to an eccentricity greater than actually observed. Since the eccentricity perturbation decreases sharply with increasing separation, this in effect means that no star approached within $\sim 3a_1$ of the BMP during the encounter. In the terminology of the previous section, we seek a clean exchange, with the BMP treated as a single particle of radius $3a_1$. We find that such an exchange is in fact the most likely outcome of the interaction for primordial binary semi-major axes $a_{MS} \sim 10$–$30$ A.U. and binary component masses $\sim 0.3$–$0.8\,M_\odot$.

Figure 3 shows the joint distribution of $a_2$ and $e_2$ for one particular choice of primordial binary parameters ($a_{MS} = 13$ A.U., each component mass $0.5\,M_\odot$). Clearly, the proposed mechanism is capable of forming triples with the desired characteristics: a nearly circular inner orbit (that remains unperturbed after the interaction) and a very eccentric outer orbit with $a_2 \sim 10^2$ A.U. The derived rate for the formation of a triple similar to B1620-26 is also consistent with the observation of one such object in the entire Galactic globular cluster system.

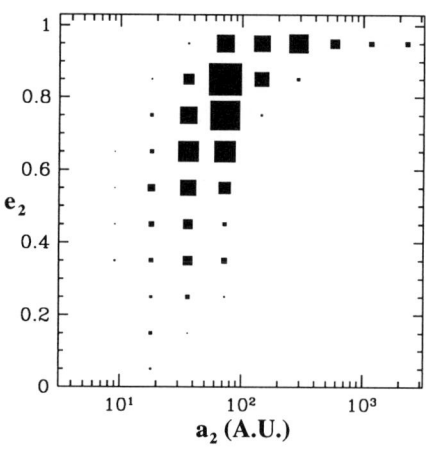

**Fig. 3:** Distribution of the semi-major axis $a_2$ and eccentricity $e_2$ of the triple's outer orbit. The size of each square dot is proportional to the differential cross-section for producing a triple system with those specific orbital parameters.

## 4.3. EXCHANGE CROSS SECTIONS FOR ARBITRARY MASSES

Finally, Heggie, Hut, & McMillan (1996) have used the STARLAB software in fully automated mode as the basis for an extensive calculation of exchange cross sections. In the limit of very hard binaries, $v_\infty \ll v_c$, the overall scaling of the exchange cross section can be shown to be

$$\sigma \sim \sigma_0 = \frac{GM_{123}a}{v_\infty^2} \left(\frac{M_{23}}{M_{123}}\right)^{1/6} \left(\frac{m_3}{M_{13}}\right)^{7/2} \left(\frac{M_{123}}{M_{12}}\right)^{1/3} \left(\frac{M_{13}}{M_{123}}\right), \qquad (6)$$

where $M_{12} = m_1 + m_2$, etc., $M_{123} = m_1 + m_2 + m_3$, and we have adopted the convention that binary component 1 is ejected following the encounter. The above expression is asymptotically valid in *all* regimes where one star's mass dominates the total, or is negligible.

By using STARLAB to determine the exchange cross sections on a grid of points in ($0 \le \mu_1 \equiv m_1/M_{12} \le 1$, $0 \le \mu_2 \equiv m_3/M_{123} \le 1$), then fitting the results to an expression of the form $\sigma = \sigma_0 f(\mu_1, \mu_2)$, we have obtained a fitting formula valid for arbitrary masses, and accurate to better than $\sim 20\%$ over most of parameter space of interest. For more details, and for the final fitting formula, see Heggie et al. (1996).

## 5. Outlook

The design of STARLAB facilitates inclusion of more detailed physical processes into our models, and this represents one obvious direction of future development of the system. In addition, binary–binary and general $N$-body scattering packages are nearing completion. While more complex, their conceptual framework is similar to that described here for the 3-body case. The entire package is available by anonymous FTP from ftp.sns.ias.edu/pub/starlab, or via http://www.sns.ias.edu/~starlab. STARLAB has been successfully installed on UNIX systems running SunOS 4, Solaris 2, HP-UX, Linux, and Dec OSF, using both native C++ compilers and the GNU g++ compiler. Real-time demonstrations of the software are available through the World-Wide Web site listed above.

This work was supported in part by NASA grant NAGW-2559 and NSF grant AST-9308005.

## References

Backer, D.C., Foster, R.S., & Sallmen, S. 1993, Nature, 365, 81
Bailyn, C.D., Rubenstein, E.P., Girard, T.M., Dinescu, D., Rasio, F.A., & Yanny, B. 1994, ApJL, 433, L89
Benz, W., & Hills, J.G. 1987, ApJ, 323, 614
Davies, M.B. 1995, MNRAS, 276, 887
Gao, B., Goodman, J.G., Cohn, H.N., & Murphy, B. 1991, ApJ, 370, 567
Heggie, D.C. 1975, MNRAS, 173, 729
Heggie, D.C., McMillan, S.L.W., & Hut, P. 1996, these proceedings; submitted to ApJ
Hills, J.G. 1992, AJ, 103, 1955
Hut, P. 1984, ApJS, 55, 301
Hut, P. 1993, ApJ, 403, 256
Hut, P. & Bahcall, J. N. 1983, ApJ, 268, 319
Hut, P., & Inagaki, S. 1985, ApJ, 298, 502
Hut, P., McMillan, S.L.W., Goodman, J.G., Mateo, M., Phinney, E.S., Pryor, C., Richer, H.B., Verbunt, F., & Weinberg, M. 1992, PASP, 105, 981
Hut, P., Makino, J., & McMillan, S.L.W. 1995, ApJL, 443, L93
Lombardi, J.C., Rasio, F.A., & Shapiro, S.L. 1995, preprint
Lyne, A., Biggs, J., Brinklow, A., Ashworth, M., & McKenna, J. 1988, Nature, 332, 45
Makino, J. 1992, ApJ, 369, 200
McMillan, S.L.W. 1996, in Binaries in Star Clusters, ed. G. Milone (San Francisco: ASP)
McMillan, S.L.W., Hut, P., & Makino, J. 1990, ApJ, 362, 522
McMillan, S.L.W., Hut, P., & Makino, J. 1991, ApJ, 372, 111
McMillan, S.L.W., & Hut, P. 1994, ApJ, 427, 793
McMillan, S.L.W., & Hut, P. 1996, submitted to ApJ
McKenna, J., & Lyne, A.G. 1988, Nature, 336, 226; erratum, 336, 698
Michel, F.C. 1994, in Millisecond Pulsars: The Decade of Surprise, ed. Fruchter et al. (San Francisco: ASP)
Rasio, F.A. 1994, ApJL, 427, L107
Rasio, F.A., McMillan, S.L.W., & Hut, P. 1995, ApJL, 438, L33
Sigurdsson, S., & Phinney, E.S. 1993, ApJ, 415, 631
Thorsett, S.E., Arzoumanian, Z., & Taylor, J.H. 1993, ApJL, 412, L33
Verbunt, F., & Hut, P. 1987, in The Origin and Evolution of Neutron Stars, eds. D. J. Helfand and J. H. Huang (Dordrecht: Reidel), p. 187

# LIVES OF HIERARCHICAL TRIPLE SYSTEMS IN CLUSTERS AND IN THE FIELD

LUDMILA G. KISELEVA
*Institute of Astronomy,*
*Madingley Road, Cambridge CB3 0HA, England*

**Abstract.** Evolution with time of isolated hierarchical triple stars, stable from a dynamical point of view, is studied using analytical and numerical approaches. The results are applied to the study of the evolutionary cycle of hierarchical systems formed by dynamical capture in open clusters (up to 10,000 stars).

## 1. Introduction

Stars show a marked tendency to form systems of different multiplicity, starting from the smallest systems, binary and triple stars, up to globular clusters with $N \sim 10^7$. Different investigators have used different methods for the identification of multiple stars, and have arrived at somewhat different conclusions, but modern observations give a frequency of binary and multiple stars in the Galactic field of up to 70% (Gliese & Jahreiss 1988, Batten, Fletcher & McCarthy 1989, Duquennoy & Mayor 1991), and between 5 and 15% of these systems are at least triple. Batten, Fletcher & McCarthy (1989) claimed that about 20% of binaries in their sample can be at least triple. Among the 50 nearest stars (G/K/M dwarfs), from van de Kamp (1971) and Henry & McCarthy (1990), are found 33 single, 13 binary and 4 triple stars. Duquennoy & Mayor (1991) gives the following ratio of single : double : triple : quadruple systems among the 164 nearest G-dwarf stars: 1.5(91 systems) : 1(62) : 0.105(7) : 0.026(2). They also pointed out that the number of triple and quadruple systems may be larger.

A particularly large fraction of triple and quadruple systems can be observed among pre-main-sequence stars in star-forming regions. For example, Ghez et al. (1993) found that triples and quadruples comprise 14%

of their sample for the Tau-Aur association. They estimate that the real frequency (taking into account the incomplete period coverage in their sample) may reach $\sim 35\%$. Simon et al. (1995) identified at least 10 binaries, two triples and one quadruple among 35 young star targets in the Ophiuchus star-forming region, and 22 binaries and 4 triples among 47 systems in the Taurus region. Of course, some fraction of these systems may be unstable, and we observe them at the stage of the distant ejection of one companion.

The majority of observed triple and higher multiplicity systems are hierarchical, i.e. a close binary has a distant component, which may also be a binary. Another possibility is that a binary with a distant component has another even more distant component still. The majority of non-hierarchical triples (except a very few special cases like, for example, the Eulerian or Lagrangian configurations) are dynamically unstable, i.e. they eventually (usually, within several crossing times) disintegrate into a bounded binary and a detached single body which can escape to infinity: for a detailed discussion, see reviews by Anosova 1986, Heggie 1988, Anosova & Orlov 1994.

Even for hierarchical triple-star systems, stability is not an easy question. There are a number of criteria to identify triple systems as stable or unstable, obtained analytically (e.g. Golubev 1967, 1968, Zare 1977, Szebehely & Zare 1977, Marchal & Bozis 1982) or numerically (e.g. Harrington 1975, 1977, Graziani & Black 1981, Donnison & Mikulskis 1992, 1995, Eggleton & Kiseleva 1995a; hereinafter EK). These criteria differ from each other, sometimes rather significantly. Part of the reason for this is that 'stability' is a difficult concept to define and authors often use different stability definitions. In this paper we adopt the stability criterion of EK and also their definition of stability: that a hierarchical triple system is stable if it persists continuously for a very long time in the *same* hierarchical configuration (which excludes exchange as well as disintegration).

Known triple systems are not so numerous in open clusters as in the field, but the statistics are increasing due to the improvement of observational techniques, and to the systematic surveys undertaken at several observatories within the last few years. There is thus growing evidence for the existence of triple and even quadruple systems in open clusters, with a variety of characteristics. These systems are usually highly hierarchical. Triple (or even higher multiplicity) systems are found in the Pleiades (Mermilliod et al. 1992), the Hyades (Griffin & Gunn 1981, Griffin et al. 1985, Mason et al. 1993), Praesepe (Mermilliod et al. 1994), M67 (Mathieu et al. 1990), and NGC 1502 (Mayer et al. 1994). The system in NGC 1502 contains an eclipsing massive binary SZ Cam ($m_1 = 13.7 M_\odot$, $m_2 = 9.7 M_\odot$, $P = 2.7d$) which is the brightest member of the cluster. The variability of the orbital period of this binary has been known for some time, but only recently new high-dispersion spectra (Mayer et al. 1994) have allowed the third body to be identified. Because of its large mass (minimum $18.6 M_\odot$) and because of observed shifts in the third-body lines, this 'third body' can possibly be a binary, and the system as a whole may be a hierarchical quadruple system with $P_{out} = 50.7y$, $e = 0.77$.

Mermilliod et al. (1994) have summarised the data for 11 main-sequence triple systems known so far in open clusters, in which one component is a spectroscopic binary. Four of these systems contain a very close binary ($P_{in} \in (2.4, 4.0)$d). Only 3 out of 11 outer orbital periods are known, and the least hierarchical system (vB 124 in the Hyades) has a period ratio $X = P_{out}/P_{in} \approx 250$.

Only one hierarchical triple system has been detected so far in globular clusters, but this is surely only the first step to the discovery of others which are likely to be present. This famous system in M4 contains the millisecond pulsar PSR B1620-26 (Backer et al. 1993, Thorsett et al. 1993; see also Rasio et al. 1995 and Hut 1995 for a discussion).

The above data indicate the importance of the numerical and analytical study of the formation and evolution of hierarchical systems in the Galactic field and in star clusters, which we discuss in the present paper. We particularly concentrate on hierarchical systems stable from the point of view of their internal dynamical evolution.

## 2. Isolated stable hierarchical triple

Dynamical stability requires that the ratio $X_0$ of outer period to inner period must be larger than a factor of $\simeq 3 - 6$, if both orbits are nearly circular and all three bodies are of comparable mass: Kiseleva, Eggleton & Anosova 1994, Kiseleva, Eggleton & Orlov 1994. More generally, for eccentric orbits, the outer periastron must be larger than inner apastron by a factor of $\simeq 2 - 16$ (EK; see also Harrington 1975). Table 1 gives, for a wide range of mass ratios, the minimum $Y_0$ necessary for stability in coplaner prograde orbits, where $Y_0 \equiv R_{peri}^{out}/R_{ap}^{in}$. This minimum $Y_0$ is not very sensitive to the two eccentricities, and so is not different by more than $\sim 20\%$ from the value for two circular orbits, which is what is shown in the Table. In Table 1, $\alpha = \log_{10} \frac{m_1}{m_2} \geq 0 S$, $\beta = \log_{10} \frac{m_1+m_2}{m_3}$, where $m_1$, $m_2$ are the masses of the components of the close binary and $m_3$ is the mass of the third body. The period ratio $X_0$ and the above distance ratio $Y_0$ are of course closely related by Kepler's law, given the two eccentricities and the two mass ratios.

Table 1 shows that for all systems with a massive distant component, i.e. $m_3 > m_1 + m_2$, $Y_0^{min} > 4$; and that for systems with $m_1 \approx m_2 \approx m_3$,

TABLE 1. Minimum initial ratio $Y_0^{min}$ of semi-major axes $a_{out}/a_{in}$ for stability of triple systems with initially doubly-circular orbits.

| $\alpha$ | 0.0 | 0.2 | 0.4 | 0.6 | 0.8 | 1.0 | 1.2 | 1.4 | 1.6 | 1.8 | 2.0 |
|---|---|---|---|---|---|---|---|---|---|---|---|
| $m_1/m_{12}$ | .50 | .39 | .28 | .20 | .14 | .09 | .06 | .04 | .025 | .016 | .01 |
| $\beta(m_3/m_{12})$ | | | | | | | | | | | |
| -2.0(100) | 15.53 | 15.62 | 15.85 | 15.90 | 16.00 | 16.05 | 16.14 | 16.15 | 16.14 | 16.17 | 16.20 |
| -1.8(63) | 13.35 | 13.49 | 13.65 | 13.77 | 13.82 | 13.87 | 13.91 | 13.94 | 13.98 | 13.98 | 14.00 |
| -1.6(40) | 11.48 | 11.64 | 11.74 | 11.88 | 11.99 | 12.04 | 12.03 | 12.10 | 12.11 | 12.11 | 12.10 |
| -1.4(25) | 9.87 | 10.01 | 10.18 | 10.29 | 10.34 | 10.42 | 10.44 | 10.44 | 10.46 | 10.48 | 10.51 |
| -1.2(16) | 8.57 | 8.65 | 8.81 | 8.86 | 8.95 | 8.99 | 9.00 | 9.05 | 9.07 | 9.07 | 9.08 |
| -1.0(10) | 7.34 | 7.42 | 7.55 | 7.66 | 7.72 | 7.76 | 7.82 | 7.83 | 7.86 | 7.86 | 7.87 |
| -0.8(6.3) | 6.31 | 6.41 | 6.51 | 6.61 | 6.67 | 6.72 | 6.76 | 6.79 | 6.79 | 6.79 | 6.81 |
| -0.6(4.0) | 5.39 | 5.51 | 5.61 | 5.70 | 5.76 | 5.82 | 5.85 | 5.87 | 5.89 | 5.90 | 5.91 |
| -0.4(2.5) | 4.61 | 4.72 | 4.82 | 4.91 | 4.97 | 5.02 | 5.06 | 5.08 | 5.09 | 5.10 | 5.10 |
| -0.2(1.6) | 3.95 | 4.03 | 4.11 | 4.20 | 4.27 | 4.31 | 4.34 | 4.36 | 4.37 | 4.38 | 4.38 |
| 0.0(1.0) | 3.37 | 3.46 | 3.47 | 3.60 | 3.63 | 3.67 | 3.70 | 3.72 | 3.73 | 3.74 | 3.74 |
| 0.2(0.63) | 3.11 | 3.12 | 3.10 | 3.22 | 3.04 | 3.05 | 3.09 | 3.12 | 3.14 | 3.15 | 3.16 |
| 0.4(.40) | 2.99 | 2.99 | 2.97 | 2.93 | 2.94 | 2.87 | 2.85 | 2.83 | 2.65 | 2.58 | 2.60 |
| 0.6(.25) | 2.88 | 2.87 | 2.87 | 2.84 | 2.81 | 2.80 | 2.73 | 2.72 | 2.71 | 2.70 | 2.69 |
| 0.8(.16) | 2.81 | 2.81 | 2.79 | 2.76 | 2.73 | 2.67 | 2.65 | 2.63 | 2.61 | 2.63 | 2.59 |
| 1.0(.10) | 2.74 | 2.74 | 2.72 | 2.70 | 2.36 | 2.33 | 2.29 | 2.26 | 2.22 | 2.23 | 2.21 |
| 1.2(.063) | 2.70 | 2.69 | 2.68 | 2.34 | 2.30 | 2.26 | 2.22 | 2.18 | 2.17 | 2.15 | 2.13 |
| 1.4(.040) | 2.67 | 2.35 | 2.33 | 2.29 | 2.25 | 2.20 | 2.18 | 2.14 | 2.07 | 2.05 | 2.04 |
| 1.6(.025) | 2.32 | 2.32 | 2.29 | 2.25 | 2.22 | 2.18 | 2.12 | 2.09 | 2.05 | 1.84 | 1.82 |
| 1.8(.016) | 2.31 | 2.31 | 2.27 | 2.22 | 2.19 | 2.15 | 2.10 | 2.08 | 2.03 | 1.83 | 1.63 |
| 2.0(.010) | 2.29 | 2.28 | 2.26 | 2.21 | 2.17 | 2.14 | 2.10 | 2.05 | 1.84 | 1.63 | 1.60 |

$3 \leq Y_0^{min} \leq 4$. Thus a strongly hierarchical structure is required for dynamical stability in these cases. In this work we consider only triple systems with $Y_0$ significantly above $Y_0^{min}$.

In dynamically stable hierarchical triples the distant component always pumps an eccentricity into the inner binary on a time scale shorter than the orbital period of the binary. In its turn, the binary also pumps an eccentricity into the outer orbit, although the inner orbit seems to be more sensitive to perturbations (Figs 1 and 2).

Fig. 1 shows that average values over time of the inner and outer eccentricities $\bar{e}_{in}$ and $\bar{e}_{out}$ increase rather smoothly as the period ratio $X_0$ decreases (but still does not approach too closely the critical value $X_0^{min}$ for stability). However, such smooth behavior of $\bar{e}_{out}(X_0)$ and $\bar{e}_{in}(X_0)$ is not universal. Kiseleva, Eggleton & Anosova (1994) found 'resonances' for some $(\alpha, \beta)$-pairs, i.e. $\bar{e}_{out}$ or $\bar{e}_{in}$ can rapidly increase and then decrease again in a narrow range of $X_0$. For yet other $(\alpha, \beta)$-pairs the resonance is 'disrup-

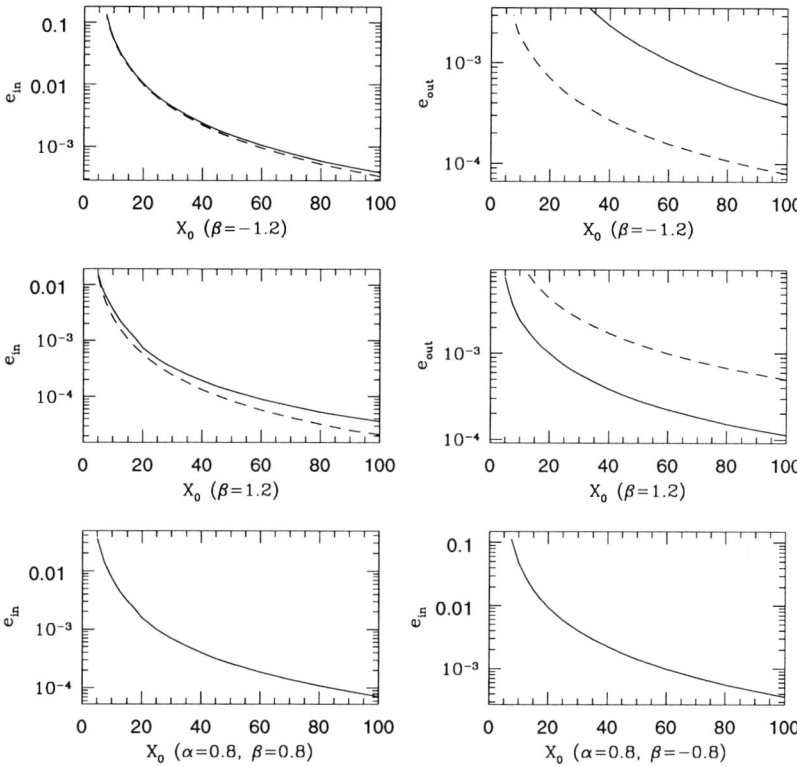

*Figure 1.* The mean values over time of the inner and outer eccentricities in hierarchical triples with coplanar initially circular orbits, as functions of the initial period ratio $X_0$. In upper and middle panels dashed lines show results for equal mass inner binaries (i.e. $\alpha = 0$) and solid lines for binaries with a rather significant mass ratio $\frac{m_1}{m_2} \approx 16$ (i.e. $\alpha = 1.2$).

tive': for some rather narrow range of $X_0$ the system breaks up, typically by ejection of a distant component, even though the system appears to be stable over very long time intervals at smaller as well as larger values of $X_0$; although of course the system will eventually break up at some smaller $X_0$ still. Fig. 1 shows that for $X_0 \geq 20$ the initial orbital elements are preserved in both binaries to better (usually much better) than 1%, so that inner and outer subsystems can be considered as rather unaffected by each other. However, Heggie (1996) has shown analytically that for non-coplanar triple systems there is a secular periodic change of the inner eccentricity which must over a sufficiently long time (which depends on both inner and outer periods, initial eccentricities, and masses) reach its maximum value $e_{\text{in}}^{\max}$. This value does not depend on any parameters of the triple system

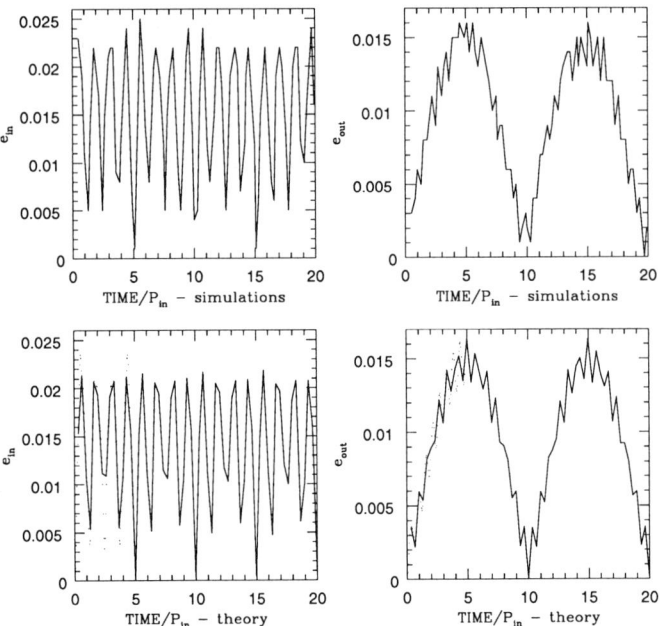

*Figure 2.* The fluctuations of inner and outer eccentricity with time for a system with $m_1=m_2=m_3$ and $X_0=10$. The upper panel shows results of numerical 3-body simulations and the lower panel gives data produced using Heggie's analytical approach (see text).

except the relative inclination $i$ between two orbital planes (other parameters affect only the time scale) If $i \approx 90°$ $e_{in}^{max} \approx 1$ and the two stars may collide or at least have a very strong tidal interaction. This affect cannot be neglected in the numerical study of triple stars in clusters which we will discuss in the next Section. Fig. 1 also show that the changes of the orbital characteristics of the inner binary in practice do not depend very much on the mass ratio $\alpha$ of the components of this binary. However, $e_{out}$ depends on $\alpha$ rather strongly.

Fig. 2 presents the behaviour in time of instantaneous $e_{in}$ and $e_{out}$ for a stable ($X_0 = 10$) triple with three components of equal masses for numerical simulations (upper panel) and the analytical approximation by Heggie (private communication) based on 1st order perturbation theory. Fig. 2 shows a resonably good aggreement between these two approaches, especially for $e_{out}$. Although the fine details of both functions differ, the amplitudes $e_{out}^{max}$ and periods of fluctuations are nearly the same. For $e_{in}$ the difference between the two approaches is more significant. The Bulirsch-Stoer integration procedure used in numerical simulations does not allow us to have more than 4-5 outputs per inner orbit, and so for uniformity we used the same sampling for the analytical curves; however the small dots on the lower panel correspond to 20 outputs per inner orbit.

## 3. Hierarchical systems in open clusters

A few numerical simulations for clusters of 500 - 10,000 stars with different fractions of primordial binaries were performed, using the N-body code NBODY4 (Aarseth 1996) on HARP in IoA. The procedure for identification and observation of dynamically formed hierarchical systems in clusters is described in Kiseleva et al. (1995): a newly formed hierarchical system (triple or quadruple in the sense that the distant component can itself be a binary) is recorded if it satisfies the EK stability criterion and at the same time the distant star forms a hard binary with respect to the centre of mass of the inner binary. Usually there are more triples than quadruples; however in some runs the fraction of quadruples may be up to 45%. Hierarchies can be destroyed because of perturbing effects of the remaining cluster stars or/and some stellar evolutionary effects.

Possible destructive effects of the secular increase of the inner eccentricity in non-coplanar triple systems, as described above, have not been taken into account so far and the inner orbit was 'frozen' as long as the hierarchy existed. However, numerical simulations show (Fig. 3) that relative inclinations $i$ of the two orbital planes of dynamically formed hierarchical systems in clusters are uniformly distributed between 0° and 180° (keeping in mind that $n(i)di \propto \sin i \, di$), and therefore the actual fraction of systems with $i \approx 90°$ is rather significant. The factor which can significantly reduce the influence of 'Heggie's effect' on the inner binary is the time scale of this effect. Some statistical properties of hierarchical systems in clusters are shown in Fig. 4. Such factors as a large ratio of semiajor axes, massive inner binary ($m_1 + m_2 \geq 2.5 M_\odot$), and relatively low-mass distant component (on average $m_3 \sim (0.2 - 0.6) m_{\text{in}}$) should increase the time interval which is required in order to reach $e_{\text{in}}^{\max}$. In this case the life-time of a hierarchy, which is typically between 1 and 3 Myrs although it can reach $\sim$10 Myrs in some cases, may not be long enough. Also the tidal circularization which time scale can be comparable or even less that one of 'Heggie's effect' can play an important role and close binary, probably, cannot even approach their $e_{\text{in}}^{\max}$. This study has just begun and we cannot yet draw any final conclusion as to how the fate of the close binary may be affected over a long time by the presence of the very distant companion. Fig. 5 presents data from the last part of the run for the model cluster with 10,000 stars and 500 primordial binaries. The left-hand panel shows the distribution of $e_{\text{in}}^{\max}$, with $e_{\text{in}}^{\max}$ calculated at the time when the hierarchy was formed. More than 15% of systems may reach $e_{\text{in}}^{\max} \in (0.9; 1)$ if they are given enough time.

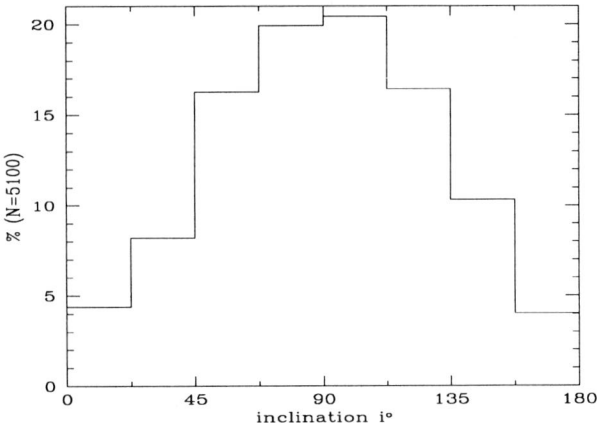

*Figure 3.* Distribution of relative orbital inclinations $i$ for hierarchical triple and quadruple systems in two runs for clusters of N = 5100 stars with 100 primordial binaries

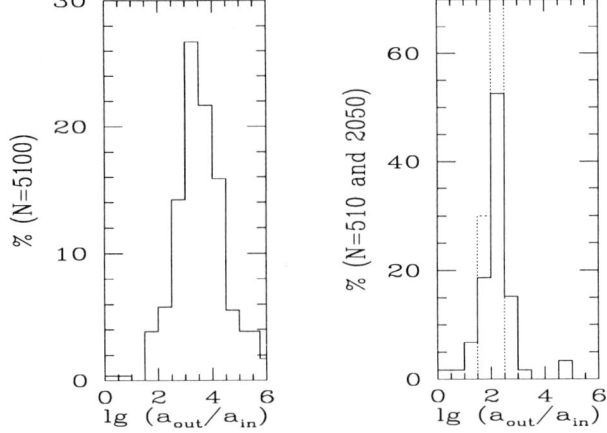

*Figure 4.* Distribution of the ratio outer/inner semi-major axis ($\log a_{out}/a_{in}$) for models with 5100 stars (left-hand panel); 510 stars with 10 primordial binaries (dots) and 2050 stars with 50 primordial binaries (solid line) on the right-hand panel

Note that at the moment of hierarchy formation *all* inner binaries had circular orbits (most likely due to tidal circularization), although about 25% of triples and 39% of quadruples contained a non-primordial inner binary. The right-hand panel shows that if systems are allowed to reach their $e_{in}^{max}$, a few percent of inner binaries would suffer collisions or very strong tidal effects at the periastrons of their orbits.

So far, we have not followed the processes of formation of hierarchical systems in clusters in detail. The most probable mechanism is the presence

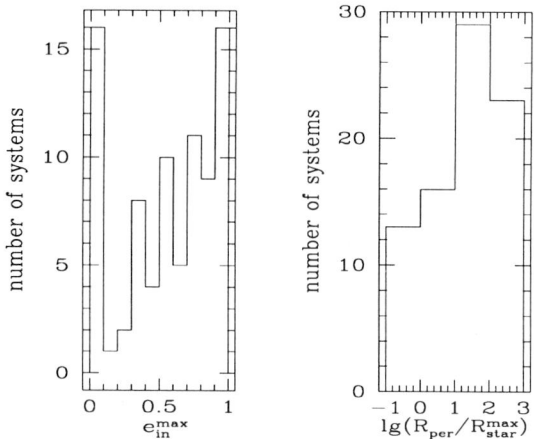

*Figure 5.* Distributions of maximum values of inner eccentricity due to the secular effect of the distant component (left), and of minimum (periastron) separations between stars in the outer binary in units of their maximum stellar radius

of primordial or, at later evolutionary stages, exchanged binaries (which can build up to 50% of all close binaries in hierarchies in some numerical runs), along with binary-binary interaction. However, our distribution of the ratio $a_{out}/a_{in}$ (Fig. 4), especially for big clusters with $N > 5000$, differs very much from the results of binary-binary scattering experiment (McMillan, Hut & Makino 1991). It appears from numerical experiments that in large clusters at a late stage of their evolution new hierarchical systems can be formed via repetitive triple-binary and triple-triple exchanges. Outer eccentricities - most systems have very eccentric outer orbits with $e_{out} \in (0.7; 0.98)$) and the periods for hierarchical systems produced in our models, are in reasonable agreement with the observations referred to above of multiples in open clusters, within the limited statistics for the latter. Let us note finally that at least one hierarchical system (in large clusters there can be up to 5 hierarchies at the same time) is present in a open cluster during about 20% of its life history and at all evolutionnary stages after the core collapse.

## Acknowledgements

My special thanks go to S. Aarseth without whom the last section of this article would not exist. I am also very grateful to D. Heggie for allowing me to use his unpublished formulae, some of which were derived specially to answer my questions. Many thanks also to P. Eggleton and R. de la Fuente Marcos for fruitful discussions and cooperation, to NATO, the Isaac Newton

Trust, and the IAU for their financial support and to IoA in Cambridge for its hospitality.

## References

Aarseth S. J. (1996) in E. F. Milone, ed., *Binaries in Clusters*, ASP series, in press
Anosova, J.P. (1986) *Astrophys. Space Sci.* **124**, 217
Anosova, J. P. and Orlov, V. V. (1994) *Celest. Mech.* **59**, 327
Backer, D. C., Foster, R. S., and Sallmen, S. (1993) *Nature* **365**, 81
Batten, A. H., Fletcher, J. M., and MacCarthy, D. G. (1989) *Publ. Dom. Astrophys. Obs.* **17**, 1
Donnison, J.R., Mikulskis, D.F. (1992) *Mon.Not.R.Astr.Soc.* **254**, 21
Donnison, J.R., Mikulskis, D.F. (1995) *Mon.Not.R.Astr.Soc.* **272**, 1
Duqennoy, A. and Mayor, A. (1991) *Astron. Astrophys.* **248**, 485
Eggleton, P. P. and Kiseleva, L. G. (1995) **Astrophys. J.**, in press
Ghez, A. M., Neugebauer, G., and Matthews, K.: 1993, *Astron. J.* **106**, 2066
Gliese, W., Jahreiss, H. (1988) *Astrophys. Space Sci.*, **142**, 49
Golubev, V. G. (1967) *Dok. Akad. Nauk USSR*, **12**, 529
Golubev, V. G. (1968) *Dok. Akad. Nauk USSR*, **13**, 373
Graziani, F. and Black, D. C. *Astrophys. J.*, **251**, 337
Griffin, R. F. and Gunn, J. E. (1981) *Observatory* **106**, 35
Griffin, R. F., Gunn, J. E., Zimmerman, B. A., and Griffin, R. E. M. (1985) *Astron. J.* **90**, 609
Harrington, R.S. (1972) *Celest. Mech.*, **6**, 322
Harrington, R.S. (1975) *Astron. J.*, **80**, 1081
Heggie, D.C. (1988) in M. Valtonen, ed., *The Few-Body Problem*, Reidel, Dordrecht, p. 213
Heggie, D.C. (1996), this Proceedings
Henry, T. J. and McCarthy, D. W. (1990) *Astrophys. J.* **350**, 334
Hut, P. (1995) in E. P. J. van den Heuvel, ed., *Compact Stars in Binaries*, in press
Kiseleva, L. G., Aarseth, S. J., Eggleton, P. P., and de la Fuente Marcos, R. (1996) in E. F. Milone, ed., *Binaries in Clusters*, ASP series, in press
Kiseleva, L. G., Eggleton, P. P., and Anosova, J. P. (1994) *Mon. Not. R. Astron. Soc.* **267**, 161
Kiseleva, L. G., Eggleton, P. P., and Orlov, V. V. (1994) *Mon. Not. R. Astron. Soc.* **270**, 936
Mason, B. D., McAlister, H. A., Hartkopf, W. I., and Bagnuolo, W. G., Jr. (1993) *Astron. J.* **105**, 220
Mathieu, R. D., Latham, D. W., and Griffin, R. F. (1990) *Astron. J.* **100**, 1859
Mayer, P., Lorenz, R., Chochol, D., and Irsmambetova, T. R. (1994) *Astron. Astrophys.* **288**, L13
McMillan, S., Hut, P., and Makino, J. (1991) *Astrophys. J.*, **372**, 111
Mermilliod, J. C., Rosvick, J., Duquennoy, A., and Mayor, M. (1992) *Astron. Astrophys.* **265**, 513
Mermilliod, J. C., Duquennoy, A., and Mayor, M. (1994) *Astron. Astrophys.* **283**, 515
Rasio, F. A., McMillan, S., and Hut, P. (1995) *Astrophys. J. Letters* **498**, L33
Simon, M., Ghezz, A. M., Leinert Ch. et al. (1995) *Astrophys. J.* **443**, 625
Szebehely, V. and Zare, K. (1977) *Astron. Astrophys.*, **58**, 145
Thorsett, S. E., Arzoumanian, Z. and Taylor, J. H. (1993) *Astrophys. J.* **412**, L33
van de Kamp, P. (1971) *Annu. Rev. Astron. Astrophys.* **9**, 103
Zare, K. (1976) *Celest. Mech.*, **14**, 73

# STELLAR ENCOUNTERS IN DENSE SYSTEMS

MELVYN B. DAVIES
*Institute of Astronomy*
*Cambridge, UK*

## 1. Introduction

The number density of stars in the solar neighbourhood is sufficiently low that encounters between two stars will be extremely rare. However, in the cores of globular clusters, and glactic nuclei, number densities are sufficiently high ($\sim 10^5$ stars/$pc^3$ in some systems) that encounter timescales can be comparable, or even less than, the age of the universe. In other words, a large fraction of the stars in these systems will have suffered from at least one close encounter or collision in their lifetime.

The cross section for two stars, having a relative velocity at infinity of $v_\infty$, to pass within a distance $R_{\rm min}$ is given by

$$\sigma = \pi R_{\rm min}^2 \left(1 + \frac{v^2}{v_\infty^2}\right) \tag{1}$$

where $v$ is the relative velocity of the two stars at closest approach in a parabolic encounter (*i.e.* $v^2 = 2G(m_1 + m_2)/R_{\rm min}$). The second term is due to the attractive gravitational force, and is referred to as gravitational focussing. In the regime where $v \ll v_\infty$ (as might be the case in galactic nuclei with extremely high velocity dispersions), we recover the result, $\sigma \propto R_{\rm min}^2$. However, if $v \gg v_\infty$ as will be the case in systems with low velocity dispersions, such as globular clusters, $\sigma \propto R_{\rm min}$.

One may estimate the timescale for a given star to undergo an encounter, $\tau_{\rm coll} = 1/n\sigma v$. For the two extreme cases mentioned above, we thus obtain

$$\tau_{\rm coll} = 7 \times 10^{10} {\rm yr} \left(\frac{10^5 pc^{-3}}{n}\right)\left(\frac{v_\infty}{10 km/s}\right)\left(\frac{R_\odot}{R_{\rm min}}\right)\left(\frac{M_\odot}{M}\right) \text{ for } v \gg v_\infty$$

$$\tau_{\text{coll}} = 7 \times 10^{12} \text{yr} \left(\frac{10^5 pc^{-3}}{n}\right) \left(\frac{100 km/s}{v_\infty}\right) \left(\frac{R_\odot}{R_{\min}}\right)^2 \text{ for } v \ll v_\infty \quad (2)$$

where $n$ is the number density of single stars of mass $m$. Thus for globular clusters, $\tau_{\text{coll}} \propto 1/R_{\min}$. For an encounter between two single stars to be hydrodynamically interesting, we typically require $R_{\min} \lesssim 3R_\odot$ (see for example, Davies, Benz & Hills 1992). For an encounter between a binary and a third, single star, we require that $R_{\min} \simeq d$, where $d$ is the semi-major axis of the binary. Even for a binary with $d \sim 1AU (\equiv 216 R_\odot)$, we see that $\tau_{\text{coll}} \ll 10^{10}$ years in the core of a dense globular cluster. Thus encounters between binaries and single stars will be important in globular clusters even if the binary fraction is $\sim 5\%$. In encounters between two binaries, we again require the two systems to pass within $\sim d$ of each other. Hence binary/binary encounters will dominate over binary/single encounters only if the binary fraction is $\geq 30\%$. The issue of the binary fraction in the core of a globular cluster is thus an important one. Unfortunately, the number is not well known.

## 2. Dynamical Evolution of Globular Clusters

Neglecting for a moment the role of binaries, a globular cluster will evolve dynamically in the following way. Stars in the halo will be heated by stars from the cluster core by two-body scattering, causing some to escape from the cluster (*i.e.* evaporate). Losing energy, the core contracts, and its velocity dispersion *increases*. In thermodynamic parlance, the system has a negative heat capacity with the velocity dispersion being equivalent to a temperature. Thus as the core contracts, the temperature gradient, and in turn the rate of energy transfer to the stars in the halo, increases. It was shown that such a system would reach an infinite central density in a finite time (Cohn 1980). This is known as the gravothermal catastrophe (Lynden-Bell & Wood 1968).

Binaries will be an important source of energy against the collapse of the core of a globular cluster. Hard binaries encountering a single, third star tend to be further hardened. In other words, the binary becomes more tightly bound, the shift in potential energy being manifested as an increase in the kinetic energies of the binary and single star, *i.e.* the stars receive kicks. This helps support a globular cluster in two ways. Firstly, the velocity dispersion of stars is increased by the kicks they receive. Further, if any stars are ejected, then the binding energy of the cluster is reduced as it contains fewer stars.

## 3. Stellar Exotica within Globular Clusters

Low-mass X-ray binaries are observed in excessive numbers in globular clusters compared to the rest of our galaxy, given the fraction of stars in globular clusters. These systems may be formed via tidal capture of a neutron star by either a main-sequence star or red giant (Fabian, Pringle & Rees 1975). Alternatively, they may be produced from primordial binaries by some evolutionary path that may include neutron stars being exchanged into a binary in some encounter. The relative frequency of these two paths will depend on the make-up of the primordial binary population and their evolution due to encounters with other binaries and single stars. It is also far from clear whether a system formed from tidal capture can survive as a binary or will form a single, merged object (Davies *et al.* 1992, and references contained within).

Millisecond pulsars (MSPs) have also been observed in large numbers in globular clusters (see, for example, Manchester *et al.* 1991). Under the standard model, MSPs are produced in Low-Mass X-ray Binaries (LMXBs) where the neutron star is spun-up as material is accreted from the Roche-filled companion. However observations seem to suggest that there are far more MSPs than LMXBs which, given their comparable expected lifetimes, poses a problem for the standard model (Kulkarni *et al.* 1990). This embarrassing profusion of MSPs coupled with the relative sparsity of LMXBs would seem to represent the nemesis of the standard model. Perhaps MSPs are formed without passing through a precursor LMXB phase, from systems containing neutron stars smothered by the remains of a main-sequence star after a collisions between two such stars, for example.

One might expect that white dwarfs experience encounters similar to neutron stars, some of which may lead to Cataclysmic Variables (CVs). As the number of LMXBs in globular clusters is large, one might expect to see copious numbers of CVs. Such objects are much harder to find, however, although CV candidates have been found in NGC 6397 by imaging the cluster in $H\alpha$ (Cool *et al.* 1995), and confirmed through spectroscopy (Grindlay *et al.* 1995). CV candidates have also been seen in NGC 6752 in the X-ray (Cool 1993), and using $H\alpha$ imaging (Bailyn *et al.* 1995).

Blue stragglers are positioned on the upper end of the main-sequence beyond the present day turn-off mass. They have been observed in many globular clusters, including: 47 Tuc (Paresce *et al.* 1991), NGC 6397 (Auriere *et al.* 1990) and M30 (Yanny *et al.* 1994, Guhathakurta *et al.* 1996). These stars may have formed from the merger of two lower-mass main-sequence stars either in an encounter between two single stars or in encounters involving binaries when two main-sequences collide and merge as part of the encounter. Such mergers may occur during an encounter between a binary and single star, as will be discussed in §6. The fraction of blue stragglers in binaries may thus be an important diagnostic for the binary fraction for globular clusters.

## 4. Encounters between Two Single Stars in Globular Clusters

We begin by considering encounters between two single stars, applicable to globular clusters where the parameter space is relatively small. The velocity dispersion is low, and essentially all stars were created in one epoch. Thus the maximum mass of primordial main-sequence stars today is well defined, as is the mass of any red giants involved in encounters today. Early 1D and 2D work was performed by Mathis (1967), De Young (1968) and Seidel and Cameron (1972). Three dimensional studies had to wait until the development of faster computers. A method known as Smooth Particle Hydrodynamics (SPH) has been applied to study stellar collisions (Benz and Hills 1987, 1992; Davies, Benz and Hills 1992, 1993, 1994; Hernquist and Goodman 1992; Rasio and Shapiro 1991, 1992). Additionally, encounters have been simulated using the PPM method (Ruffert and Müller 1991). The results of these studies can be summarised with three numbers: the capture radius, $R_{capt}$, which is the minimum distance required for a given pair of stars to form a bound system assuming a relative speed at infinity of 10km/s; the merger radius, $R_{merg}$, which is the largerst value of the minimum distance of an encounter which produces a single merged object; and a typical value for the amount of material ejected from the system, $M_{lost}$.

Below we review the current understanding of encounters between various stellar species, considering encounters involving main-sequnece stars and red giants of turn-off mass ($0.8M_\odot$). We discuss encounters between two main-sequence stars (MS/MS encounters), encounters between a main-sequence star and a more compact star, such as a white dwarf or neutron star (MS/CO encounters), and encounters between a red giant and smaller stars (RG/CO encounters).

### 4.1. MS/MS ENCOUNTERS

As alluded to in §3, collisions between two main-sequence stars remains a viable route to produce the observed blue stragglers. Simulations yield values of $R_{capt} \sim 3R_{ms}$, and provide a lower limit of $R_{merg} \sim 2R_{ms}$. The mass lost from the system on the initial impact is small, typically $M_{lost} \leq 0.01 M_{ms}$. Early work suggested that the merged stars would be well mixed (Benz and Hills 1987). More recently simulations, using a more centrally concentrated model for the main-sequence stars, seem to suggest that the material in the cores will not (at least initially) mix with the envelope gas (Lombardi, Rasio and Shapiro 1995). The subsequent evolution of the merged objects remains an open question: will they expand significantly up the Hyashi track to come into thermal equilibrium? Will they become deeply convective and thus mix the material at some later time? Some clue to their subsequent evolution may be found in observations of blue stragglers in binaries: if we see such objects in hard, but eccentric, binaries then we may deduce that the stars didn't expand very much after formation. Conversely, if we see a blue straggler in a wide, but circular binary, this may suggest that the star had a much larger radius at an earlier time.

### 4.2. MS/CO ENCOUNTERS

Simulations yield values of $R_{\text{capt}} \sim 3.5 R_{\text{ms}}$, and provide a lower limit of $R_{\text{merg}} \sim 1.8 R_{\text{ms}}$. The mass lost from the system on the initial impact is larger than seen in MS/MS encounters, with $M_{\text{lost}} \sim 0.1 M_{\text{ms}}$. In the case of encounters involving a neutron star, the subsequent evolution of merged systems is extremely unclear. It has been speculated that a small amount of material will be accreted by the neutron star, spinning it up, possibly to millisecond periods, whilst the enveloping gas prevents it from being visible as an X-ray object. Such a path way might help explain the apparent birthrate problem between LMXBs and MSPs, providing a sufficient number of smothered neutron star systems can be produced relative to clean binaries. In the case of encounters involving white dwarfs, any merged system seems likely to evolve into a red giant.

### 4.3. RG/CO ENCOUNTERS

Simulations yield values of $R_{\text{capt}} \sim 2.5 R_{\text{rg}}$, and provide a lower limit of $R_{\text{merg}} \sim 2 R_{\text{rg}}$. The mass lost from the system on the initial impact $\sim 0.1 M_{\text{rg}}$. Thus the vast majority of encounters would seem to produce merged systems, containing the red-giant core (essentially a white dwarf) and the impactor engulfed in a common envelope of gas. In such a system, the white dwarf and impactor will spiral together as the common gaseous envelope is ejected. The final separation of the two stars may be estimated by simply equating the binding energy of the envelope with the change in binding energy of the two stars, within some efficiency $\alpha_{\text{ce}}$ (typically taken to be 0.2-0.4 [Taam & Bodenheimer 1989]), Such a common envelope phase will be of great use in producing hard binaries in globular clusters. For systems containing two compact objects, inspiral will occur due to angular momentum loss from gravitational radiation. A system will merge in $\sim 10^{10}$ years if the initial separation (after a common envelope phase) $\leq 3 R_\odot$ (Landau & Lifshitz 1962), as will often be the case for post-CE binaries.

## 5. Encounters between Two Single Stars in Galactic Nuclei

Less computational work has been done on encounters that will occur in galactic nuclei. The relevant parameter space is much larger than for globular clusters, with a wide range of relative velocities being important. As the relative velocity increases, encounters involving larger stars become progressively more important as the effect of gravitational focussing is reduced. In other words, encounters involving red giants may be more important than encounters involving main-sequence stars if the velocity dispersion $\geq$ 600km/s, as will be the case in the very centres of some active galaxies.

Some work considering encounters between main-sequence stars has been performed (Benz and Hills, 1987, 1992; Lai *et al.* 1993; and Davies 1995 [in preparation]).

## 6. Encounters between Binaries and Single Stars

Encounters between binaries and single stars treating the stars as point masses have been considered in numerous works (Hut 1994, and references contained therein). Some allowance has also been made for the finite size of the stars in three-body simulations (see for example, Sigurdsson & Phinney 1994). Cleary & Monaghan (1990), and Davies, Benz & Hills (1993, 1994) also considered the hydrodynamical effects by performing some encounters using a smoothed particle hydrodynamics (SPH) code.

Encounters between binaries and single stars lead to three main outcomes: the incoming star replaces one of the original components forming a new, detached, binary in a so-called *clean exchange*, the incoming star simply hardens the binary without any exchange occuring – a so-called fly-by, or a merger between two of the stars occurs, where the product of the merger may or may not remain bound to the third star.

Before discussing the various cross sections, we will mention what has become known as Heggie's Law, namely that *hard binaries get harder, and weak binaries are broken up* (Hut 1983). In this context, hard means a binary that is sufficiently tightly bound that the orbital speeds of the two stars in the binary are much larger than the relative speed, at infinity, of the system and the third star. The dividing line between these two regimes is a function of the masses of the stars, and the relative velocity of the binary and the single stars it encounters. For example, with solar-mass stars, and a velocity dispersion of 10km/s (typical for a globular cluster), the water shed occurs at a semi-major axis of $\sim$ 10AU. In a somewhat dense cluster such as 47 Tuc, we would expect surviving binaries to be hardened today from this separation to values $\sim 0.2 - 0.5$AU (Hut, McMillan & Romani 1992).

The cross section for a single star to pass within a distance $R_{\min}$ of the center of mass of a binary is given by $\sigma = \pi R_{\min}^2 (1 + V_c^2/V_\infty^2)$, where $V_\infty$ is the relative speed at infinity, and $V_c$ is the relative speed at which the system has zero total energy and is given by $V_c^2 = F(M_1, M_2, M_3)/d = GM_1M_2(M_1 + M_2 + M_3)/M_3(M_1 + M_2)d$, where $M_1$ is the mass of the primary, $M_2$ the mass of the secondary and $M_3$ is the mass of the incoming, single star. In order for an exchange encounter to occur where the impacting star replaces one of the binary components, one might expect $R_{\min} \sim d$, where $d$ is the separation of the two components in the original binary. For a binary of separation $\sim 10$ AU, $V_c \simeq 10$km/s (a typical value for the velocity dispersion in a globular cluster). Hence for hard binaries in globular clusters (where $d \sim 0.1 - 0.5$AU), $V_c \gg V_\infty$, and the exchange cross section can be written as

$$\sigma_{\rm ex} = k_{\rm ex}(q_1, q_2) \pi d^2 \cdot \frac{GM_1 F(q_1, q_2)}{d} \cdot \frac{1}{V_\infty^2} \qquad (3)$$

where $q_1 = M_2/M_1$, $q_2 = M_3/M_1$, and $F(q_1, q_2) = q_1(1+q_1+q_2)/q_2(1+q_1)$. The constant $k_{\rm ex}(q_1, q_2)$ has to be determined through numerical simulations. Once we have the value of $k_{\rm ex}$ for a range of values of $q_1$ and $q_2$, we are able to compute the exchange cross section for any reasonable binary-single star encounter (Davies 1995).

Hut and Inagaki (1985) found that the cross section for any two stars to pass within some minimum distance can be written in the following form,

$$\sigma(R < R_{\min}) = k_{\rm rmin}(q_1, q_2,) \pi d^2 \frac{GM_1 F(q_1, q_2)}{d} \cdot \frac{1}{V_\infty^2} \left(\frac{R_{\min}}{d}\right)^\gamma \qquad (4)$$

where both $k_{\rm rmin}$ and $\gamma$ can be found through simulations of encounters, with $\gamma \sim 0.5$ for encounters involving stars of equal masses. A form suggested for the differential exchange cross section is (Heggie 1975, Hut 1984)

$$\frac{d\sigma}{d\Delta} = 3.5 k_{\rm ex}(q_1, q_2) \pi d^2 \cdot \frac{GM_1 F(q_1, q_2)}{d} \cdot \frac{1}{V_\infty^2} (1+\Delta)^{-4.5} \qquad (5)$$

where $\Delta = \delta E_{\rm bin}/E_{\rm bin}$, the relative shift in the binding energy of the binary. The values predicted from the above equation are in close agreement with those obtained from three-body runs. On average an encounter will harden a binary by 20-40%, the change in potential energry being seen in a boost to the binary's kinetic energy. We can thus compute the kick velocity given to a binary for each encounter. Very few binaries with separations $\sim 0.2$AU will receive kicks sufficient to eject them from a typical globular cluster ($V_{\rm kick} \sim 40$km/s). Rather, it is more interesting to consider a kick velocity

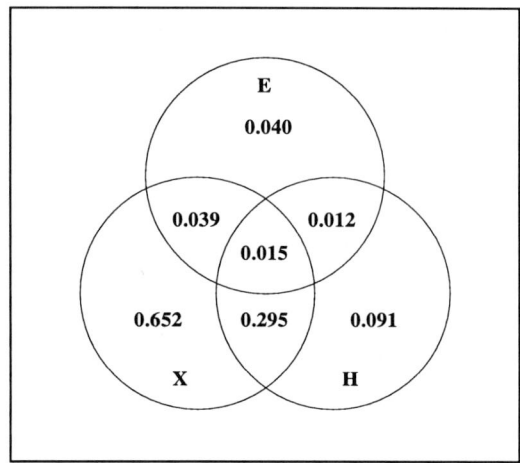

*Figure 1.* Relative cross sections for the various combinations of exchange (X), ejection (E), and the necessity to include hydrodynamical effects (H), calculated from a series of 7000 three-body simulations of encounters between a binary of separation $60R_\odot$ and eccentricity $e = 0.5$ containing an $0.8M_\odot$ main-sequence star and a $1.4M_\odot$ neutron star, and a $1.4M_\odot$ neutron star. The cross sections are renormalized in such a way that the total cross section for processes leading to exchange is unity.

$\sim 20$km/s which would be sufficient to propel the binary into an orbit that takes it far outside the core of a typical globular cluster.

We can compute the cross section for the following three processes: ejection from the globular cluster core, exchange, and strong hydrodynamical encounters (*i.e.* two stars pass within $3R_{\rm ms}$). A convenient way to illustrate the fruit of these calculations is in a Venn diagram. In such a diagram, the three circles represent the three processes, and, for example, the number placed in the region common to all three circles is the cross section for exchange *and* ejection *and* hydrodynamical encounters, renormalized in such a way that the cross section for exchange is unity. Such a Venn diagram is illustrated in Figure 1 for encounters between an eccentric binary ($e = 0.5$, with a semi-major axis, $d = 0.2$AU) containing stars of mass $1.4M_\odot$ and $0.8M_\odot$, and third star of mass $1.4M_\odot$. We see from this figure that some 30% of encounters which were labelled as exchanges from the three-body runs would have involved a hydrodynamic encounter between two of the stars, likely forming a merged object either bound or unbound to the third star. The remaining exchanges would have been *clean*, meaning we expect a binary to be produced without any hydrodynamic processes playing an important part. A very small fraction of encounters would have resulted in the ejection of the binary from a typical globular cluster core. In an equivalent diagram for a harder binary, the relative cross section for hydrodynamic events would be larger, with fewer clean encounters occurring.

We are thus left with the following picture. Binaries are fed into the core of a globular cluster and are either broken up immediately or harden progressively, receiving larger kicks at each succesive encounter, as the binary becomes more tightly bound. Eventually, the binary will be sufficiently tight that some form of merger is likely, unless the system contains two compact objects, in which case the system may well be ejected from the globular cluster. This evolutionary scenario is considered in more detail in McMillan *et al.* 1992, Davies (1995), and Davies & Benz (1995).

## 7. Encounters between Two Binaries

Encounters between two binaries will be more important than encounters between binaries and single stars if the binary fraction is larger than $\sim 10\%$, which it might be in the cores of some globular clusters. Relatively little work has been carried out on binary-binary encounters (see Bacon *et al.* 1995, and references contained therein). Such encounters may lead to a relatively large number of mergers, and triple systems.

## 8. Summary

Collisions happen in globular clusters and galactic nuclei. In globular clusters

- Encounters involving binaries may be important.
- Collisions may produce the stellar exotica that we see (LMXBS *etc.*).
- At least half of encounters between two single stars produce merged objects.
- Wide binaries $\Longrightarrow$ broken up.
- Hard binaries $\Longrightarrow$ harder (Heggie's Law)
- Massive stars exchange into binaries.
- Really hard binaries $\Longrightarrow$ stellar *exotica*
- The relative importance of binary/single and binary/binary encounters is a function of binary fraction. Binary/binary encounters will generally only be important if binary fraction $\geq 30\%$.
- Better modelling of globular clusters and binary/binary cross sections are required and will come in the near future.

In galactic nuclei

- Many more simulations of stellar encounters are required.
- Very high velocity, head on encounters between main-sequence stars lead to their complete disruption.

• Grazing encounters between two main-sequence stars lead to some mass loss but not complete disruption and leave the two stars unbound.

## Acknowledgements

The support of a Royal Society URF is gratefully acknowledged. I thank the IAU for providing financial aid to attend IAU174.

## References

Auriere M., Ortolani S., Lauzarel G., 1990, Nature, 344, 638
Bacon D., Sigurdsson S., Davies M. B., 1995, MNRAS, submitted
Bailyn C. D., Grindlay J. E., 1987, ApJ, 316, L25
Bailyn C. D., Grindlay J. E., Garcia M., 1990, ApJ, 357,L35
Bailyn C. D., et al. 1995, in preparation
Benz W., Bowers R. L., Cameron A. G. W., Press W. H., 1989, ApJ, 348, 647
Benz W., Hills J. G., 1987, ApJ, 323, 614
Benz W., Hills J. G., 1992, ApJ, 389, 546
Cannon R. C., 1993, MNRAS, 263, 817
Cleary P. W., Monaghan J. J., 1990, ApJ, 349, 150
Cohn, H., 1980, ApJ, 242, 765
Cool A., 1993, PhD thesis, Harvard University
Cool A., Grindlay J., Cohn H., Lugger P., Slavin S., 1995, ApJ, 439, 695
Davies M. B., 1995, MNRAS, 276, 887
Davies M. B., Benz W., 1995, MNRAS, 276, 887
Davies M. B., Benz W., Hills J. G., 1992, ApJ, 401, 246
Davies M. B., Benz W., Hills J. G., 1993, ApJ, 411, 285
Davies M. B., Benz W., Hills J. G., 1994, ApJ, 424, 870
de Young D. S., 1968, ApJ, 153, 633
Fabian, A., Pringle, J. E., Rees, M. J., 1975, 172, 17p
Grindlay J., Cool A., Callanan P., Cohn H., Lugger P., Bailyn C., 1995, ApJ, in press
Guhathakurta, P., Yanny B., Schneider D. P., Bahcall J. N., 1996, ApJ, submitted
Hartwick F. D. A., Hesser J. E., 1972, ApJ, 175, 77
Hernquist L., Goodman J., 1991, ApJ, 378, 637
Hut P., 1994, ApJS, 55, 301
Hut P., McMillan S., Romani R. W., 1992, ApJ, 389, 527
Kulkarni, S. R., Narayan, R., Romani, R. W., 1990, ApJ, 356, 174
Landau L., Lifschitz E., 1962, Quantum Mechanics (London: Wiley)
Lai D., Rasio F. A., Shapiro S. L., 1993, ApJ, 412, 593
Lombardi J. C., Rasio F. A., Shapiro S. L., 1995, ApJ, submitted
Lynden-Bell, D., Wood, R., 1968, MNRAS, 138, 495
Manchester, R. N., Lyne, A. G., Robinson, C., D'Amico, N., Lim, J., 1991, Nature, 352, 219
Mathis J. S.,, 1967, ApJ, 147, 1050
Nemec J. M., Harris H. C., 1987, ApJ, 316, 172
Paresce F., et al, 1991, Nature, 352, 297
Rasio F. A.,, Shapiro S., 1991, ApJ, 377, 559
Ruffert M., Müller E., 1990, A&A, 238, 116
Seidl F. G. P., Cameron, A. G. W. 1972, AP Space Sci, 15, 44
Sigurdsson S., Phinney E. S. P., 1995, ApJS, 1995, 99, 609
Taam R. E., Bodenheimer P., 1989, ApJ, 337, 849
Yanny B., Schneider D. P., Bahcall J. N., Guhathakurta, P., 1994, ApJ, 435, L59

# SPH CALCULATIONS OF COLLISIONS BETWEEN MAIN-SEQUENCE STARS

FREDERIC A. RASIO
*Department of Physics, M.I.T. 6-201, Cambridge, MA 02139, USA*

**Abstract.** The hydrodynamics of collisions and mergers of main-sequence stars is discussed in the light of recent 3-D calculations using the smoothed particle hydrodynamics (SPH) method. Theoretical models for the formation of blue stragglers are reviewed in the context of recent comparisons between the observed properties of blue stragglers in dense globular clusters and the predictions of those models.

## 1. Introduction

Close dissipative encounters and direct physical collisions between stars occur frequently in dense star clusters. The dissipation of kinetic energy in close stellar encounters can have a direct influence on the dynamical evolution of a cluster, since it encourages secular core collapse. At the same time, however, mass loss due to accelerated stellar evolution in merger products tends to *unbind* the parent system (Spitzer 1987; Statler et al. 1987; Goodman & Hernquist 1991). Observational evidence for stellar collisions and mergers in globular clusters is provided by the existence of large numbers of blue stragglers in these systems. Blue stragglers are main-sequence stars that appear above the turnoff point in the color-magnitude diagram of a cluster. They have long been thought to be formed through the merger of two lower-mass stars, either in a collision or following binary coalescence (Leonard 1989; Livio 1993; Stryker 1993; Bailyn 1995). Clear indication for a collisional origin of blue stragglers in dense globular clusters has come from recent observations of cluster cores by the Hubble Space Telescope. Large numbers of blue stragglers were found to be concentrated in the cores of the densest clusters, such as M15 (De Marchi & Paresce 1994; Guhathakurta et al. 1995) and M30 (Yanny et al. 1994).

Collisions can happen directly between two single stars only in the cores of the densest clusters, but even in somewhat lower-density clusters they can also happen indirectly, during resonant interactions involving wide primordial binaries (Leonard 1989; Sigurdsson & Phinney 1995; Davies & Benz 1995). The existence of dynamically significant numbers of primordial binaries in globular clusters is now well established observationally (Hut et al. 1992; Cote et al. 1994; see also the article by Pryor in this volume). In dense cluster cores, close binaries, perhaps formed by tidal capture, may be quickly destroyed by interactions with other stars or binaries, also leading to collisions (Goodman & Hernquist 1991; see also the articles by Mardling and by Shara in this volume).

Following early numerical work in 2-D (e.g., Shara & Shaviv 1978), Benz & Hills (1987, 1992) performed the first 3-D calculations of direct collisions between two main-sequence stars. An important conclusion of their pioneering study was that stellar collisions could lead to thorough mixing of the fluid. In particular, they pointed out that the mixing of fresh hydrogen fuel into the core of the merger remnant could reset the nuclear clock of a blue straggler, allowing it to remain visible for a full main-sequence lifetime $t_{MS} \sim 10^9$ yr after its formation.

In subsequent work it was generally assumed that the merger remnants resulting from stellar collisions were nearly homogeneous. Blue stragglers would then start their life close to the zero-age main sequence, but with an anomalously high helium abundance coming from the hydrogen burning in the parent stars. In contrast, little hydrodynamic mixing was expected to occur during the much gentler process of binary coalescence, which could take place on a stellar evolution timescale rather than on a dynamical timescale (Mateo et al. 1990; Bailyn 1992; but see Rasio 1993, 1995, and Rasio & Shapiro 1995).

On the basis of these ideas, Bailyn (1992) suggested a way of distinguishing observationally between the two possible formation processes. The helium abundance in the envelope of a blue straggler, which reflects the degree of mixing during its formation process, can affect its observed position in a color-magnitude diagram. Blue stragglers made from collisions would have a higher helium abundance in their outer layers than those made from binary mergers, and this would generally make them appear somewhat brighter and bluer.

A detailed analysis was carried out by Bailyn & Pinsonneault (1995) who performed stellar evolution calculations for blue stragglers assuming various initial chemical composition profiles. To represent the collisional case, they assumed chemically homogeneous initial profiles with enhanced helium abundances, calculating the total helium mass from stellar evolution models of the parent stars. For the dense cluster 47 Tuc they concluded that the observed luminosity function and numbers of blue stragglers were then consistent with a collisional origin.

## 2. The SPH Method

The vast majority of recent 3-D calculations of stellar interactions (collisions, binary coalescence, common envelope evolution, tidal disruption, etc.) have been done using the smoothed particle hydrodynamics (SPH) method (see Monaghan 1992 for a recent review). Since SPH is a Lagrangian method, in which particles are used to represent fluid elements, it is ideally suited for the study of hydrodynamic mixing. Indeed, chemical abundances are passively advected quantities during the dynamical evolution. Therefore, the chemical composition in the final fluid configuration can be determined after the completion of a calculation simply by noting the original and final positions of all SPH particles and by assigning particle abundances according to an initial profile.

A straightforward derivation of the basic SPH equations can be obtained from a Lagrangian formulation of hydrodynamics (Gingold & Monaghan 1982). Consider for simplicity an ideal fluid undergoing adiabatic evolution. The Euler equations of motion,

$$\frac{d\mathbf{v}}{dt} = \frac{\partial \mathbf{v}}{\partial t} + (\mathbf{v} \cdot \nabla)\mathbf{v} = -\frac{1}{\rho}\nabla p, \qquad p = A\rho^\gamma, \qquad (1)$$

can be derived from a Lagrangian principle with

$$L = \int \left\{ \frac{1}{2}\dot{\mathbf{x}}^2 - u[\rho(\mathbf{x})] \right\} \rho \, d^3x. \qquad (2)$$

Here $p$ is the pressure, $\rho$ is the density, $A \propto \exp(s)$ is a function of the specific entropy $s$ (assumed here to be constant in space and time), and $u[\rho] = p/[(\gamma - 1)\rho] = A\rho^{\gamma-1}/(\gamma - 1)$ is the specific internal energy of the fluid.

The basic idea in SPH is to use the discrete representation

$$L_{SPH} = \sum_{i=1}^{N} m_i \left[ \frac{1}{2}\dot{\mathbf{x}}_i^2 - u(\rho_i) \right] \qquad (3)$$

for the Lagrangian, where the sum is over a large but discrete number of small cells, or "particles," covering the volume of the fluid. Here $m_i$ is the mass of a particle, $\mathbf{x}_i$ is its position, and $\dot{\mathbf{x}}_i$ is its velocity. For expression (3) to become the Lagrangian of a system with a finite number $N$ of degrees of freedom, we need a prescription to compute the density $\rho_i = \rho(\mathbf{x}_i)$ at the position of a given particle, as a function of the masses and positions of neighboring particles. In SPH, this is done by introducing a local average,

$$\rho_i = \sum_j m_j W_{ij}, \qquad W_{ij} = W(|\mathbf{x}_i - \mathbf{x}_j|; h), \qquad (4)$$

where $W(r; h)$ is a smoothing kernel, normalized to unity and of width $\sim h$. A very common choice for the smoothing kernel is the cubic spline

$$W(r; h) = \frac{1}{\pi h^3} \begin{cases} 1 - \frac{3}{2}\left(\frac{r}{h}\right)^2 + \frac{3}{4}\left(\frac{r}{h}\right)^3, & 0 \leq \frac{r}{h} < 1, \\ \frac{1}{4}[2 - \left(\frac{r}{h}\right)]^3, & 1 \leq \frac{r}{h} < 2, \\ 0, & \frac{r}{h} \geq 2. \end{cases} \qquad (5)$$

(Monaghan & Lattanzio 1985). With the prescription (4) for the density, we can now obtain the equations of motion for all the particles. Deriving the Euler-Lagrange equations from $L_{SPH}$ we get

$$\frac{d\mathbf{v}_i}{dt} = -\sum_j m_j \left(\frac{p_i}{\rho_i^2} + \frac{p_j}{\rho_j^2}\right) \nabla_i W_{ij}. \qquad (6)$$

The expression in the right-hand side of equation (6) is a sum over neighboring particles (within a distance $\sim h$ of $\mathbf{x}_i$) representing a discrete approximation to the pressure force $[-(1/\rho)\nabla p]_i$ acting on the particle at $\mathbf{x}_i$. Typically, a full implementation of SPH for astrophysical problems would add to equation (6) a treatment of self-gravity (e.g., using one of the many grid-based or tree-based algorithms developed for N-body simulations) and an artificial viscosity term to allow for entropy production in shocks. In addition, we have assumed here in deriving equation (6) that the smoothing length $h$ is constant in time and the same for all particles. In reality, individual and time-varying smoothing lengths $h_i(t)$ are almost always used, so that the local spatial resolution can be adapted to the (time-varying) density of SPH particles (see Nelson & Papaloizou 1994 for a rigorous derivation of the equations of motion in this case).

The following energy and momentum conservation laws are satisfied *exactly* by the simple SPH equations of motion given above

$$\frac{d}{dt}\left(\sum_{i=1}^{N} m_i \mathbf{v}_i\right) = 0, \qquad (7)$$

and

$$\frac{d}{dt}\left(\sum_{i=1}^{N} m_i [\frac{1}{2}v_i^2 + u_i]\right) = 0, \qquad (8)$$

where $u_i = p_i/[(\gamma - 1)\rho_i]$. Note that energy and momentum conservation in SPH is independent of the number of particles $N$.

## 3. Discussion of Recent Results

Lombardi, Rasio, & Shapiro (1995a,b) have studied collisions between main-sequence stars, and, in particular, the question of mixing during mergers, by performing a new set of numerical hydrodynamic calculations using SPH. This new work improves on the previous study of Benz & Hills (1987) by adopting more realistic stellar models, and by performing numerical calculations with increased spatial resolution. Benz & Hills (1987) used an early version of the SPH method and performed their calculations with a small number of particles ($N = 1024$). They also represented all stars by simple $n = 1.5$ polytropes. Unfortunately, $n = 1.5$ polytropes have density profiles that are not steep enough to represent main-sequence stars close to the turnoff point. Turnoff main-sequence stars have very shallow convective envelopes and are much better modeled by $n = 3$ polytropes (which have much more centrally concentrated density profiles).

The new SPH calculations of Lombardi et al. (1995a,b) are done using $N = 3 \times 10^4$ particles, and the colliding stars are modeled as composite polytropes (with $n = 3$ for the radiative interior and $n = 1.5$ in the convective envelope), which provide accurate representations of the density profiles in the entire mass range of interest for globular clusters (cf. Rappaport, Verbunt, & Joss 1983; Ruciński 1988). This is particularly important for collisions between two stars of different masses, which in general will also have different internal structures (and, for this reason, the later calculations of Benz & Hills 1992, done for two $n = 1.5$ polytropes with a mass ratio of $1/5$, are of very limited applicability).

Stars close to the main-sequence turnoff point in a cluster are in fact the most relevant ones to consider for stellar collision calculations. Indeed, as the cluster evolves via two-body relaxation, the most massive stars will tend to concentrate in the dense cluster core, where the collision rate is highest (see, e.g., Spitzer 1987). In addition, collision rates can be increased dramatically by the presence of a significant fraction of primordial binaries in the cluster, and the more massive stars will preferentially tend to be exchanged into such a binary, or collide with another star, following a dynamical interaction between two binaries or between a binary and a single star (Sigurdsson & Phinney 1995).

The main new results of Lombardi et al. (1995a,b) can be summarized as follows. Typical merger remnants produced by collisions are far from chemically homogeneous. In the case of collisions between two nearly identical main-sequence stars close to the turnoff point, the amount of hydrodynamic mixing during the collision is minimal. In fact, the final chemical composition profile is very close to the initial profile of the parent stars. For two turnoff stars, this means that the core of the merger remnant is still mostly helium and that the object may not be able to remain on the main sequence for a very long time, since it is born with very little hydrogen left to burn at the center. In the case of collisions between two stars of very different masses, the chemical composition profiles of the merger remnants can be rather peculiar. For example, it often happens that the maximum helium abundance does not occur at the center of the remnant. These results are illustrated in Figure 1, which compares interior profiles of the final configurations following collisions between two stars with a mass ratio of 1/2 and between two identical stars, at various impact parameters.

At a qualitative level, these results can be understood very simply in terms of the requirement of convective stability of the final configurations. If entropy production in shocks could be neglected entirely (which may not be too unreasonable for the low-velocity collisions occurring in globular clusters), then one could predict the final composition profile simply by observing the composition and entropy profiles of the parent stars. Convective (dynamical) stability requires that the specific entropy $s$ increase monotonically from the center to the surface ($ds/dr > 0$, cf. Fig. 1) in the final hydrostatic equilibrium configuration. Therefore, in the absence of shock-heating, fluid elements conserve their entropy and the final composition profile of a merger remnant could be predicted simply by combining mass shells in order of increasing entropy, from the center to the outside. Many features of the results follow directly.

For example, in the case of a collision between two identical turnoff stars, it is obvious why the composition profile of the merger remnant remains very similar to that of the parent stars, since shock-heating is significant only in the outer layers of the stars, which contain a very small fraction of the total mass. The low-entropy, helium-rich material is concentrated in the deep interior of the parent stars, where shock-heating is negligible, and therefore it remains concentrated in the deep interior of the final configuration. For two stars of very different masses, the much lower-entropy material of the lower-mass star tends to concentrate in the core of the final configuration, leading to the unusual composition and temperature profiles seen in Figure 1(a). In essence, the smaller-mass star simply sinks in and settles at the center of the merger, while the higher-entropy, helium-rich material has been pushed out.

Sills, Bailyn & Demarque (1995) were the first to explore the consequences of blue stragglers being born unmixed. Using detailed stellar structure calculations, they compared the predicted colors ($U - B$ and $B - V$) of initially unmixed blue stragglers with observations. They concluded that

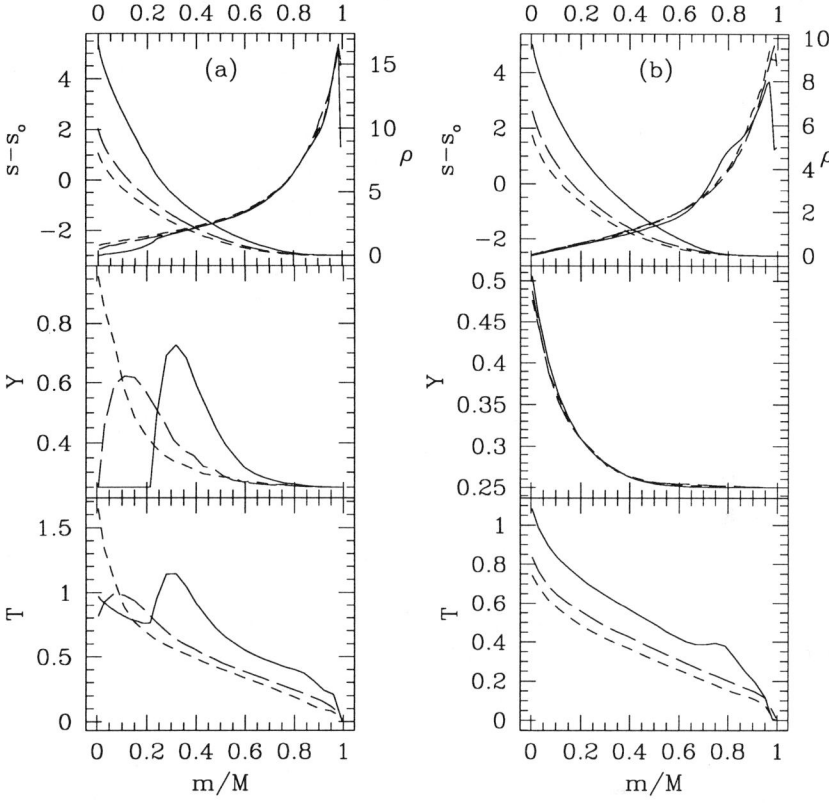

*Figure 1.* Final interior profiles for several merger remnants. The density $\rho$, specific entropy $s$ (up to a constant $s_o$), fractional helium abundance $Y$, and temperature $T$ are shown as a function of interior mass fraction. The masses of the two colliding stars were $M_1 = 0.8\,M_\odot$ and $M_2 = 0.4\,M_\odot$ in (a), and $M_1 = M_2 = 0.6\,M_\odot$ in (b). The solid, long-dashed and short-dashed curves correspond to initial trajectories with increasing periastron separations, $r_p/(R_1 + R_2) = 0$, 0.25, and 0.5, respectively ($R_1$ and $R_2$ are the initial stellar radii). The units are defined by $G = M_{TO} = R_{TO} = 1$, where $M_{TO} \simeq 0.8\,M_\odot$ and $R_{TO} \simeq 1\,R_\odot$ (i.e., the mass and radius of a turnoff main-sequence star). The specific entropy $s$ increases monotonically from the center to the outside, as required for convective stability (except in the outermost few percent of the mass, which have not yet reached hydrostatic equilibrium). Note the peculiar composition and temperature profiles in (a).

some blue stragglers have observed colors that *cannot be explained* using unmixed initial models. Initially homogeneous models, however, can reproduce all the observations. In addition, unmixed models have much shorter main-sequence lifetimes than homogeneous models (because the cores of unmixed models have a very limited supply of hydrogen to burn), and therefore may be incompatible with the observed *numbers* of blue stragglers. Thus we now have some indications from the observations that *significant mixing must take place* during the blue straggler formation process, and this is in apparent conflict with the latest results of hydrodynamic calculations. In addition, spectroscopic measurements of surface rotation rates of blue stragglers indicate that they are slow rotators (Mathys 1987, 1991), also in contradiction with the predictions of dynamical merger calculations. These problems are not peculiar to stellar collisions. Binary coalescence on a dynamical timescale also produces rapidly rotating merger remnants, and since the binary coalescence process is less dissipative than direct collisions (i.e., it creates less shock-heating), we expect even less hydrodynamic mixing in this case (Rasio & Shapiro 1995).

To resolve these apparent conflicts, it may be necessary to take into account processes that have not yet been incorporated into the theoretical models. Hydrodynamic calculations are of course limited to following the evolution of mergers on a dynamical timescale ($t_{dyn} \sim$ hours for main-sequence stars) but are not capable of following processes taking place on a thermal timescale ($t_{th} \sim 10^7$ yr). The final configurations obtained at the end of hydrodynamic calculations (such as the ones illustrated in Fig. 1) are very close to hydrostatic equilibrium, but are generally *far from thermal equilibrium*. This is evident simply from the typical size of the merger remnants: the 95% mass radius at the end of the dynamical phase is typically several times the radius of a main-sequence star of the same total mass. Thus the object will need to contract (on its Kelvin time) before it can become a main-sequence star. In addition, the interior profiles of the merger remnants show clear evidence of departures from thermal equilibrium. A temperature gradient inversion (as seen in Fig. 1a) is a particularly clear sign. In addition, in the case of significantly off-axis collisions, it is found that the rapidly rotating final configurations are not barotropic (i.e.. the angular velocity is not simply constant on cylinders centered on the rotation axis), and therefore they must in general be out of thermal equilibrium (see, e.g., Tassoul 1978).

As the merger remnant contracts to the main sequence and evolves towards thermal equilibrium, many processes can lead to additional mixing of the fluid. These include convection (which is well known to occur during the evolution of ordinary pre-main-sequence stars), and, for rapidly rotating configurations, meridional circulation. These processes can lead not only to mixing, but also to loss of angular momentum and rapid spin-down through magnetic breaking (Leonard & Livio 1995). Even in regions where the convective (dynamical) stability criterion $ds/dr > 0$ is satisfied, *local thermal instabilities* (i.e., secular instabilities) can still occur. The small vertical oscillations (at the local Brunt-Väisälä frequency $\Omega_{BV} \propto [ds/dr]^{1/2}$)

of a fluid element in such a region have amplitudes that grow unstably on a timescale comparable to the local radiative damping time (see, e.g., Kippenhahn & Weigert 1990). In a thermally unstable region, mixing will occur on this timescale. For example, in a region where there is a positive molecular weight gradient ($d\mu/dr > 0$) stabilized by a positive temperature gradient ($dT/dr > 0$), as seen in Figure 1 (solid lines in Fig. 1a), fingers of helium-rich material will tend to develop and penetrate the hydrogen-rich material below. Detailed calculations of the thermal relaxation phase ("pre-main-sequence blue straggler" evolution) incorporating a treatment of all relevant mixing processes will be necessary in order to develop a complete theoretical understanding of blue straggler formation.

At the same time, more extensive comparisons with the latest observational data will be necessary. The first study completed by Sills, Bailyn & Demarque (1995) focused on a small number of bright blue stragglers in the core of one particular cluster, NGC 6397. Preliminary results from another study, by Ouellette & Pritchet (1996), looking at a much larger sample of blue stragglers in the globular cluster M3, apparently lead to rather different conclusions. For the blue stragglers observed by HST in the core of M3 (Guhathakurta et al. 1994), this study finds better agreement with *unmixed* initial models. This is in direct contradiction with the results of Sills et al. (1995) if one assumes that blue stragglers in the cores of dense clusters are all formed by the same mechanism. On the other hand, Ouellette & Pritchet (1996) also find that the distribution of blue stragglers in the low-density outer region of M3 (Ferraro et al. 1993) agrees better with the predictions of *fully mixed* models. Clearly the final word on this question has not been said, but with the rapidly increasing quantity and quality of observational data on blue stragglers, and with several groups now working on improving the theoretical calculations and making detailed comparisons with observations, we can look forward to rapid progress in this area over the next few years.

## Acknowledgments

I am very grateful to Piet Hut, Eiichiro Kokubo, Junichiro Makino, and Daiichiro Sugimoto for their hospitality in Tokyo. I thank Sverre Aarseth, Rosemary Mardling, and Frank Verbunt for many stimulating exchanges. I also thank Jamie Lombardi for his help in preparing the manuscript. Many of the results presented here were obtained from computations performed at the Cornell Theory Center, which receives major funding from the NSF and IBM Corporation, with additional support from the New York State Science and Technology Foundation and members of the Corporate Research Institute.

# References

Bailyn, C. D. 1992, ApJ, 392, 519
Bailyn, C. D. 1995, ARAA, 33, 133
Bailyn, C. D., & Pinsonneault, M. H. 1995, ApJ, 439, 705
Benz, W., & Hills, J. G. 1987, ApJ, 323, 614
Benz, W., & Hills, J. G. 1992, ApJ, 389, 546
Cote, P., Welch, D. L., Fischer, P., Da Costa, G. S., Tamblyn, P., Seitzer, P., & Irwin, M. J. 1994, ApJS, 90, 83
Davies, M. B., & Benz, W. 1995, MNRAS, in press
De Marchi, G., & Paresce, F. 1994, ApJ, 422, 597
Ferraro, F. R., Fusi Pecci, F., Cacciari, C., Corsi, C., Buonanno, R., Fahlman, G. G., & Richer, H. B. 1993, AJ, 106, 2324
Gingold, R. A., & Monaghan, J. J. 1982, J. Comp. Phys., 46, 429
Goodman, J., & Hernquist, L. 1991, ApJ, 378, 637
Guhathakurta, P., Yanny, B., Bahcall, J. N., & Schneider, D. P. 1994, AJ, 108, 1786
Guhathakurta, P., Yanny, B., Schneider, D. P., & Bahcall, J. N. 1995, AJ, in press
Hut, P., McMillan, S., Goodman, J., Mateo, M., Phinney, E. S., Pryor, C., Richer, H. B., Verbunt, F., & Weinberg, M. 1992, PASP, 104, 981
Kippenhahn, R., & Weigert, A. 1990, Stellar Structure and Evolution (Springer-Verlag)
Leonard, P. J. T. 1989, AJ, 98, 217
Leonard, P. J. T., & Livio, M. 1995, ApJL, 447, 121
Livio, M. 1993, in ASP Conf. Ser. Vol. 53, Blue Stragglers, ed. R. A. Saffer (San Francisco: ASP), 3
Lombardi, J. C., Jr., Rasio, F. A., & Shapiro, S. L. 1995a, ApJ, 445, L117
Lombardi, J. C., Jr., Rasio, F. A., & Shapiro, S. L. 1995b, ApJ, submitted
Mateo, M., Harris, H. C., Nemec, J., & Olszewski, E. W. 1990, AJ, 100, 469
Mathys, G. 1987, A&AS, 71, 201
Mathys, G. 1991, A&A, 245, 467
Monaghan, J. J. 1992, ARAA, 30, 543
Monaghan, J. J., & Lattanzio, J. C. 1985, A&A, 149, 135
Nelson, R. P., & Papaloizou, J. C. B. 1994, MNRAS, 270, 1
Ouellette, J., & Pritchet, C. 1996, in Binaries in Clusters, ed. E. Milone (ASP Conf. Series), in press
Rappaport, S., Verbunt, F., & Joss, P. C. 1983, ApJ, 275, 713
Rasio, F. A. 1993, in ASP Conf. Ser. Vol. 53, Blue Stragglers, ed. R. A. Saffer (San Francisco: ASP), 196
Rasio, F. A. 1995, ApJ, 444, L41
Rasio, F. A., & Shapiro, S. L. 1995, ApJ, 438, 887
Ruciński, S. M. 1988, AJ, 95, 1895
Shara, M. M., Drissen, L., Bergeron, L. E., & Paresce, F. 1995, ApJ, 441, 617
Shara, M. M., & Shaviv, G. 1978, MNRAS, 183, 687
Sigurdsson, S., Davies, M. B., & Bolte, M. 1994, ApJ, 431, L115
Sigurdsson, S., & Phinney, E. S. 1995, ApJS, 99, 609
Sills, A. P., Bailyn, C. D., Demarque, P. 1995, ApJL, in press
Spitzer, L. 1987, Dynamical Evolution of Globular Clusters (Princeton: Princeton Univ. Press)
Statler, T. S., Ostriker, J. P. & Cohn, H. N. 1987, ApJ, 316, 626
Stryker, L. L. 1993, PASP, 105, 1081
Tassoul, J. 1978, Theory of Rotating Stars (Princeton: Princeton Univ. Press)
Yanny, B., Guhathakurta, P., Schneider, D. P., & Bahcall, J. N. 1994, ApJ, 435, L59

# FREQUENCY OF STELLAR COLLISIONS IN THREE-BODY HEATING

DAVID F. CHERNOFF AND XIAOLAN HUANG
*Department of Astronomy*
*Cornell University*
*Ithaca, NY 14853*

**Abstract.** The probability for collisional interaction of three body binaries is calculated as a function of the physical radius and mass of the stellar objects and the depth of the cluster potential well. For typical cluster parameters, there is a significant chance of physical collision for objects as small as white dwarfs. One consequence of the collisions is to lower the amount of heat produced from hardening a binary, thereby diminishing the efficiency of the three-body heating mechanism.

## 1. Introduction

When a cluster undergoes core collapse stellar interactions tend to harden the binaries and some of these encounters lead to physical collisions. In this paper, we explore the frequency of stellar collisions involving binaries formed as a result of three-body encounters (3B binaries, hereafter). Such binaries are born when the core reaches high densities, interact predominantly with stars in the core and release heat at a rate that roughly balances the transport of energy to the outer regions of the cluster by gravitational relaxation (for a review Goodman 1989). Some of the encounters eject stars and/or binaries from the cluster center; some lead to physical collisions. Generally, it has been assumed that finite size effects are unimportant for the case of 3B binary heating, an assumption explored here in some detail. By way of contrast, it is well-known that finite size effects are crucial to understanding the role of two-body binaries (binaries formed in dissipative encounters of two single stars, hereafter 2B-binaries). For example, the cross section for the formation of 2B binaries is the subject of ongoing work to disentangle the events that lead to collisions and mergers from those that actually produce binaries (Kochanek 1992, Mardling 1995ab). Collisional processes will limit the ability to extract binding energy since a strong encounter with a free star often yields a complicated hydrodynamical interaction instead of a more tightly bound binary (Davies, Benz and Hills 1993, 1994). Clearly, the efficiency of cluster heating is altered by these collisional effects.

Under typical cluster conditions, newly formed 3B binaries have large semi-major axes compared to the stellar radii of main sequence or compact constituents. Let $x = E_b/m_s\sigma_s^2$ be the binary hardness, where $E_b$ is the binary energy, $m_s$ is the mass of a single star, $\sigma_s$ is the one-dimensional velocity dispersion of the singles and we assume that the translational degrees of freedom of binaries and singles are in thermal equilibrium. Then $x = 4.4(\mathrm{AU}/a)([m_1 m_2/m_s]/M_\odot)(10\text{ km/s}/\sigma_s)^2$ where $a$ is the semi-major axis and $m_1$ and $m_2$ are the mass elements of the binary. 3B binaries form at small $x$ and harden until the recoil from a strong encounter liberates them from the cluster potential. For an average energy change of $0.4E_b$, ejection occurs for hardness $x_c = 2.5(m_1 + m_2)(m_1 + m_2 + m_s)\Phi(0)/m_s^2\sigma_s^2$ where $\Phi(0)$ is the depth of the potential in the core. The corresponding semi-major axis is $a_c/\mathrm{AU} = 2.1 \times 10^{-2} y(10\text{ km/s}/\sigma_s)^2(14\sigma_s^2/\Phi(0))$, where $y = ([6m_1 m_2 m_s/(m_1+m_2)(m_1+m_2+m_s)]/M_\odot)$, or $a_c = 4.6 R_\odot$ for the fiducial parameters. The experiments described in this paper show that *repeated* resonant encounters raise the probability for collision dramatically so that white-dwarf 3B binary systems will often collide before they harden enough to be ejected. (A more detailed report, Huang and Chernoff 1995, is in preparation.)

## 2. Binary History

Assume a background of single stars with the following properties: homogeneous in space, steady-state in time, Maxwellian in velocity. Place a binary in the sea of stars and follow its history, assuming that the density is low enough that all changes may be described in terms of pairwise encounters between the binary and a single star. Each interaction alters the essential parameters of the binary (the semi-major axis, the eccentricity, the orientation of the plane of the orbit, and so forth). Denote the sequence of parameter states that exist between the pairwise encounters as a "binary history."

A specific realization of a binary history involves generating the initial conditions for an encounter, integrating the three-body problem to find the final state and repeating. From the vast set of possible encounters, only those with pericenters $\leq \zeta a$ in the two-body limit are considered. Here $a$ is the current semi-major axis and $\zeta$ is a constant, chosen to be large enough to include all "important" encounters. The mean rate of encounters is

$$\frac{dN}{dt} = \int d^3v f(v) \pi (\zeta a)^2 |v - v_B| \left[1 + \frac{2GM_t}{\zeta a |v - v_B|^2}\right] \quad (1)$$

where $M_t$ is the total mass of the stars involved and $v_B$ is the binary velocity. Specific encounters are realized as a Poisson process subject to the above mean rate. All initial conditions (impact parameter, velocity of approach at infinity, orientation angles) are drawn from the appropriate probability distributions. The outcome of each encounter is calculated numerically by a three-body integration and the endstate is one of three types: flyby, exchange or ionization. In addition, the outgoing stellar velocity, binary center of mass velocity, closest distance of approach, and other quantities are calculated. A new encounter is realized and the process repeated until (1) a collision occurs, (2) the binary escapes, or (3) the binary is ionized.

The value of $\zeta$ was fixed as follows. The rate coefficients for the average energy change, the mean square energy change, the average angular momentum change, and the mean square angular momentum change were calculated for hard binaries of a range of $x$ and for $\zeta = 4, 8, 12, 16$ and $20$. The error incurred for $\zeta = 12$ was less than 1% in all quantitites. In addition, the mean and mean square energy changes at selected $x$ agreed with the results of Heggie and Hut (1993) to within the statistically dominated errorbars.

## 3. Steady-State

Assuming a steady-state homogeneous, single star population, Goodman and Hut (1993, hereafter GH93) found the steady-state binary distribution $f(x)$. At small $x$ the 3B binary creation rate is high but the probability of ionization is large; at large $x$ the creation rate is small but the survival probability is high. Eventually, for $x \gtrsim x_c$ recoil ejects binaries from the potential well of the cluster. The solid lines in Figure 1 display $f(x)$ for three potential depths; there are many soft binaries and they are closely approximated by a thermal distribution (dotted line); the distribution function is flat at large $x$, corresponding to a constant net flux until ejection becomes possible. Binaries at large $x$ are immortal, i.e. they have virtually no chance of being ionized.

A direct computation of the steady-state distribution has been carried out as follows. A number of binary histories $(16,000)$ were created. The hardness at birth was distributed according to the 3B binary creation rate but confined to the range $x > 0.3$; the eccentricity distribution was thermal. (The range in $x$ accounts for roughly 40% of the immortal binaries in the

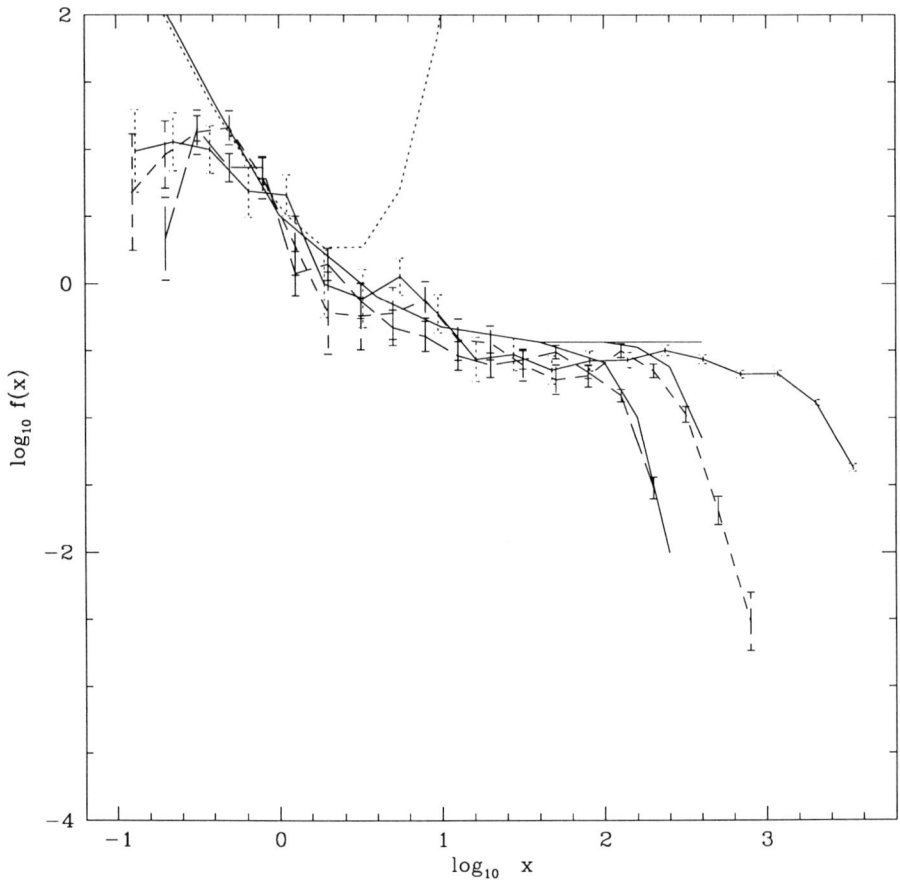

*Figure 1.* The solid line shows the steady-state distribution funciton found by Goodman and Hut (1993) for three different escape criteria ($\Phi(0)/\sigma_s^2 = 6$, 14 and $\infty$). The lines through points with error bars are the same results based on the binary history (long-dashed, short-dashed and solid correspond to $\Phi(0)/\sigma_s^2 = 6$, 14 and 128, respectively). The dotted line gives the distribution in thermal equilibrium.

GH93 solution but it is impractical to decrease the lower limit of $x$ by much, since the probability of survival for a binary with $x = 0.3$ is $\sim 2 \times 10^{-3}$ and decreases rapidly. Tests briefly described below show that it is possible to "renormalize" the distribution function to include the effect of births with $x < 0.3$.) The time span of the simulation was long enough to realize the steady-state binary distribution. The result is illustrated in Figure 1 (see points with errorbars) and agrees in shape quite well with GH93 for

$x \gtrsim 1$. The discrepancy at small $x$ is a consequence of the truncated range of the birth function. The fraction of binaries which are born with a given $x$ that go on to become immortal binaries agrees well with previous analyses (Hut 1985, GH93). About 1% of the histories yielded immortal binaries (on average, $\sim 360$ encounters per history) and the rest were destroyed ($\sim 170$ encounters).

The results of the binary history may be used to examine how quickly memory of the initial conditions of a binary's birth are lost. For example, a population of newly formed binaries with $0.6 < e < 0.7$ and $1 < x < 2$ evolves a thermal distribution of eccentricity in the time necessary for $\sim 42$ encounters (within $12a$) for a binary with $x = 1$. (The number of encounters for individual members of the population ranges greatly.) Detailed analysis (see Huang and Chernoff 1995) shows that the memory of the specific initial conditions ($x$ and eccentricity) is quickly lost so that the statistical properties of the binary population that have hardened to $x \gtrsim 3$ are nearly independent of the detailed form of the binary birth distribution. Since the issues of interest in this paper concern the binary behavior as $x$ grows large, we have simply renormalized the birth rate to yield the immortal birth rate from GH93.

## 4. Close Encounters

Consider the steady-state distribution of immortal binaries. At any instant, there will be a range of values of $x$ and range of values of $d_{min}$, the closest approach suffered by any star which is or was a member of the binary. Figure 2a is a realization of the steady-state distribution of the values of $(x, d_{min})$. The evolution of the points as they age can be understood fairly simply: the hard binaries tend to move to larger values of $x$ (but not monotonically) and the value of $d_{min}$ always decreases. Note that the objects that eventually disrupt are not included. The general flux in Figure 2a is down and to the right. Figure 2b is an analogous plot for the distribution of $(V_{max}, d_{min})$ where $V_{max}$ is the maximum recoil the binary has ever suffered; all points move monotonically down and to the right. The rate of loss by ejection is the flux across the line $V_{max}$ equals the escape velocity. A vertical line is included in the diagram for illustrative purposes. Likewise, the collision criterion corresponds to the flux across a horizontal line at a given value of $d_{min}$.

It is striking that so many of the points in the steady-state snapshot have undergone close encounters. The collision probability for an immortal binary in a sea of similar stars may be inferred from the two dimensional distribution. Figure 3 illustrates the probability as a function of the size (in dimensionless units). For fiducial parameters of $\sigma_s = 10$ km/s, $r_{min} = 2R_*$,

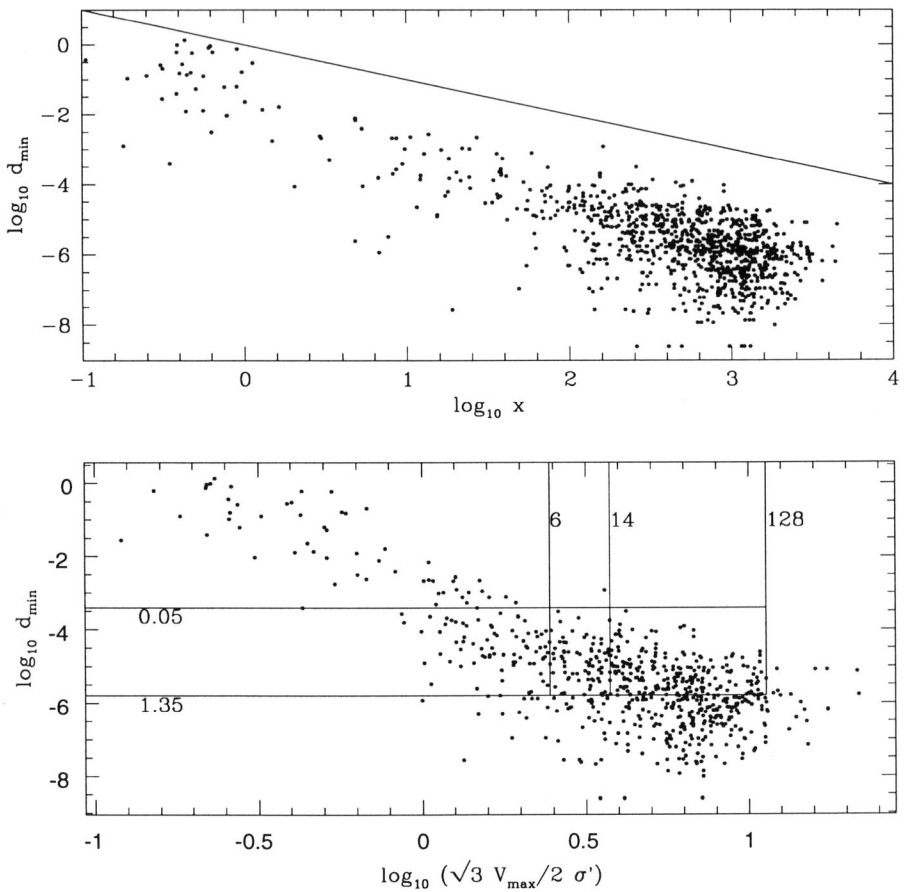

*Figure 2.* Top: The steady-state distribution of hardness and distance of closest approach for immortal 3B binaries (with $x = 3Gm/4a\sigma'^2$, $\sigma'^2 = \sigma_s^2 + \sigma_B^2$, and $d_{min} = 4d\sigma'^2/3Gm$). Binaries are born in range $x > 0.3$. All points must lie below the solid line which is the condition $a = d_{min}$. Bottom: The steady-state distribution of maximum binary recoil ($\sqrt{3}V_{max}/2\sigma'$) and distance of closest approach for immortal 3B binaries. The vertical lines mark the escape velocity in clusters with $\Phi(0)/\sigma_s^2 = 6$, 14 and 128; the horizontal lines mark $d_{min}$ for 0.05 and $1.35 M_\odot$ white dwarfs.

then the abscissa is marked to indicate the location of neutron stars with $1.4 M_\odot$, $R_* = 10^6$ cm at $-7.7$ and and of main sequence stars with $1.0 M_\odot$, $R_* = R_\odot$ at $-2.7$. In addition, white dwarfs span a range from $-5.8$ to $-3.4$ depending upon mass (heavy ones with $1.35 M_\odot$, $R_* = 10^{-3} R_\odot$ lie at the small end, light ones with $0.05 M_\odot$, $R_* = 10^{-2} R_\odot$ at the large end).

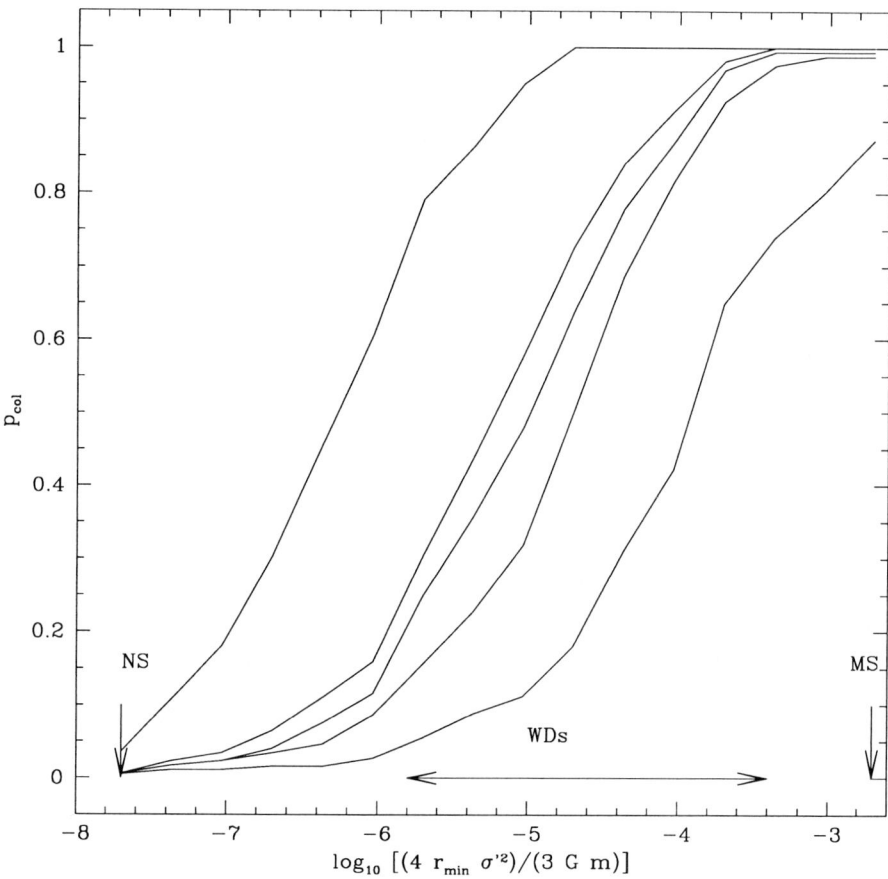

*Figure 3.* The probability of collision of an immortal binary as a function of the physical distance of separation $r_{min}$ that yields a collision. The arrows mark the location of neutron stars (NS), white dwarfs (WDs) and main sequence stars (MS) for fiducial parameters (see text). The multiple lines correspond to different depths of the central potential $\Phi(0)/\sigma_s^2 = 2, 6, 11, 14$ and $128$ (right to left).

A detailed analysis of the factors that lead to the remarkably small distances of closest approach shows that three independent physical effects contribute. First, as is well-known, resonant encounters increase the collision probability as the three stars form a long-lived interacting three-body system. The cross section for two stars to pass within $0.01a$ when a hard, circular binary interacts with a third star is of order $a^2$ times the gravitational focusing term. Hence, strong encounters immediately imply elevated collisional cross sections. Cross section for close encounters have been derived by Hut and Inagaki (1985) and Sigurdsson and Phinney (1993); our own calculations of these cross sections are in good agreement with the previous studies.

Second, 3B binary hardening involves a large number of successive encounters. The repetitive nature raises the probability that at least one collision will occur as the binary ages (and is analogous to the effect of resonant interactions on collision probability). Although only a fraction of the simulated encounters with pericenter less than $\zeta a$ are strong (roughly $3/\zeta$), these significantly increase the cumulative probability of a collision for a binary of a given hardness.

Third, the value of the binary eccentricity appears to play some role, albeit a subdominant one, in enhancing collisions. The cross section for close encounters depends upon $e$: the probability for an $e = 0.7$ (0.99) binary to suffer a collision is about 1.6 (10) times that of a circular binary. The 3B binaries have a nearly thermal eccentricity distribution and this raises the collision probability over estimates made based on purely circular calculations. A related but distinct physical effect is that most of the simulated encounters are weak, angular momentum changing events. Thus, the binary eccentricity undergoes some diffusion in the interval between the strong, resonant encounters. This diffusion provides a small enhancement of the collision probability because the binary explores some very eccentric configurations.

Taken together, these factors appear to account for the high propensity for collisions to occur during the hardening process.

## 5. Effective Heating

When collisions occur the net heat extracted from an immortal binary is altered. Figure 4 illustrates the average change in binary binding energy per immortal binary until ejection or collision. All subsequent consequences of the collision have been ignored in so far as the heating is concerned (i.e. changes due to altered stellar evolution, mass ejected to infinity by the collision, etc.). Also, no attempt has been made to differentiate the cluster heating via bound versus ejected single stars; the inferred heating is simply the average change in binary binding energy. The total heating (solid line) decreases as the finite size of the objects and the collision probability increases. Two contributions to the total are independently noted: the average heating by binaries ultimately ejected and by binaries that collide. The results are presented to allow one to estimate easily the steady-state heating rate as the generation rate of immortal binaries ($\dot{n} = 0.75 n^3 G^5 m^5 / \sigma_s^9$ GH93) times the mean heating per binary ($dE/dt = \dot{n}\overline{\Delta E}$).

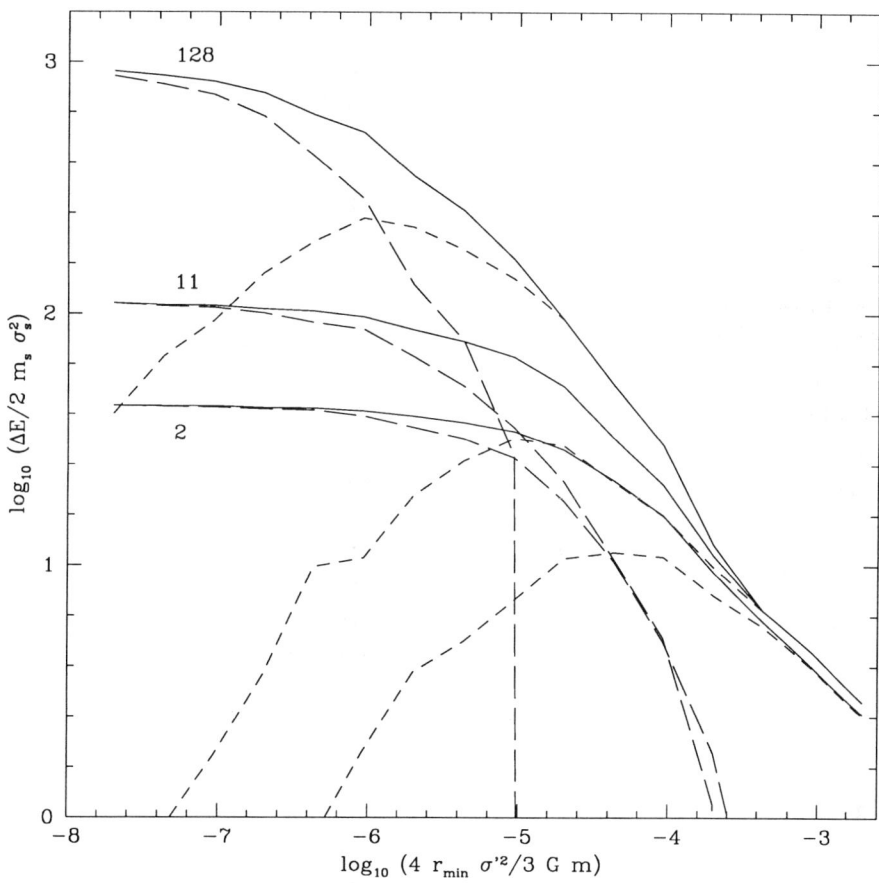

*Figure 4.* The change in binding energy per immortal binary is given as a function of the physical distance that yields a collision. The solid lines are the total heating; the heating of ejected binaries (long dashes) and of binaries that collide (short dashes) are independently noted. Groups of three lines are marked by the depth of the potential in units of $\Phi(0)/\sigma_s^2$.

## 6. Conclusions

The steady-state distribution of binding energy, eccentricity, distance of closest approach, and maximum recoil velocity for 3B binaries has been calculated. This generalizes previous calculations of the distribution of binding energy for hardening binaries (Goodman and Hut, 1993) and estimates of the collisional probability based on averaged cross sections. The collision probability (meaning the probability that two point masses pass within a distance of two physical radii) is significant for objects as small as white dwarfs in typical cluster potentials with typical velocity dispersions. The heating efficiency is diminished when collisions are included. When 3B binaries are responsible for heating post collapse cores the results provide a direct relationship between the stellar collision rate, the stellar heating rate and the rate of energy transport through the cluster as a whole.

## 7. Acknowledgements

We acknowledge support from the IAU Symposium organizers, NASA NAGW-2224, NSF AST-86-57467 and AST-91-19475.

## References

Davies, M.B., Benz W. and Hills J.G. (1994), *Ap. J.* **424**, 870.
Davies, M.B., Benz W. and Hills J.G. (1993), *Ap. J.* **411**, 285.
Goodman, J. and Hut, P. (1993), *Ap. J.* **403**, 271.
Heggie, D. C. and Hut, P. (1993), *Ap. J. Supp.* **85**, 347.
Hut, P. (1985) in *Dynamics of Star Clusters, IAU Symposium No. 113*, eds. J. Goodman and P. Hut, Reidel, Dordrecht, p. 231.
Hut, P. and Inagaki, S. (1985), *Ap. J.*, **298**, 502.
Goodman, J. (1989) in *Dynamics of Dense Stellar Systems*, ed. D. Merritt, Cambridge University Press, New York, p. 183.
Kochanek, C.S. (1992), *Ap. J.*, **385**, 604.
Mardling, R. A. (1995a) *Ap. J.* **450**, 722.
Mardling, R. A. (1995b) *Ap. J.* **450**, 732.
Sigurdsson, S. and Phinney, E. S. (1993), *Ap. J.*, **415**, 613.
Takahashi, K. and Inagaki, S. (1991), *P. A. S. J.*, **43**, 589.

# TIDAL CAPTURE IN STAR CLUSTERS

ROSEMARY A. MARDLING
*Mathematics Department,*
*Monash University, Melbourne, Australia*

**Abstract.** We review the tidal capture process and in particular the chaotic orbital evolution which follows capture. We discuss the formation of low-mass X-ray binaries in globular clusters via tidal capture and speculate on the possibility that some field low-mass X-ray binaries were formed this way in *open* clusters which have since dispersed, or in existing old open clusters which are not accessible to observation because of obscuration by dust or because they are indistinguishable from the rich background of galactic stars.

## 1. Introduction

Tidal capture is a mechanism for binary formation which involves transferring the excess energy of unbound orbital motion to the tides of two stars which pass each other at a distance of a few stellar radii so that a bound system can result. The process was originally suggested by Fabian, Pringle & Rees (1975) to explain the origin of the X-ray sources being observed for the first time in globular clusters. It was suggested that these sources were neutron stars accreting from low-mass main sequence companions, *ie.*, low-mass X-ray binaries (LMXBs). The idea appears to explain the excess of such systems. Since the X-ray luminosity to mass ratio is at least 100 times higher in globular clusters than in the galactic disk (Katz 1975), it is natural to assume that a process peculiar to globular clusters is responsible. The high stellar densities and low velocity dispersions found in globular clusters greatly favour the tidal capture mechanism, but it should be noted that *open* clusters have velocity dispersions an order of magnitude lower than globular clusters, so that although the stellar densities are also much lower, rare tidal captures may take place frequently enough to explain some of the field sources observed. This will be discussed in Section 3.1.

Considerable doubt has been cast recently on the ability of the tidal capture process to produce long-lived binaries, that is, binaries which avoid tidal disruption or self-ionization. Such conclusions have been drawn from heuristic discussions of the evolution following capture (with careful attention to energy deposition - McMillan, McDermott & Taam 1987, Ray, Kembhavi & Antia 1987), or studies of the dynamical evolution of extremely close binaries for at most a few orbits after capture (see, for example, Gingold & Monaghan 1980, Rasio & Shapiro 1991, Benz & Hills 1992, Kochanek 1992). Such studies have been governed by the fact that in order to follow the evolution numerically after capture, the timestep is restricted by the dynamical timescales of the stars. For a typical capture, a star may oscillate more than a million times per orbit, at least for the first few orbits following capture, and this has severely limited the number of orbits accessible to study.

Recent work by the author (Mardling 1995a,b) has overcome this problem. By employing a normal mode analysis devised by Gingold & Monaghan (1980), and taking advantage of the fact that the orbit and the tides interact for only part of a very long period orbit, one is able to study thousands of orbits following capture. It is found that for a short time, the orbital evolution is chaotic (large tides accompanied by unpredictable large changes in eccentricity) and after this phase (given that the binary survives) the orbital evolution becomes "normal" in that tides are small and the binary circularizes slowly. We will elaborate on this process in Section 2. Section 3 discusses the formation of LMXBs via tidal capture, including field objects. Section 4 briefly discusses the role of tidal capture binaries in cluster dynamics.

## 2. Tidal Capture

In this section we summarize the main elements of the tidal capture mechanism. A more extensive review of this subject may be found in Mardling (1996a).

### 2.1. STELLAR SEPARATIONS FOR CAPTURE

Tidal capture becomes possible when two stars pass each other so closely that their tides are able to absorb the excess energy of unbound orbital motion (Figure 1). For typical globular cluster velocity dispersions ($\sim$ 10 km s$^{-1}$), this means passing within about 3 stellar radii for equal mass

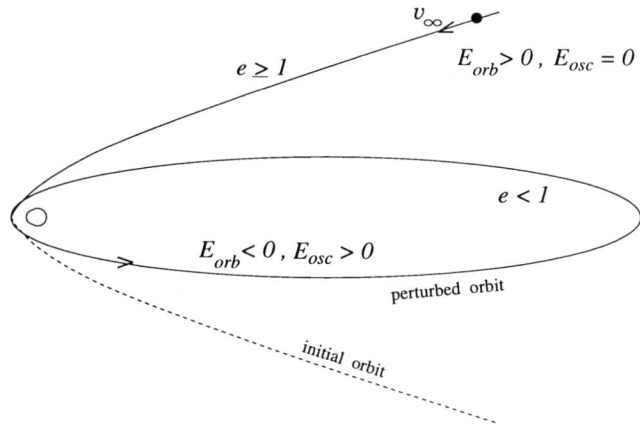

*Figure 1.* Tidal capture formation of a binary. $E_{\rm orb}$ and $E_{\rm osc}$ are the orbital binding energy and the tidal or oscillation energy respectively, and $e$ is the orbital eccentricity.

stars where one is compact.[1] This distance increases with increasing mass ratio of compact to extended star, as well as with decreasing relative velocity at infinity (less energy to absorb). The process is enhanced by gravitational focussing, for example, the impact parameter for the equal mass system described above is about 1 AU. The situation is quite different in galactic nuclei where the velocity dispersions are much higher.[2] In this case, gravitational focussing is ineffective and stars need to collide in order to absorb all the energy of unbound motion.

## 2.2. CAPTURE RATES

Given two species of stars with number densities $N_1$ and $N_2$, masses $M_1$ and $M_2$ living in an environment where the velocity dispersion is $\sigma$, the capture rate between them is given by (see, for example, Mardling 1996a)

$$\Gamma_{12} = 6.8 \frac{N_1}{10^4 \text{ pc}^{-3}} \frac{N_2}{10^4 \text{ pc}^{-3}} \frac{M_1+M_2}{M_\odot} \frac{R_p^{\min}}{R_\odot} \frac{10 \text{ km s}^{-1}}{\sigma} \text{Gyr}^{-1}\text{pc}^{-3}. \quad (1)$$

It should be noted that this formula includes *collisions*; in practise the rate at which long-lived binaries are formed by tidal capture will be a fraction of $\Gamma_{12}$ ($\sim 25\%$?). We apply this formula to capture of neutron stars in Section 3.1.

---

[1] This maximum periastron separation is for stars with convective envelopes; stars with radiative envelopes must pass even closer.
[2] The core of the nucleus of M33 is an exception; here the velocity dispersion is estimated to be as low as 21 km s$^{-1}$ (Kormendy & McClure 1991).

## 2.3. CHAOTIC ORBITAL EVOLUTION

If a Lagrangian is employed to derive the equations of motion for a system, one gains the advantage of a conserved energy. In order to examine the tide-orbit coupling in a binary system, a suitable Lagrangian consists of terms describing the orbit, terms describing the fluid motion of the stars, and a term describing the interaction between the two. Gingold & Monaghan (1980) contructed such a Lagrangian and used the normal modes of a polytrope and their associated orthogonality conditions to derive a Lagrangian which depends only on the orbital variables and the mode amplitudes (which describe the tides). The equations of motion derived from this consist of a Kepler orbit perturbed by the tides, and forced simple harmonic oscillators representing the orbital forcing of the tides. This system of equations is constrained by an energy and an angular momentum integral and it is this which makes the system self-consistent and allows one to examine in detail the tide-orbit interaction over many orbits.

For most binary orbits specified by the mass ratio, the orbital eccentricity and, say, the orbital period or the semi-major axis or the periastron separation, the solutions to these equations are periodic in that energy is exchanged between the tides and the orbit over a few (or sometimes many) orbits. Except for resonant orbits, that is, for orbits where the orbital period is nearly an integral number of oscillation periods of the most energetic mode, the tides remain small for periodic orbits. In contrast to this regular kind of behaviour, there exists a range of eccentricities and periastron separations for which the solutions are *chaotic*, that is, they exhibit extreme sensitivity to initial conditions. In addition, the orbit can (at least partially) circularize in the absence of dissipation, and this implies large tides.

Chaotic orbits are highly eccentric and generally extremely close at periastron. Capture orbits are clearly candidates for chaotic behaviour and in fact, one can show that *all* capture orbits are chaotic (Mardling, 1996b).

## 2.4. DISSIPATION AND LONG-TERM EVOLUTION

The tides raised during the chaotic phase following capture are generally extremely energetic, and it follows that nonlinear dissipative processes will be important. These in turn will limit the tidal energy, although the extent to which this is so is a matter of debate (Mardling 1995b, Kumar & Goodman 1996). The response of the stars to the huge tidal energies has also generated much discussion. In general, it is necessary to know where in the star the tidal energy is deposited. If it is in the extreme outer layers or in the core, the system will not be much affected, while if it is somewhere in between, the star may expand and cause the system to enter a common envelope phase, possibly destroying the binary (Podsiadlowski 1996).

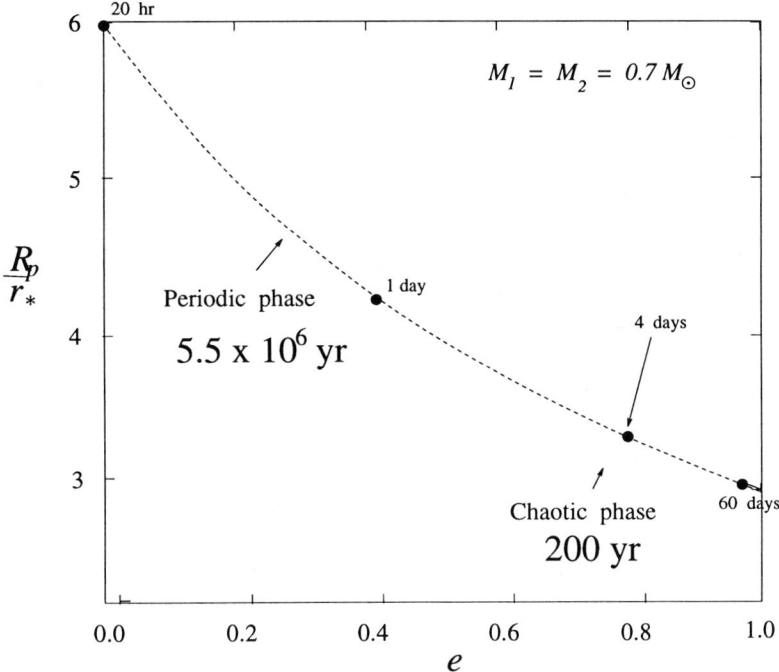

Figure 2. Chaotic and periodic circularization of an equal mass binary captured at 3 stellar radii. The times shown in days are the orbital periods at these points. The dashed curve represents constant orbital angular momentum.

Given that the system survives the short violent chaotic phase, it will eventually become quiescent in the sense of normal tidal circularization. This involves the system permanently crossing the boundary between chaotic and periodic behaviour after it has dissipated an amount of energy associated with the initial conditions at capture. For example, an equal mass system with an initial periastron separation of 3 stellar radii will leave the chaotic phase with an eccentricity of 0.8, and will take at least another few million years to circularize, depending on the linear damping timescale. This is in contrast to previous estimates of the circularization time which were of the order of a few years (McMillan, McDermott & Taam 1987).

Figure 2 illustrates the two phase circularization process by plotting the periastron separation against eccentricity for such a system. In this case, the nonlinear damping timescale is about 30 yr and the linear damping timescale is 1000 yr. The nonlinear damping timescale may well be shorter than this, but whenever it is longer than the orbital period, chaotic behaviour will persist. In this case, the chaotic phase lasts for about 200 years after which the binary takes more than 5 million years to finish circularizing.

## 3. Low-Mass X-Ray Binaries via Tidal Capture

The number of LMXBs per unit mass in our globular cluster system is at least two orders of magnitude higher than in the galaxy generally. This naturally leads to the suggestion that a process peculiar to globular clusters is responsible for this population of LMXBs, and that the field population is formed some other way. Fabian, Pringle & Rees (1975) proposed tidal capture as a viable mechanism in globular clusters, while Hut, Murphy & Verbunt (1991), faced with the suggestion that tidal capture binaries do not survive the circularization process (McMillan, McDermott & Taam 1987, Ray, Kembhavi & Antia 1987), suggested that globular cluster LMXBs are formed via exchange of neutron stars into primordial binaries. Mardling (1995b) has suggested that the exchange process generally does not produce binaries with periods as short as those observed.[3] For example for solar mass stars, the cross sections for collisions and non-collisions are about equal for binaries with periods around 20 days, with collisions dominating for smaller period binaries (McMillan 1995) and clean exchanges tending to leave binaries with around the same period as the original binary. Since the companion in a neutron star binary thus formed will have a mass at or below the turn-off mass of the cluster, the mass ratio of the binary will be such that mass transfer will be stable once the secondary evolves. Thus this scenario cannot rely on a common envelope phase to shrink the binary. On the other hand, one might expect to observe some long period globular cluster LMXBs with giant components; there is no evidence for this so far. Magnetic breaking and gravitational radiation will also be ineffective for such orbital periods (at least for low to moderate eccentricities).

Some of these binaries will become harder through flyby interactions with other cluster stars (Davies 1995), but it is not clear whether such a process is capable of producing all the LMXBs observed. In addition, if LMXBs *are* the progenitors of millisecond pulsars, the process may need to be even more efficient.

It should be pointed out that in spite of the discussion above, it is still possible for this and other dynamical interactions (such as binary-binary interactions) to produce short period mass transfer binaries.

If the tidal capture process is viable, it should be capable of producing LMXBs directly. For example, if a turn-off mass 0.8 $M_\odot$ main sequence star

---

[3] Although at present only three globular cluster LMXBs have known orbital periods, there is evidence to suggest that the other LMXBs also have short periods (Verbunt 1995).

captures a neutron star at, say, 3.5 stellar radii, it will leave the chaotic phase with an eccentricity of around 0.9 and slowly circularize to a separation of 7 stellar radii. This is assuming there is no mass loss during the evolution, an assumption which almost certainly does not hold during the chaotic phase. With mass loss the orbit will widen at periastron and this may in fact reduce the length and intensity of the chaotic phase, although as the mass ratio of the system changes, so does the chaos boundary in a way which competes with the increasing orbital separation. On the other hand, the tides can absorb a significant fraction of the orbital angular momentum during the chaotic phase (the amount being proportional to the tidal energy) so that substantial mass loss can also mean substantial angular momentum loss. Throughout the whole circularization process, tidal forces will act to spin up the star. Magnetic breaking, effective in the low-mass main sequence stars of interest, will act against this spin-up so that orbital angular momentum will be continually lost during circularization. Note that old low-mass main sequence stars in globular clusters have negligible spin, precisely because of magnetic breaking, so that the spin-up process after capture could result in significant angular momentum loss from the system by the time of circularization (recall that the circularization time predicted by the model discussed in this paper is at least four orders of magnitude longer than previous estimates).

In the absence of angular momentum loss the binary described above, once circularized, will have an orbital period of around 24 hours. In order for mass tranfer to start, the secondary will either have to evolve (in this case to a radius of about 1.8 $R_\odot$) or the orbit will have to shrink via magnetic breaking (or both). Stars less massive than the turn-off mass of the cluster must rely on the latter process. Of the three globular cluster LMXBs with known orbital periods, two are tidal capture candidates. The object in M15 which has an orbital period of 17.1 hours (Ilovaisky et al. 1993) must contain a slightly post main sequence donor, while the object in NGC 6441 with an orbital period of 5.7 hours (Sansom et al. 1993) must contain a main sequence star and may have started mass transfer after a period of orbital shrinkage via magnetic breaking (enhanced in the later stages by gravitational radiation).

3.1. THE FIELD POPULATION OF LMXBS

The main formation scenario for the field population involves common envelope evolution (van den Heuvel 1983), and there are many evolutionary constraints which a binary must satisfy in order to evolve to an LMXB via this channel. On the other hand, these objects are rare so that any model for their formation must necessarily reflect this. Webbink & Kalogera (1994)

TABLE 1. CAPTURE RATES FOR NEUTRON STARS

|     | $N_1$  | $N_2$ | $M_1/M_\odot$ | $M_2/M_\odot$ | $R_p/R_*$ | $\sigma$ (km s$^{-1}$) | $\Gamma_{12}$ |
|-----|--------|-------|---------------|---------------|-----------|------------------------|---------------|
| (a) | 30,000 | 900   | 0.8 (MS)      | 1.4           | 4         | 10                     | 13            |
| (b) | 300    | 3     | 1.0 (MS)      | 1.4           | 5         | 1                      | 0.007         |
| (c) | 100    | 3     | 1.0 (giant)   | 1.4           | 2(?)      | 1                      | 0.01          |

list nine constraints for LMXB formation via common envelope evolution and there are many uncertainties involved in each one, making statistical estimates of the number of LMXBs formed via this process difficult.

Here we propose that at least some of the field LMXBs were produced via tidal capture in *open* clusters which have since dispersed, or in existing old open clusters which are not accessible to observation because of obscuration by dust or which are indistinguishable from the rich background of galactic stars. An obvious problem with this scenario is cluster retention of neutron stars. We appeal to the fact that globular clusters manage to retain a significant number of neutron stars as evidenced by their large population of LMXBs and millisecond pulsars, despite having escape velocities of only a few tens of kilometers per second. While the escape velocities of open clusters are lower still, the process which allows globular clusters to retain some neutron stars might be expected to operate in open clusters as well.

Table 1 uses equation 1 to estimate the capture rates of neutron stars for various hypothetical clusters, where $N_2$ is the number density of neutron stars. Case (a) gives the capture rate of neutron stars by turn-off mass main sequence stars in a globular cluster like 47 Tuc; these numbers are taken from Davies & Benz (1995). Case (b) might represent an old open cluster containing 10,000 stars, one which is two orders of magnitude less massive than 47 Tuc. We take $N_1$ to be simply a hundredth of the corresponding value for case (a). Since the neutron star content of such a cluster is unknown, we take $N_2$ to be a third of a hundredth of the corresponding value for case (a). Case (c) considers capture of neutron stars by giants of radius $\sim 50 R_\odot$ in the same open cluster as case (b). While these numbers are highly speculative, the results serve to illustrate that *given* such neutron star densities, tidal capture of neutron stars in *open* clusters must be considered seriously. For instance, although the capture rate for case (b) is low, this represents 7 captures per 1000 clusters per Gyr, and given the large number of such clusters which exist in the galaxy over a Gyr, as well as the low birthrate and longevity of LMXBs, it is not unreasonable to suggest that some LMXBs are produced in this way.

$N$-body simulations of open clusters containing 10,000 stars (Aarseth, this volume) have revealed several two-body hyperbolic encounters with closest approach separations less than a few stellar radii. None resulted in tidal capture because in each case the relative velocity was too high. Nonetheless, given that only a few such calculations have been performed to date, these results lend some support to the above estimate for capture rates in open clusters.

## 3.2. CATACLYSMIC VARIABLES

There has been considerable debate in recent years about the existence or otherwise of cataclysmic variables (CVs) in globular clusters. Grindlay & Cool (this volume) argue that the abundant soft X-ray sources are CVs, while Shara (this volume) argues that more should be observed in outburst if these sources are indeed CVs. One might expect that if tidal capture is efficient at making LMXBs, then it should be efficient at making CVs. The tidal capture process is still not well understood, particularly the stellar response to the chaotic phase, and it may well be that a better understanding will lead to a resolution of this problem.

## 4. Tidal Capture Binaries and Cluster Dynamics

Before the discovery of giants in the outer regions of some globular clusters (Pryor et al. 1989), it was believed that such clusters were devoid of primordial binaries (Gunn & Griffin 1979). At the same time, it was known that binaries are necessary for core support against collapse and for reexpansion of the core, and tidal capture binaries were shown to be capable of performing these roles (Statler, Ostriker & Cohn 1987). Assumed to be extremely hard circular binaries, they were found to be an *indirect* heat source in that they tended to be ejected from the core.

More recent calculations have included the effect of primordial binaries on cluster evolution (McMillan, Hut & Makino 1990, 1991 and Heggie & Aarseth 1992) and these are found to be a much more efficient direct energy source for the core. $N$-body calculations now include more realistic circularization times for tidal capture binaries (Aarseth, this volume, Mardling & Aarseth, 1996), but it will be necessary to wait until larger values of $N$ are achieved before the role of tidal capture binaries in globular cluster dynamics becomes apparent (note that the circularization times for such binaries are comparable to the central relaxation timescales for many globular clusters).

## Acknowledgements

I thank Dr Jun Makino and Professor Daiichiro Sugimoto for inviting me to Tokyo, and my colleagues and friends Drs Sverre Aarseth, Fred Rasio, Steve McMillan and Piet Hut for getting me there. Thanks also to Dr Frank Verbunt for introducing me to Noh theater.

## References

Brandt, N. and Podsiadlowski, Ph.: 1995, *Mon. Not. R. Astron. Soc.*, **274**, 461
Davies, M. B. : 1995, *Mon. Not. R. Astron. Soc.*, in press
Davies, M. B. and Benz, W.: 1995, *Mon. Not. R. Astron. Soc.*, in press
Fabian, A. C., Pringle, J. E. and Rees, M. J.: 1975, *Mon. Not. R. Astron. Soc.*, **172**, 15P
Gingold, R. A. and Monaghan, J. J.: 1980, *Mon. Not. R. Astron. Soc.*, **191**, 897
Gunn, J. E. and Griffin, R. F.: 1979, *Astronom. J.*, **84**, 752
Heggie, D. C. and Aarseth, S. J.: 1992, *Mon. Not. R. astr. Soc.*, **257**, 513
Hut, P., Murphy, B. W. and Verbunt, F.: 1991, *Astron. & Astrophys.*, **241**, 137
Ilovaisky, S. A., Aurière, M., Koch-Miramond, L., Chevalier, C., Cordoni, J.-P. & Crowe, R. A. 1993, *Astron. & Astrophys.*, **270**, 139
Katz, J. I.: 1975, *Nature*, **253**, 698
Kormendy, J. and McClure, R. D.: 1993, *Astronom. J.*, **105**, 1793
Kumar, P. and Goodman, J.: 1996, submitted
Mardling, R. A.: 1995a, *Astrophys. J.*, **450**, 722
Mardling, R. A.: 1995b, *Astrophys. J.*, **450**, 732
Mardling, R. A.: 1996a, to appear in *Evolutionary Processes in Binary Stars*, eds R. A. M. J. Wijers, M. B. Davies and C. A. Tout, Kluwer, Dordrecht, Holland
Mardling, R. A.: 1996b, in preparation
Mardling, R. A. and Aarseth, S. J.: 1996, in preparation
McMillan, S. L. W., 1995, private communication
McMillan, S. L. W., Hut, P. and Makino, J.: 1990, *Astrophys. J.*, **362**, 522
McMillan, S. L. W., Hut, P. and Makino, J.: 1991, *Astrophys. J.*, **372**, 111
McMillan, S. L. W., McDermott, P. N. and Taam, R. E.: 1987, *Astrophys. J.*, **318**, 261
Podsiadlowski, Ph.: 1996, *Mon. Not. R. Astron. Soc.*, in press
Pryor, C., McClure, R. D., Hesser, J. E. and Fletcher, J. M.: 1989, in *Dynamics of Dense Stellar Systems*, ed. D. Merritt, (New York: Cambridge University Press), 175
Ray, A., Kembhavi, A. K. and Antia, H. M.: 1987, *Astron. & Astrophys.* **184**, 164
Sansom, A. E., Dotani, T., Asai, K. & Lehto, H. J. 1993, *Mon. Not. R. astr. Soc.*, **262**, 429
Statler, T. S., Ostriker, J. P. and Cohn, H. N.: 1987, *Astrophys. J.*, **316**, 626
van den Heuvel, E. P. J.: 1983, in *Accretion-driven Stellar X-ray Sources*, eds W. H. G. Lewin and E. P. J. van den Heuvel, Cambridge University Press, p. 303
Verbunt, F.: 1995, private communication
Webbink, R. F. and Kalogera, V.: 1994, in *Evolution of X-Ray Binaries*, eds S. S. Holt and C. S. Day, AIP Press, New York, p 321

# THE FORMATION OF GLOBULAR CLUSTERS AND OF THE STARS WITHIN THEM

D.N.C. LIN
*Lick Observatory, Univ. of California, Santa Cruz, CA 95064*
AND
S.D. MURRAY
*Lawrence Livermore Nat'l Lab., L-23, Livermore, CA 94550*

We propose that proto-globular cluster clouds form in a collapsing protogalactic cloud as a consequence of thermal instability. The clouds are photoionized and heated by nearby massive stars. Most are not self-gravitating, but are confined by the residual hot gas in the protogalactic cloud. Their masses evolve as they undergo cohesive collisions with each other and erosion due to interaction with the residual halo gas. Collisions may also trigger thermal instability and fragmentation within protocluster clouds. The resulting cloudlets are pressure confined, and fall toward the center of the protocluster cloud due to inverse buoyancy. Their mass distribution is also regulated by coagulation and erosion. While most cloudlets have substellar masses, the largest become self-gravitating, and collapse to form protostellar cores without further fragmentation. The initial stellar mass function is established as these cores capture additional residual cloudlets. Energy dissipation from the mergers ensures that the cluster will remain bound in the limit of low star formation efficiency. Dissipation also promotes the formation and retention of the most massive stars in the cluster center.

## 1. Introduction

Globular clusters contain the oldest stars in the Galaxy. Investigations of the formation of these clusters and the stars within them can therefore provide important clues regarding the origin and evolution of the Galaxy. Our main objective here is to provide an overview of some recent developments in the theory of cluster formation, and their implications in terms of a comprehensive unified theory.

## 2. Fragmentation of Protogalactic Clouds

Our basic conjecture is that globular clusters formed within infalling protogalactic clouds (PGC's) (see, e.g. Fall & Rees 1985). During the collapse of a PGC, density inhomogeneities and velocity variations lead to shocks which heat the gas to the virial temperature of the galactic halo (Binney 1977; Rees & Ostriker 1977; White & Rees 1978). In order for a PGC to collapse, its cooling time scale, $\tau_c$, must be shorter than the dynamical time scale, $\tau_d$, on which it can contract. For systems with masses comparable to the Galaxy, this condition is satisfied when the characteristic length scale of the PGC, $D < 100$ kpc (Blumenthal et al. 1984).

Subsequent fragmentation of the PGC requires the growth of density inhomogeneities on a time scale, $\tau_g < \tau_d$. If the PGC is cold, gravitational instability causes perturbations with initial amplitude $\delta_0$ to become nonlinear when the system collapses by a factor $\sim \delta_0^{2/3}$ (Hunter 1962). Even if $D = 100$ kpc, the initial perturbations must be nearly nonlinear for gravitational instability to trigger fragmentation of the PGC at a few kpc. Such a requirement is difficult to satisfy on scales much smaller than $D$.

Thermal instability can, however, lead to the rapid growth of perturbations, even in the limit of infinitesimal $\delta_0$, when $\tau_c < \tau_d$ (Field 1965). At the virial temperature of the PGC, the dominant cooling mechanisms are bremsstrahlung and recombination processes (Dalgarno & McCray 1972) for which $\tau_c$ increases with temperature. In this case, the temperature difference between cooler perturbed regions and the background is amplified and a pressure gradient across the interface is established. The resulting gas flow from the hot background towards the cooler perturbed regions further acts to decrease $\tau_c$ rapidly from $\sim \tau_d$ at the onset of thermal runaway to $\ll \tau_d$.

Fragmentation of a cloud requires instabilities for which the growth time scale, $\tau_g$, increases with wavelength, $\lambda$. For thermal instability associated with local cooling, $\tau_g$ is independent of $\lambda$. During the growth of thermal instability, however, pressure balance is maintained on scales smaller than that ($l_s$) traveled by sound on a time scale $\tau_c$. Compact ($< l_s$), cool dense regions contract as entropy is lost at a much faster rate than the background. Since $\tau_c$ increases with temperature, the interface separating cool regions and the background retreats at an accelerating pace, leading to the growth of Rayleigh-Taylor instability, for which $\tau_g$ is an increasing function of the wavelength (Burkert & Lin in preparation). The acceleration of the interface increases as the temperature in the perturbed region decreases. The long wavelength disturbances become stabilized when their growth time scales become long compared with that over which the acceleration of the interface is modified. Nevertheless, they grow to nonlinearity due to the Richtmyer-Meshkoff instability. As the temperature in the perturbed regions continues to plummet, $l_s$ decreases rapidly to below the length scale of the perturbed regions, and so most parts of the perturbed region cool without any significant change in density. Pressure balance is, however, enforced and Rayleigh-Taylor instability grows within $l_s$ from the interface.

Bremsstrahlung, recombination, and atomic hydrogen emission decrease rapidly in efficiency below $\sim 10^4$ K. In even a metal-free PGC, however, non-equilibrium recombination leads to the formation of a small amount of $H^-$ ions which combine with neutral $H$ to form $H_2$. Radiative emission by $H_2$ reduces the gas temperature to $\sim 10^2$ K (Murray & Lin 1990). If $[Fe/H] > -3$, lower temperatures ($\sim 10$ K) are attainable due to cooling by heavy elements (e.g. CI, CO, and grains). Equilibrium temperatures $\sim 10^4$ K may be maintained in the presence of external heat sources (see § 3). Once the final temperature is attained, pressure balance is re-established on all scales. The PGC becomes a two-phase medium with density contrast inversely proportional to the temperature difference. The residual halo gas (RHG) in the background remains at the virial temperature, with a density such that its thermal energy is lost on a time scale $\sim \tau_d$. Energy loss from the RHG can occur through both radiative cooling and thermal conduction between the RHG and the cool, dense clouds (McKee & Cowie 1977). At the distance of $\sim 10$ kpc, the energy balance implies $nT \sim 10^3 - 10^4$ where $n$ and $T$ are the halo number density and temperature.

Most of the cool dense clouds are pressure confined by the RHG, and so have similar $nT$. The RHG also exerts a drag on the motion of the clouds as they are accelerated by the gravity of the Galactic halo. The terminal speed of clouds with size L is $V_t \sim (f_n L/D)^{1/2} V_k$ where $f_n$ is the density ratio of the clouds to the RHG, and $V_k$ and $D$ are the velocity dispersion and size of the halo, respectively. The motion of the clouds through the RHG also leads to mass loss due to Kelvin-Helmholtz instability (Murray *et al.* 1993), whose growth time scale ($\tau_{KH}$) is a few times $L/V_t$. Because $\tau_{KH}$ increases with $\lambda$, the KH instability leads to fragmentation. The break down of the clouds increases their collective area filling factor and so increases their collision frequency. A balance between disruption and coagulation establishes an equilibrium size distribution.

## 3. Formation of Protocluster Clouds

A lower limit on their size distribution is set by the clouds' evaporation due to conduction from the hot RHG. The central density of these clouds increases with their mass. In the high mass limit, the clouds' self-gravity suppresses the Kelvin-Helmholtz instability. But at a critical mass $M_c \sim T^2/(nT)^{1/2} M_\odot$, thermal pressure can no longer support the weight of the envelope (Bonner 1956), and the clouds undergo inside-out collapse (Shu 1977). During the collapse, although the Jean's mass decreases with density, it is larger than the mass contained inside any radius. The collapse is stable and does not lead to fragmentation without any further unstable cooling. Thus, contrary to the opacity-limited fragmentation scenario (Hoyle 1953; Low & Lynden-Bell 1976), $M_c$ represents the minimum mass for isothermal collapsing clouds (Tsai, in preparation).

In a metal-free environment, $T \sim 10^2$ K and $M_c \sim 10 - 10^2 M_\odot$. The resulting massive stars are, however, copious sources of UV radiation. A population of $\sim 10^4$ O5 stars is adequate to photoionize the entire PGC out to 100 kpc. Photoionization raises $T \sim 10^4$ K and $M_c \sim 10^6 M_\odot$. Small (a few $M_\odot$) heated clouds are stable and star formation is quenched. As the massive stars evolve off the main sequence, the UV flux diminishes, cooling again leads to $T \sim 10^2$ K in sheltered regions, and spontaneous star formation is resumed. This self-regulated star formation scenario has three implications: 1) Stars formed in a metal-poor environment are massive and short-lived, consistent with their rarity today. 2) The elemental abundance distribution are produced by type II supernovae, consistent with that observed among stars with [Fe/H]< $-1$ (Wheeler et al. 1989). 3) The self-regulated star formation rate naturally yields [Fe/H] $\sim 0.1$ on the collapse time scale $\tau_d \sim 1$ Gyr, consistent with the halo stars.

The magnitude of $M_c$ for the warm ($\sim 10^4$ K) clouds is comparable to the mass of globular clusters. We refer to these warm, massive clouds as protocluster clouds (PCC's). PCC's with mass $(M) < M_c(\sim 10^6 M_\odot)$ are confined by the RHG. These PCC's are completely photoionized by a single O5 star at a distance greater than their size (typical a few pc). They would persist for a significant fraction of $\tau_d$ if the accretion of smaller clouds can compensate for their mass loss due to stripping by the RHG. To verify that the PCC's were pressure-confined, we first estimate their $nT$ from the current properties of globular clusters, averaged over their half-mass radius $(r_h)$. We use these because $n$ and the velocity dispersion at $r_h$ do not change significantly during post-formation evolution. Extrapolation to the stage prior to star formation is, however, highly uncertain. If, after their formation, the stars undergo collapse and virialization from rest, the clouds' initial radii $(r_i)$ would be $\sim 2 r_h$. Larger $r_i$ would be expected if star formation requires dissipative collisions and coagulation of substellar fragments (see § 4). But $r_i$ is unlikely to be larger than the tidal radii of the PCC's, which are typically a few times larger than $r_h$. Thus, the initial density of the PCC's may be 1-3 orders of magnitude smaller than the average cluster density at $r_h$ today. Based on the present velocity dispersion of the clusters, we infer the initial temperature of the PCC's to be $\sim 10^4$ K, comparable to that expected if they were photoionized. From these estimates, we infer $nT \sim 10^{2-5}$, and that $nT \propto D^{-3}$ where $D$ is the cluster's distance to the Galactic center. In accordance with the pressure confinement scenario, both the magnitude and the spatial dependence of PCC's $nT$ are consistent with those expected for the RHG (Murray & Lin 1992). ¿From these results, and the cluster metallicities, we can also estimate the cooling time scale ($\tau_{cc}$) and dynamical time scale ($\tau_{dc}$), of the PCC's. The ratio $\tau_{cc}/\tau_{cd}$ increases from $\sim 10^{-4}$ near the Galactic bulge to $\sim 1$ at $\sim 100$ kpc. In most PCC's, $\tau_{cc} \ll \tau_{cd}$ and thermal equilibrium is only possible in the presence of external UV photons with a flux comparable to that required by self regulated star formation in the halo.

## 4. Induced Star Formation

A constraint on the duration of the star formation epoch can be imposed from the metal homogeneity (primarily in [Fe/H]) among stars within individual clusters. Based upon the color of the red giant branch, upper limits can be placed on the spread, $\Delta$ [Fe/H], ranging from 0.1 for metal poor clusters to 0.01 for metal rich clusters (Sandage & Katem 1977; Richer & Fahlman 1984). In M92, $\Delta$ [Fe/H]$< 0.2$, and [Fe/H]$= -2.2$ (Stetson & Harris 1988). With a mass $3 \times 10^5 M_\odot$, the total content of all metals in the cluster is $\sim 30 M_\odot$, representing the production of only a few supernovae. If star formation in M92 occurred over more than a few million years, supernovae would increase both [Fe/H] as well as $\Delta$ [Fe/H] to values much larger than observed.

The small $\Delta$ [Fe/H] implies that PCC's are well mixed and pre-enriched prior to a rapid burst of star formation. The most efficient mixing process within the PCC's is large scale turbulence (Murray & Lin 1990), which may be excited by the Kelvin-Helmholtz instability at the interface with the RHG. Since shock dissipation damps supersonic motion efficiently, the mixing time scale is typically a few $\tau_{dc} \gg \tau_{cc}$. External heating by nearby massive stars is then required to maintain a stable thermal equilibrium for the PCC's. Such an equilibrium is only stable, however, if the PCC's are not strongly self-gravitating.

In PCC's with marginal self gravity, mass loss due to ram pressure stripping by RHG is stabilized. Mass loss may also be compensated by coagulation in a collisional equilibrium. Collisions also introduce perturbations which, with a sufficiently large amplitude, could increase the cooling efficiency (Murray & Lin 1989). If the new state is thermally unstable, runaway cooling would lead to $T < 100$ K. Once the PCC's re-establish pressure equilibrium, their $n$ increases by the same factor by which $T$ is reduced. Density enhancement increases the recombination rate and opacity, which blocks the penetration of UV photons and external heating.

Coagulation may also increase the masses of the PCC's to a point where they become self-gravitating, at which point the heating and cooling equilibrium is no longer stable (Murray & Lin 1992). With sufficiently large column density, self-shielding may also act to prevent the heating of the PCC's interior. All of these effects indicate that there is a small range of mass over which the PCC's become thermally unstable and begin to collapse.

During collisions, PCC's dissipate much of the kinetic energy of their Galactic orbits. Today, the Galactic distance of most globular clusters is smaller than that of the Sun. Thus, while most of the clusters are members of a spheroidal population, their distribution implies that, like the disk, their progenitors have much lower specific energy than the PGC, as would result from collisional dissipation. The collisions which triggered the onset of thermal instability may also have lead to the separation of PCC's from dark matter as well as any older stars which may have formed within them and contributed to their prior enrichment.

The onset of thermal and subsequent dynamical instabilities induce the fragmentation of PCC's into small cloudlets analogous to the formation of PCC's within a PGC. The rapid growth of thermal instability results in a strong pressure imbalance between the cooling cloudlets and a tenuous, warm ($10^4$ K) residual gas (WRG). A pressure wave is therefore driven into the cloud, which is subject to Rayleigh-Taylor instabilities, leading to fragmentation of the PCC's into cold, dense, low-mass cloudlets, confined by the thermal pressure of WRG which is, in turn, in balance with that of the RHG. With the clusters' present metallicity, the cloudlets cool to $\sim 10$ K so that $M_c \sim 0.1 M_\odot$. The cloudlets' velocity dispersion is comparable to their sound speed, so that the system is dynamically cold. The inverse buoyancy of the cloudlets in the WRG therefore causes the system of cloudlets to collapse under their collective gravity.

During their infall, the cloudlets lose mass through ram pressure stripping and gain mass through coagulation. In a collisional equilibrium, a power-law size distribution is established. At the upper limit of the mass spectrum, the cloudlets' masses exceed $M_c$ and they collapse to form protostellar cores. Since no additional fragmentation is expected, the minimum stellar mass is $\sim M_c$. Protostellar cores can continue to acquire additional mass as they merge with residual cloudlets. The observed number of stars in the range $m$ to $m + dm$ is usually approximated as $\frac{dN_*}{dm} \propto m^{-(1+x)}$. Such a power-law mass distribution is expected in idealized coagulation models (Nakano 1966; Kwan 1979; Silk & Takahashi 1979). Approximate asymptotic solutions of the coagulation equations indicate $x \approx 0$ if the collisional cross-sections are given by the geometric cross-sections of the particles, whereas $x \approx 1$ if encounters are strongly gravitationally focussed.

These values of $x$ encompass much of the observed range for both open and globular clusters (Salpeter 1955; Miller & Scalo 1979; Scalo 1986; Francic 1989; Capaccioli, Ortolani, & Piotto 1991).

Numerical simulations of the coagulation of cloudlets in PCC's (Murray & Lin 1996) shows that the geometric cross section determines the collisional frequency between uncollapsed cloudlets leading to a relatively flat mass distribution. But the protostellar cores have compact sizes and their capture rate of residual cloudlets is strongly affected by gravitational focusing. In this case, the growth time scale is a decreasing function of the mass. A few massive cores rapidly grow prior to any significant mass increase among most other cores, leading to a steep initial mass function (IMF) for the protostellar cores. Collisions between the residual cloudlets and protostellar cores lead to the growth of the cores' mass and the dissipation of their relative motion, while global instabilities increase their velocity dispersion (Aarseth et al. 1988). In the final stage of the PCC's collapse, the velocity dispersion increases more rapidly than does the mass of protostellar cores, and the collisional frequency is once again determined by the geometrical cross section of the cloudlets, leading to a flattening of the IMF during the collapse. The final slope of the IMF is determined by whether or not most cloudlets have undergone collisions to form protostellar cores before gravitational focusing ceases to be important.

The growth of the most massive cores is terminated when their UV emission heats and ionizes nearby cloudlets. The most massive stars require many dissipative mergers, and so are preferentially formed in the cluster center. This expectation is in contrast to the implication of opacity-limited fragmentation scenario, in which the Jean's criterion suggests that the high mass stars are preferentially formed in low density environments.

After the first generation of stars are formed, the residual gas must be removed from the clusters to avoid self-contamination. Extrapolating from the present stellar population, a Salpeter IMF would imply at least several tens of early type stars within a typical globular cluster. The UV flux from these massive stars is sufficiently intense to photoionize the residual gas throughout a PCC (Tenorio-Tagle et al 1986). In the shallow potential of PCC's, the resultant internal heating leads to the formation of an expanding ionization wave which is efficient in clearing out the residual gas within a few $10^6$ yr (Noriega-Crespo et al. 1989).

Unless more than half of the gas in PCC's is converted into stars, the removal of the gas on a time scale comparable to the dynamical time scale rapidly reduces the depth of the cluster's potential. In most cases, the disposal of residual gas would then lead to the disruption of the cluster (Lada al. 1984). If, however, young stars form through the coagulation of low mass cloudlets, energy dissipation resulting from the mergers leads to the resulting star clusters having radii much smaller than the original PCC's. It is therefore more likely that the newly formed clusters will remain gravitationally bound even in the limit of inefficient star formation. Numerical simulations show an order of magnitude reduction in the half mass radius of the star cluster formed through coagulation of a system of collapsing cloudlets (Murray & Lin 1996).

## 5. Comparison with present-day cluster and star formation

In the construction of the above comprehensive model for the formation of globular clusters and the stars within them, we used the observed dynamical and chemical properties of globular clusters as constraints. This model has many implications on the observational properties of young clusters. However, young globular clusters are only found in distant galaxies such as NGC 1275 (Holtzman et al. 1992), where they cannot presently be resolved. However, recent observations find that most stars in the Galaxy today form in clusters in which the time scale for star formation is quite short (Lada et al. 1991; Lada 1992). The central density of some young clusters are comparable to that of some globular clusters. In the Trapezium cluster, all the stars appear to have an age $< 10^6$ yr (Prosser et al. 1994) which is $\sim 2\tau_{dc}$. These time scales are consistent with the above model, in which star formation proceeds through a sequence of initial gas fragmentation, coagulation, protostellar collapse, and the clearing of residual gas. (There are exceptions such as IC 348, where star formation has persisted for many $\tau_{dc}$, Lada & Lada, 1995). It is therefore of value to compare the implications of our model with the dynamical properties of nearby sites of ongoing star formation, to provide both supporting evidence and constraints upon the above picture of star formation.

Molecular clouds are clumpy on all scales (Scalo 1985; Zinnecker et al. 1993). Complex cloud substructure may also be inferred from the large dispersion, over a small field, in the observed extinction of the stars in the background cluster IC 5146 (Lada et al. 1994). (The extinction is equivalent to the surface density of the intervening clouds). These observations suggest that fragmentation occurs in the clouds prior to the gravitational collapse of individual protostellar cores as we have postulated above.

Magnetic fields, neglected in our cluster and star formation scenario, are observed to be important in regulating the structure of star forming regions today. In these regions, the velocity dispersions, which is correlated with the length scale of the substructures (Larson 1981), are often larger than the sound speed inferred from the transition temperature, but are comparable to the Alfvén speed (Heiles et al. 1993; Caselli & Myers 1995). In some regions, the dispersive motion of the clouds may be regulated by the interstellar magnetic fields. There are also, however, magnetic supercritical regions, where the magnetic field can no longer balance gravity. These are the regions where massive stars and small clusters are formed. The lack of polarization in the densest cores of molecular clouds suggest that magnetic fields are excluded from these regions (Goodman *et al.* 1995). The decoupling of the field from the protostellar clouds is equivalent to the loss of thermal support during a cooling instability, and may also lead to complex substructures (Terquem & Lin in preparation).

The mass function of dark cloudlets in star forming complexes such as Ophiuchus, Taurus, Orion (Scalo 1985), and L1630 (Lada *et al.* 1991) has a very similar power-law distribution. It is considerably flatter than that of the stellar IMF. The extrapolated collisional time scale for the small cloudlets is comparable to a few local dynamical time scales, and so their size distribution could arise naturally from a collisional equilibrium. The relatively flat mass spectrum is then consistent with that obtained from numerical simulations of the coagulation processes among protostellar cores (Murray & Lin 1996), and implies that the physical cross section determines the merger rate among the cloudlets. In contrast, gravitational focusing is more important for the capture of residual cloudlets by the collapsed stellar cores, and so the stellar IMF is steeper.

Shortly after the formation of protostellar cores, they are surrounded by the residual gas and appear as embedded sources. IR observations indicate that the embedded sources are strongly clustered (Greene *et. al.* 1994). These clusters are much more centrally condensed than the host cloud complex. Furthermore, the brightest embedded sources are usually found at the center of the clusters (Lada 1992). The luminosity segregation is consistent with the concept that protostellar cores form through dissipative mergers of small cloudlets, such that the most massive stars preferentially form in regions of high density where the collision frequency is greatest.

Finally, in older star forming regions, the young stellar objects emerge as T Tauri stars. In the Orion Nebula, T Tauri stars also appear to be clustered. The higher luminosity T Tauri stars are more centrally condensed than those with low luminosities (Prosser *et al.* 1994). In these regions, there is not sufficient time for post-formation dynamical evolution toward mass segregation. The observations are consistent with the more massive stars forming in dense, central regions. Since most of their kinetic energy is dissipated during the coagulation, these massive stars remain near PCC's center after the residual gas is cleared.

## References

Aarseth, S.J., Lin, D.N.C., & Papaloizou, J.C.B. 1988, ApJ, 324, 288
Binney, J. J. 1977, ApJ, 215, 483

Blumenthal, G. R., Faber, S. M., Primack, J. R., & Rees, M. J. 1984, *Nature*, 311, 517
Bonner, W. B. 1956, MNRAS, 116, 356
Capaccioli, M., Ortolani, S., & Piotto, G. 1991, A&A, 244, 298
Caselli, P. & Myers, P.C. 1995, ApJ, 446, 665
Dalgarno, A., & McCray, R. A. 1972, ARAA, 10, 375
Fall, S. M., & Rees, M. J. 1985, ApJ, 298, 18
Field, G. B. 1965, ApJ, 142, 531
Francic S. P. 1989, AJ, 98, 888
Goodman, A.A., Jones, T.J., Lada, E.A., Myers, P.C. 1995, ApJ 448, 748
Greene, T.P., Wilking, B.A., André, P., Young, E., & Lada, C.J. 1994, ApJ, 434, 614
Heiles, C., Goodman, A.A., & McKee, C.F. 1993 in *Protostars and planet III*, eds. E. H. Levy & J. I. Lunine (Tucson: Univ. Arizona Press), 279
Holtzman, J. A., et al. 1992, AJ, 103, 691
Hoyle, F. 1953, ApJ, 118, 513
Hunter, C. 1962, ApJ, 136, 594
Kwan, J. 1979, ApJ, 229, 567
Lada, C. J., Lada, E. A., Clemens, D. P., & Bally, J. 1994, ApJ, 429, 694
Lada, C. J., Margulis, M., & Dearborn, D. 1984, ApJ, 285, 141
Lada, E. A. 1992, ApJL, 1992, 393, L25
Lada, E. A., DePoy, D. L., Evans, N. J., & Gatley, I. 1991, ApJ, 371, 171
Lada, E.A. & Lada, C.J. 1995, AJ, 109, 1684
Larson, R. B. 1981, MNRAS, 194, 809
Lin, D. N. C., & Murray, S. D. 1992, ApJ, 394, 523
Low, C., & Lynden-Bell, D. 1976, MNRAS, 176, 367
McKee, C. F., & Cowie, L. L. 1977, ApJ, 215, 213
Miller, G. E., & Scalo, J. M. 1979, ApJS, 41, 513
Murray, S. D., & Lin, D. N. C. 1989, ApJ, 339, 933
—— 1990, ApJ, 357, 105
—— 1992, ApJ, 400, 265
—— 1996, ApJ, submitted
Murray, S. D., White, S. D. M., Blondin, J. M., & Lin, D. N. C. 1993, ApJ, 407, 588
Nakano, T. 1966, *Prog. Theor. Phys.*, 36, 515
Noriega-Crespo, A. Bodenheimer, P. Lin, D. Tenorio-Tagle, G. 1989, MNRAS, 237, 461
Prosser, C. F. et al. 1994, ApJ, 421, 517
Rees, M. J., & Ostriker, J. P. 1977, MNRAS, 179, 541
Richer, H. B., & Fahlman, G. G. 1984, ApJ, 277, 227
Salpeter, E. E. 1955, ApJ, 121, 161
Scalo, J.M. 1985, in *Protostars and Planets II*, eds. D.C. Black & M. Matthews, (Univ. of Arizona Press), 201
Scalo, J. M. 1986, *Fundam. Cosmic Phys.*, 11, 1
Sandage, A. & Katem, B. 1977, ApJ, 215, 62
Shu, F. 1977, ApJ, 214, 488
Silk, J. & Takahashi, T. 1979, ApJ, 229, 242
Stetson, P. B. & Harris, W. E. 1988, AJ, 96, 909
Tenorio-Tagle, G. Bodenheimer, P. Lin, D. Noriega-Crespo, A. 1986, MNRAS, 221, 635
Wheeler, J. C., Sneden, C., & Truran, J. W. 1989, ARAA, 27, 279
White, S. D. M., & Rees, M. J. 1978, MNRAS, 183, 341
Zinnecker, H., McCaughrean, M., & Wilking, B. A. 1993, in *Protostars and Planets III*, eds. E. Levy & J. Lunine, (Univ. of Arizona Press), 429

# DYNAMICS OF GALACTIC NUCLEI CONTAINING MASSIVE REMNANT STARS

H. M. LEE
Pusan University
Department of Earth Sciences, Pusan 609-735, Korea

**Abstract.** We have examined the dynamical evolution of stellar system containing massive remnant stellar component. If individual mass of remnant stars is much heavier than that of normal stars which comprise most of the mass in the cluster, remnant stars quickly form a subsystem within the core of cluster of ordinary stars. The subsystem evolves on its own relaxation time scale which is very short. However, the post collapse expansion driven by the three-body binary heating becomes very slow because the expansion energy of the compact subcluster can be easily absorbed by surrounding cluster. The gravitational radiation can lead to the merger of binaries when binaries become very hard. A central seed black hole might form if repeated merger becomes very efficient. Otherwise, relatively stable two-component phase of central compact cluster of remnant stars surrounded by larger cluster of low mass stars would last for a long time.

## 1. Introduction

The stellar evolution inevitably produces remnant stars such as white dwarfs, neutron stars and black holes. In the old stellar systems dominated by low mass stars, these remnant stars can play important roles in dynamical evolution. For example, the concentrated dark mass indicated by spectroscopic observations in some nearby galaxies are interpreted as central black holes, but compact cluster of remnant stars can be an alternative explanation (Goodman & Lee 1989).

The dynamical evolution of spherical stellar systems have been studied in great detail using various numerical techniques. The core collapse is a natural consequence of the gravothermal instability existing in collisionless stellar systems. The time to reach the core collapse is a several multiple of initial half-mass relaxation time.

The relaxation time scale is difficult to define for galactic nuclei star cluster since there is no clear boundary. The central relaxation time scale can be a well define quantity, but most galactic nuclei are not resolved even with the HST (Lauer et al. 1995). Thus the physical conditions are very uncertain, but galactic nuclei are thought to have too long relaxation time to undergo core collapse within Hubble time.

However, the time scale for core collapse in units of relaxation time can become significantly shorter than single component case if more than one mass components are present (e.g. Inagaki 1985). In multi-mass clusters, the energy exchange between different mass groups leads to the segregation of mass. This process happens in a time scale shorter than core collapse time of a single component cluster. As a result of mass segregation, the central parts are dominated by heavier component. Since the velocity dispersion of the heaver component is smaller than the virial value, the central relaxation time scale becomes very short. Thus the core collapse of the heavy component becomes very fast. If the mass ratio between heavy and light components is very large, the whole process takes place in much shorter time scale than relaxation time scale of the cluster.

White dwarfs and neutron stars may be quite abundant but they can not accelerate the dynamical evolution significantly because the individual mass of these components is not much larger than the mass of normal stars. Black holes are not thought to be very abundant, but typical mass would be much larger than ordinary stars. Even a small fraction of mass in the form of black holes would be able to influence the dynamical evolution significantly. The mass of the black holes formed by stellar evolution is uncertain, but 10 $M_\odot$ appears to be a good representative value.

In the present paper, we discuss various aspects of dynamical evolution of dense stellar systems comparable to galactic nuclei containing small amount of massive remnant stars.

## 2. Dynamical Evolution of Two-Component Clusters

We restrict our discussion to the two-component star clusters for simplicity: one component representing old population of ordinary stars, and the other component representing remnant stars. The individual mass and abundance of remnant stars are taken as free parameters. Let's denote $m_*$ and $m_D$ as individual masses of normal and degenerate stars, respectively. The total mass stored in normal and degenerate stars are denoted by $M_*$ and $M_D$, respectively. We assume that $m_D > m_*$ and $M_* \gg M_D$. Many of our discussions are only dependent on $m_D/m_*$ and $M_D/M_*$, but we assume

TABLE 1. Minimum $t_c/t_{rh,0}$ for Two-Component Clusters

| $m_D/m_*$ | $(t_c/t_{rh,0})_m$ | $(M_D/M_*)_m$ |
|---|---|---|
| 2.0 | 7.8 | 0.2 |
| 4.0 | 2.6 | 0.1 |
| 8.0 | 0.95 | 0.08 |
| 16.0 | 0.41 | 0.06 |

that $m_* = 0.7$ M$_\odot$ when conversion to physical units is necessary.

The time for mass segregation is typically order of $t_r \times (m_*/m_D)$ where $t_r$ is the relaxation time of the stellar system. However, the exact amount of time for mass segregation is difficult to define. But core collapse of the cluster of remnant stars gives well defined epoch and the core collape takes place in a very short time once the remnant stars are sufficiently segregated. Therefore, core collapse time can be a good measure for time scale for mass segregation. The collapse time is strongly dependent on the ratio of $m_*/m_D$, but it also varies with $M_D/M_*$. In Table 1, we have shown the minimum $t_c/t_{rh,0}$ for given $m_D/m_*$, where $t_c$ is the collapse time and $t_{rh,0}$ is the initial half-mass relaxation time. In computing these numbers, we have used multi-mass Fokker-Planck equation and have assumed Plummer model as initial models. We have ignored any collisional effects in these calculations. Note that the collapse time of the single component cluster is $15.8 t_{rh,0}$ (Cohn 1980). The last column of this table is the value of $M_D/M_*$ giving minimum $t_c/t_{rh,0}$, but it is not an accurately determined number because the minimum is rather broad. From this table, it is clear that the dynamical evolution can be accelerated significantly by the presence of massive remnant stars.

The cluster of normal stars is essentially unaffected because the time scale for the dynamical evolution of the entire system is much longer than the collapse time. Thus the degenerate stars formed a subsystem within the core of ordinary star clusters. The half-mass radius of the degenerate component becomes smaller than the core radius of the normal component. Until the core collapse, the core radius of normal star component does not vary. Thus a compact cluster of degenerate stars can reside within the central core of much broader distribution of normal stars. If we determine the mass distribution based on the measured velocity distribution, there would be an indication of central point mass. As long as the half-mass radius of degenerate cluster remains to be much smaller than the resolution of the observations, it is impossible distinguish from a point mass.

Our assumption of neglecting the effects of collisions breaks down as the density of degenerate component cluster becomes very large. We now consider the collisional effects on the dynamics of the compact cluster.

## 3. Further Evolution of the Subsystem

Quinlan & Shapiro (1989; abbreviated as QS hereafter) studied the dynamical evolution of very dense stellar systems composed of neutron stars. As the density of the central degenerate component grows, the situation becomes similar to that considered by QS. The physical processes studied by QS include the formation of binaries via three-body and two-body processes. In addition, tidal interactions between degenerate and normal stars would take place in two-component clusters. We now examine these processes.

### 3.1. FORMATION AND EVOLUTION OF THREE-BODY BINARIES

The rate of formation of hard binaries can be expressed (Hut & Goodman 1993) as

$$\frac{dn_B}{dt} = 126 G^5 m^5 n^4 v^{-9}, \qquad (1)$$

where $n$ is the number density and $v$ is the three-dimensional velocity dispersion. Compared to a single component case with ordinary stars only, the binary formation rate is greatly enhanced because the degenerate component has large $n$ and $m$ and small $v$. The binaries become energy source by releasing binding energy to the cluster through the interactions with surrounding stars. If the cluster is isolated so that there exists a finite escape velocity, the binaries are eventually ejected when they reach certain hardness, where the hardness is defined as

$$x \equiv \frac{3}{2} \frac{Gm}{av^2}. \qquad (2)$$

Here $a$ is the semi-major-axis of the binary orbit. The maximum hardness before the ejection depends on the central potential depth in single component case, but typically a few hundreds. However, the situation in two-component cluster is somewhat different in two aspects. First, the potential depth felt by degenerate component is much deeper because the velocity dispersion of the degenerate component is smaller. Thus the binaries can reach larger hardness than in the single component cluster. Second, the gravitational radiation energy loss could be important for binaries of compact stars if the orbital separation becomes smaller.

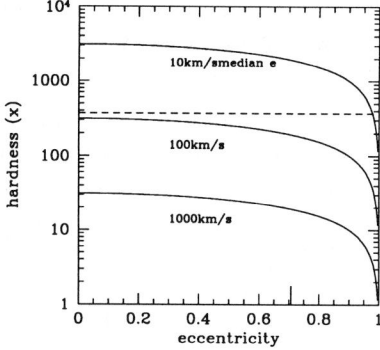

*Figure 1.* The border lines for gravitational radiation merger and ejection.

As binaries become tighter, typical recoil energy per collision increases but typical time interval between two successive collisions decreases while the time required for gravitational radiation to bring two object together decreases. A merger would take place before the binary is ejected if the gravitation radiation merger time scale becomes shorter than typical collision time scale.

The dynamical properties of a binary can be characterized by the hardness and the eccentricity ($e$) of the binary orbit. In Fig. 1, we have shown the the borderlines for a few processes in ($e, x$) plane. A typical binary is formed near $x = 3$ and wonders upward as a result of encounters with other objects. While the hardness changes mostly by strong encounters ($r_p \sim a$, where $r_p$ is the pericentral distance of the binary-single relative orbit), eccentricities can be easily changed by week encounters. A thermal distribution of $f(e) \propto e$ can be established.

In computing the location of the border lines, we have assumed 10 M$_\odot$ black holes as degenerate stars (see Lee 1995 for details). The broken horizontal line near $x = 300$ is the typical hardness for binaries to escape as a result of single strong encounter. This is somewhat larger than typical value of $x \sim 100$ for a single component cluster because the potential depth measured by the velocity dispersion of degenerate stars becomes deeper.

The solid curves are the locations on which gravitational radiation merger time scale equals to the typical strong collision time. Therefore, binaries cannot cross this line: if the ejection line lies above this line, a binary would become mergers within the cluster. These lines are drawn for three different values of velocity dispersion because the physical orbital separation for a given hardness depends on the velocity dispersion. Thus the binaries in clusters with small velocity dispersion would be ejected (except for binaries with $e$ close to 1). If there were some population of massive degenerate stars in globular clusters, they would have been ejected as suggested by Kulkarni, McMillan & Hut 1993.

Since the gravitational radiation merger time scale is very short for binaries with highly eccentric orbits, such binaries would be quickly depleted. Even in clusters with relatively small velocity dispersion, a significant fraction of binaries can become mergers. If the velocity dispersion is greater than 100 km/s, virtually all the binaries would become mergers. A typical binary can release energy of order of a few hundred $kT$ until merger. Therefore the binaries of remnant stars formed by three-body processes are also an efficient heat source to the cluster. The core collapse can be stopped by the heating effects of binaries. Through the formation of binaries and subsequent merging, a population of more massive black holes grows. Eventually a single massive black can be formed by accreting many black holes if one waits for more than 'collision time' (Lee 1993).

If the velocity dispersion of the cluster were very large (say, 500 km/s or greater), the amount of heating per binary reduces significantly. The binaries may be unable to stop and reverse the core collapse. Thus one eventually reaches regime studied by QS who concluded that the formation of a single massive black hole is inevitable in the dense stellar systems composed of only neutron stars. By adopting more massive objects than neutron stars and two-component models, the growth to a single massive black hole would be even easier.

Although the degenerate binaries are efficient heat source for galactic nuclei with velocity dispersion of $\sim$ 100 km/s, there is a possibility that the efficiency could be significantly lower. The strong encounters between binaries and singles often involve complex motion among three objects. Any two of three objects can approach to very small distance, and the gravitational radiation energy loss can be very large. Merger can occur through this process for binaries with relatively large orbital separation. Some preliminary numerical calculations of binary-single scattering experiments are made and about 10 % of typical interactions for binaries with $e = 0.67$ (a median value for thermal distribution of eccentricity) and $x = 100$ at velocity dispersion of 100 km/s (Kong & Lee, this volume). More extensive computations are necessary in order to assess the importance of the gravitational radiation during the triple interactions.

## 3.2. CAPTURE BY GRAVITATIONAL RADIATION

If the energy loss by gravitational radiation exceeds the initial orbital kinetic energy, a binary can be formed. The rate of binary formation by this process is larger than that by three-body processes in low density phase.

However, binary formation is important only at high densities for both mechanisms. As the central density becomes very high, the three-body process dominates the two-body process unless the velocity dispersion is very large. For a typical condition during the post collapse expansion, the number of objects in the core becomes around 100. Then the velocity dispersion should be greater than 350 km/s in order for the two-body binary formation rate to exceed the three-body rate.

The gravitational radiation capture binaries have very large eccentricities and very small pericentral distances. Thus the gravitational radiation merger time scale is much shorter than any other time scales (QS; Lee 1994). Therefore they do not provide any energy to the cluster. It can only be considered as a process forming more massive black hole.

### 3.3. NORMAL-DEGENERATE TIDAL INTERACTIONS

Since the central parts of the entire cluster is dominated by degenerate component, close encounters between a degenerate and a normal star can occur. Close interactions between normal stars are much rarer. In stellar systems with relatively low velocity dispersion (order of 10 km/s) a tight binaries can be formed via tidal capture. If the velocity dispersion is large, the close encounters usually lead to the disruption of the normal component.

Some of the disrupted material will be unbound from the degenerate star just after the disruption because the stellar debris would have typical expansion velocity of order of several hundreds km/s. The maximum amount of escaping material will be 50% of the mass of the disrupted star. The remaining material will be orbiting around the degenerate star.

If the bound material is accreted onto the degenerate star, there will be strong radiation in the form of X-rays. The luminosity depends on the accretion rate which is quite uncertain. If the stellar debris forms a disk around the degenerate star, the disk mass is significant fraction of the central object. In such a massive disk, the viscous time scale is very short. The accretion rate can easily exceed the Eddington limit by a few orders of magnitude. Thus it is highly uncertain what the fate of the massive disk would be. It is possible that the majority of the material is blown out by strong radiation produced by the rapid accretion of matter. If this is the case, the entire energy released followed by a disruption of a normal star is be only a small fraction (i.e. $< 10^{-3}$) of the rest mass energy of the star. The entire duration of bright phase will be very short.

On the other hand, if the accretion rate is maintained at around the Eddington limit but there is no loss of disk mass, the entire duration to accrete the whole stellar debris will be relatively long. For example, 0.5 $M_\odot$ mass can be accreted in in approximately $2 \times 10^6$ years onto a 10 $M_\odot$

black hole if the accretion rate is limited by Eddington rate. The bolometric luminosity is $3.5 \times 10^5$ $L_\odot$ during this period. Thus the total luminosity in X-ray will be fairly large if there exist several systems having massive disks at a given time.

Because of these uncertainties, it is too early to dismiss the possibility of having a compact cluster of remnant stars in the central parts of galactic nuclei as a model for the dark central object on the basis of absence of strong X-ray emission. We now turn to numerical simulations of dynamical evolution of two-component clusters.

## 4. Fokker-Planck Models for the Dynamical Evolution

Here we present some of the model calculations based on the integration of multi-component Fokker-Planck equation. We assume spherical symmetry and isotropic velocity dispersion. The dynamical effect of binaries formed via three-body processes is included by adding a heating term to the Fokker-Planck equation. We also include the fact that the binaries formed via two body and three body processes eventually become mergers.

The heating of binaries enables the reversal of the collapse of the degenerate component. The density at the reversal is very large and there are only order of $10 \sim 100$ objects. Like the solutions for globular clusters, the central density exhibits rapid oscillation. However, as shown in Fig. 2 for the post collapse evolution of core radius of the degenerate component and half-mass radii of normal and degenerate components, the core of degenerate component does not expand in time scale of half-mass relaxation time of its own. The amplitude of core radius oscillation is very large, but the tendency for general expansion is not seen in this figure. In addition the half-mass radius of the degenerate component remains nearly constant, unlike the evolution of the single component case where $r_c \propto r_h \propto t^{3/2}$ after core collapse. This is due to the fact that the degenerate stellar system is embedded in the much larger stellar system of normal stars. The surrounding stellar system simply absorbs the expansion energy of the degenerate component. Since the energy content of the surrounding stellar system is very large, it can be considered as a heat reservoir.

If such a configuration of compact cluster of remnant stars surrounded by much larger cluster of normal stars is stable, concentration of dark mass in some of nearby galaxies can also be interpreted as the compact cluster of degenerate stars. There are two possibilities that this scenario cannot explain situations in our galaxy and nearby galaxies such as M31 and M32. First, the tidal interactions between a normal and a degenerate stars is estimated to be rather frequent. For a typical run with $M_{tot} = 10^7$ $M_\odot$, $M_D = 0.01 M_{tot}$, $m_D = 10$ $M_\odot$, we obtain roughly $10^{-3}$ encounters per year.

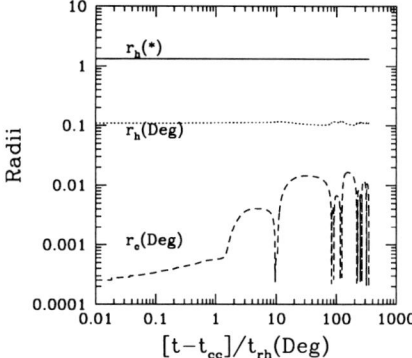

*Figure 2.* Core radius of normal component $[r_c(*)]$ and half-mass radii of normal $[r_h(*)]$ and degenerate $[r_h(Deg)]$ components.

Since the duration of active accretion phase can be up to $10^6$ years after the disruption of a normal star if the accretion rate is regulated by Eddington limit (see §3.3), there will be order of 1000 X-ray emitting sources at any given time. This clearly exceeds the observational limit. On the other hand, if most of the massive disk is blown away by strong radiation produced by accretion of small mass, the X-ray luminosity could be lower by several orders of magnitudes.

Second, the merger of degenerate binaries via gravitational radiation could significantly alter the population of degenerate component. The formation rate is sufficiently small in low density phase of the oscillation so that it takes more than Hubble time to transform majority of degenerate stars into more massive population. However, the conditions in the core can become rather extreme during the high density phase of core oscillations. The entire mass of the core can be transformed into a single black hole. The changes in black hole mass in rapidly oscillating system has to be studied more carefully to decide the long term stability of highly segregated two component clusters.

## 5. Summary

The presence of massive degenerate component in a stellar system mainly composed of low mass stars can change the course of dynamical evolution significantly. The segregation of mass takes place in a short time scale. The central part of the stellar system is dominated by degenerate stars. The subsystem of degenerate component evolves on its own time scale while the surrounding system of normal stars remains nearly static because the relaxation time scale for this component is much longer than that of degenerate component.

As the core density of the degenerate system grows, various physical interactions occur. The binaries formed via three-body processes can release significant of energy to the cluster before they become mergers by gravitational radiation. If the potential depth of the stellar system is shallow (i.e., velocity dispersion is small) most of the binaries will be ejected. However, under the conditions of galactic nuclei where the escape velocity exceeds several hundreds km/s, binaries would become mergers well before they will reach the hardness for ejection.

The post collapse evolution driven by the binary heating is characterized by gravothermal oscillations, but the half-mass radius of the degenerate component stays nearly static because the expansion energy of the degenerate component is absorbed by much larger cluster of normal stars. The two-component phase may be dynamically unstable in the long run if the growth of the individual mass of degenerate stars grows rapidly. Typical time scale for the runaway growth of the mass is collision time, but it is difficult to define in a rapidly oscillating system.

The X-ray emission due to the accretion of debris of a tidally disrupted star onto the degenerate component can pose serious observational constraint on the presence compact cluster of remnant stars in quiet nuclei of galaxies. However, the physics after the disruption is sufficiently uncertain to draw any firm conclusion based on this argument.

## Acknowledgements

This work was supported in part by the Korea Science and Engineering Foundation under grant No. 941-0200-001-2 and in part by the Basic Science Research Institute, Ministry of Education, project No. BSRI-95-2413.

## References

Brown, G. E., & Bethe, H. A.: 1994, *Ap. J.*, 423, 659.
Cohn, H., 1980, *Ap. J.*, **242**, 765
Goodman, J., & Lee, H. M., 1989: *Ap. J.*, 337, 84.
Hut, P., & Goodman, J., 1993: *Ap. J.*, 403, 271.
Inagaki, S., 1985, in Goodman J.,, Hut P., eds, Dynamics of Star Clusters, Reidel, Dordrecht, p. 189.
Kulkarni, S. R., Hut, P., & McMillan, S. L. W., 1993, *Nature*, 364, 421.
Lauer, T., et al, *A. J.*, submitted.
Lee, M. H., 1993, *Ap. J.*, 418, 147.
Lee, H. M., 1995, *MNRAS*, 272, 605.
Quinlan, J., & Shapiro, S. L., 1989, *Ap. J.*, 356, 483. (QS)

# DARK MATTER IN GLOBULAR CLUSTERS

DOUGLAS C. HEGGIE
*University of Edinburgh*
*Department of Mathematics and Statistics, King's Buildings,*
*Edinburgh EH9 3JZ, U.K.*

AND

PIET HUT
*Institute for Advanced Study*
*Princeton, New Jersey 08540, U.S.A.*

**Abstract.** We first review reasons why dark matter is an interesting issue in connection with star clusters. Next we consider to what extent the presence of dark matter is consistent with their dynamics and structure. We review various model-dependent and model-independent methods which have been applied to two well studied clusters, NGC 6397 and 47 Tuc. We suggest that about half of the mass in each object is still unobserved, possibly in the form of a mixture of low-mass stars and white dwarfs.

## 1. Introduction

It is remarkable how many of the major problems in astronomy are influenced by what we know about globular clusters. Nevertheless, the dark matter problem is an exception. The question whether globular clusters contain significant amounts of dark matter has not often emerged in the literature, however much it may have been discussed informally, and to the best of our knowledge it has never previously been reviewed.

In this paper we first attempt to set out the background to the problem, explaining why the search for dark matter in the context of globular clusters is an interesting problem. Next we consider ways in which such a search may be carried out, with particular emphasis on techniques based on dynamical modelling. We shall see that this topic has reached a new and exciting point of development, thanks to the depth to which the mass function of globular clusters can now be examined observationally. Nevertheless important uncertainties still remain, and direct attention to both theoretical and observational problems for the coming years.

## 2. The Importance of Dark Matter in Globular Clusters

There are several *a priori* reasons why the search for dark matter may be taken seriously.

1. In terms of mass, globular clusters are the next step down from the smallest stellar systems which definitely contain dark matter – the dwarf spheroidals (see Ashman 1992, Pryor 1992).
2. The presence or absence of dark matter in globulars would be a significant piece of evidence in the study of galaxy formation. It would clarify the relation between globulars and their host galaxy, and between globulars and other small stellar systems.
3. Renewed interest in the topic is timely because of the flood of new results on deep mass functions, and the observation of white dwarfs, in several globular clusters (cf. the papers by Piotto and by Fahlman, these proceedings).
4. Hypothetical black hole stellar remnants were invoked by Larson (1984) to account for anomalously high-velocity stars in some clusters.
5. Upper limits on the dark matter content of globulars are needed in order to assess the feasibility of searches by such techniques as microlensing (Griest, pers. comm.).
6. Though the issue seems controversial, there is a possibility that certain types of dark matter might affect the evolution of stars (Renzini 1987, and references in Heggie et al. 1993).
7. The typical masses of globular clusters have a special significance in reasonably standard cosmological models (Peebles 1984, Rosenblatt et al. 1988, West 1993), and Peebles suggested that globulars might contain a reasonably uniform dark matter component. Other theories (e.g. Silk & Stebbins 1993, Moore & Silk 1995) are also relevant to the structure and content of globular clusters, but give a less clear picture of the expected distribution of the dark matter.

## 3. Searching for Dark Matter in Globular Clusters

### 3.1. SEARCH STRATEGIES

How one searches for dark matter depends on what it is, and in principle one may consider all the usual suspects: black holes, neutron stars, white dwarfs, low-mass stars, wimps, etc (see Carr 1994 for a review, especially on baryonic varieties). Certain types of dark matter could be detected from the resulting gamma-ray emission (cf. references in Heggie et al. 1993). On the more conventional side, low-mass stars and brown dwarfs may be detected by a variety of means. Detection and spectroscopy from the ground in the IR may be feasible with 8m class telescopes (Fusi Pecci et al. 1994). For estimates of the usefulness of space-based observations, and of microlensing, see the paper by Longaretti et al. in this volume, and references therein.

In the remainder of this paper we consider the traditional *model-building strategy*. It is the counterpart in globular clusters of the classical Oort technique which revealed the possibility of missing mass in the solar neighbourhood (Oort 1932). In principle one should construct a model of a cluster which is consistent with all relevant dynamical data, including (i) the surface brightness profile; (ii) star counts; (iii) radial velocities; (iv) proper motions; (v) pulsar "spin-up" (see the paper by Kulkarni in these proceedings); and (vi) dynamical evolution. Finally the mass of the model may be compared with that of visible matter in order to determine the amount of dark matter.

In practice dynamical models are often constructed from a subset of the above list of data. For example, the commonest models are constructed from the surface brightness profile only. Without kinematic data, however, these do not constrain the mass sufficiently. (The most obvious illustration of this assertion is a single-component model, to which any amount of dark stars of the same mass could be added without altering the surface brightness profile; see also Longaretti et al., this volume.) Though radial velocity data is rather commonly used, it is unusual for data on proper motions to be taken into account, Leonard et al. (1992) being a notable exception.

An example of a cluster in which this technique has revealed the presence of dark matter is M71. Richer & Fahlman (1989) obtained a mass-to-light ratio $(M/L)_V \simeq 0.57$ from star counts out to radius 3.4', compared with a global value in the range $1 - 1.4$ obtained by dynamical modelling of kinematic data. This result indicates that at least about 50% of the mass of this cluster resides in unobserved components, and Richer & Fahlman concluded, after further modelling, that it was most likely to consist of stars with masses below that of the faintest stars observable in their study, i.e. less than about $0.33 M_\odot$.

What makes a reexamination of such investigations timely is that it now seems possible to push star counts down to masses closer to $0.1 M_\odot$ in some clusters, as the following examples illustrate. And another dark matter candidate which can now be counted convincingly is the population of white dwarfs, or at least the brightest ones. It has already been shown (see the paper by Fahlman, these proceedings) that their numbers correspond nearly to what would be expected from the evolution of stars originally slightly more massive than the present turnoff. Therefore they may be expected to contribute substantially to any missing mass.

The following two examples are meant to illustrate the advances in modelling to which these new observational data should lead. In addition, however, it is our aim to illustrate the variety of modelling techniques which are now available. We shall try to show that each has its strengths and weaknesses, and that use of several techniques for the same cluster helps to guard against the risk of drawing model-dependent conclusions.

3.2. EXAMPLE: NGC 6397

As the frequency of its appearance in the papers in this volume show, this cluster has almost replaced M15 as the classic example of a post-collapse cluster, one reason being its relative proximity. Like several post-collapse clusters, it exhibits population gradients (Djorgovski et al. 1991). According to Aguilar et al. (1988) it is one of the most fragile galactic globular clusters.

Anisotropic multi-mass King models for NGC 6397 can be found in Meylan & Mayor (1991), who fitted to radial velocities and the surface brightness profile. An isotropic model was constructed by King et al. (1995), who fitted to star counts and the surface brightness profile, but no kinematic data. This last study furnishes a beautiful example of mass segregation in observational data, and the fit to the surface brightness profile is excellent. There is also a single-component King model given by Da Costa (1979).

As already mentioned, deep star counts and kinematic data are important for establishing the existence of dark matter, and so we have constructed a multi-mass isotropic King model which fits the projected radial velocity dispersion profile of Meylan & Mayor (their Table 2), as well as their $V$-band surface brightness profile and the deep star counts of Paresce et al. (1995a). (Note, incidentally, that there exists some disagreement between different groups [King, pers. comm., and Piotto, this volume] in the counts of the stars of lowest mass.) We used the same mass bins as Meylan & Mayor, and our best fitting model has the global mass function given in Table 1. Other parameters of the model, in standard notation, are $W_0 = 12.7$, $\sigma^2 \equiv 1/(2j^2) = 12.0 \mathrm{km}^2 \mathrm{s}^{-2}$, $r_c = 0.19\mathrm{pc}$, and $r_t = 26\mathrm{pc}$. We have checked that the resulting model is consistent with the star counts of King et al. at two radii. The fit to the surface brightness profile is of comparable quality (judged by $\chi^2$) to that of the models of Meylan & Mayor.

At this point it is worth pausing to consider the limitations and advantages of multi-mass King models. Their principal advantage (and it is a formidable one, which explains their great popularity) is speed. Among their limitations, however, are

– *Neglect of anisotropy, rotation, flattening* All three pose considerable

TABLE 1. Global Mass Function for a New Model of NGC 6397

| Mass range ($M_\odot$) | Mean mass ($M_\odot$) | Mass ($M_\odot$) |
|---|---|---|
| $[5, 100]_{hr}$ | 1.400 | 400 |
| $[2.5, 5]_{wd}$ | 1.093 | 5600 |
| $[1.5, 2.5]_{wd} + [0.63, 0.79]$ | 0.726 | 3400 + 9600 |
| $[0.79, 1.5]_{wd} + [0.50, 0.63]$ | 0.578 | 23700 + 5500 |
| $[0.40, 0.50]$ | 0.447 | 3400 |
| $[0.32, 0.40]$ | 0.355 | 2900 |
| $[0.25, 0.32]$ | 0.282 | 3400 |
| $[0.20, 0.25]$ | 0.224 | 2300 |
| $[0.16, 0.20]$ | 0.178 | 1800 |
| $[0.13, 0.16]$ | 0.141 | 1000 |
| | Total cluster mass | 63000 |

Note: the notation for the mass range follows that of Meylan & Mayor exactly. Thus the first bin contains heavy remnants of stars with initial mass in the stated range. Similarly $wd$ denotes white dwarf remnants. Other ranges refer to main sequence stars.

modelling problems, and even though anisotropy is often included, it is far from clear on dynamical grounds that the usual recipe (i.e. Michie-King models) is at all appropriate.

- *Approximate dynamical evolution* The lowered Maxwellian distribution was introduced as an approximate solution of the one-component Fokker-Planck equation. For a long time, however, it has been possible to solve this equation by direct numerical methods (see below).
- *Lack of primordial binaries* These effectively give rise to a small population of bright objects more massive than the turnoff mass. They are usually ignored, though the work of King et al. is a notable exception.
- *Problems of population gradients* The implication is that the surface brightness profile is sampling different populations at different radii, and in principle this may affect modelling based on the surface brightness profile.
- *Poor statistical methodology* It has been claimed (Merritt & Tremblay 1994) that astronomy is one of the last disciplines to hold out against the trend towards non-parametric statistics, and the use of parametrised models such as multi-mass King models introduces unquantified biases and other undesirable deficiencies.

Many of these drawbacks can be put right, but usually at the cost of computational speed. Dynamical evolution can be handled better by means of Fokker-Planck calculations, and in fact Drukier (1994) has carried out an excellent study of this cluster using this technique. To some extent his preferred models were guided by ground-based faint star counts which have now been superseded, but there is little doubt that modest adjustment of the parameters of his models would be sufficient to restore good agreement.

The limitations of the Fokker-Planck method include

- *Time-consuming computations*
- *Neglect of anisotropy, rotation and flattening*, though it is now becoming possible to include anisotropy efficiently (see the papers by Takahashi and Giersz in these proceedings). Also there has been a modest recent revival of interest in the Fokker-Planck modelling of rotating clusters.
- *Omission of disk shocking*, though this could easily be included at a satisfactory level of approximation.
- *Omission of primordial binaries*, despite their known importance in core bounce and post-collapse evolution (see Hut, these proceedings). They could be included, but the necessary cross sections are not well known, especially for unequal masses, and the level of approximation would be rough. Monte Carlo methods (see Giersz, these proceedings) offer the best prospect here.

TABLE 2. Inferred data for NGC 6397

| Source | Tidal Radius pc | Total Mass $10^5 M_\odot$ | Neutron Stars % by mass | White Dwarfs % by mass |
|---|---|---|---|---|
| Da Costa (1979) | 24 | – | – | – |
| Meylan & Mayor (1991) | 66 ± 14 | 1 ± 0.1 | 2 | 25 |
| Drukier (1994) | 19 ± 2 | 0.66 ± 0.05 | 4 | ? |
| This paper | 26 | 0.63 | 0.6 | 52 |

Note: where necessary we have assumed $1\text{pc} = 1'.6$.

To return to NGC 6397, Table 2 summarises the main data from the various models which are relevant to the dark matter problem, though insufficient details of the model of King et al. were available to us. A comparison between the data derived from different models or different selections of observational data helps to assess the the magnitude of the systematic errors in these estimations. The high mass of the model of Meylan & Mayor, for example, may stem from their use of anisotropy, which extends the halo. In any case $r_t$ is difficult to determine for this cluster because of the high background density (Drukier et al. 1993).

The interesting number in this table is the proportion of white dwarfs, which we think may be higher than was previously believed. Our model also contains a lower proportion of low-mass stars (in the last six bins in Table 1) than in the models of Meylan & Mayor. At the time when the latter models were constructed the main sequence mass function was poorly constrained.

### 3.3. EXAMPLE: 47 TUC

It can be claimed that conclusions such as this are still too model-dependent, and could be relaxed if some of the specific choices of multi-mass King models (e.g. the choice of distribution function) were altered. One of the aims of non-parametric methods is to avoid this pitfall. Of the four clusters studied non-parametrically by Gebhardt & Fischer (1995), we select 47 Tuc, being one of those for which recent deep star counts have become available. Though it would be desirable to construct new King models taking account of this data, we have not yet done so, and it is interesting to see what conclusions can be drawn from the above non-parametric study alone.

The dynamical status of 47 Tuc is a little controversial, though it is commonly assumed to be a high-concentration cluster approaching core collapse. Without doubt it is amongst the most massive clusters.

The advantages of non-parametric methods have already been touched upon, and it is worth listing their possible defects. These include

- *Neglect of anisotropy, rotation and flattening*
- *Neglect of dynamical and physical constraints* The method makes no assumption about the form of distribution function of the population used to trace the potential, even where it may be well constrained by theory. The method also makes no assumption that the mass density is positive, and in fact the results inferred for 47 Tuc imply that a negative mass-to-light ratio is acceptable at the 90% confidence level in one range of radius. This defect may be related with the previous item (concerning isotropy), as Richstone has pointed out (pers. comm.) that the problem with $M/L$ is avoided if some anisotropy is introduced.
- *Effect of population gradients.* In fact Guhathakurta et al. (1992) have reported the existence of a population gradient in 47 Tuc. As with multi-mass King models fitted to the surface brightness, however, it is not known how important the effect may be.

From the results of Gebhardt & Fischer we compute that the mass within a sphere of $7'$ projected radius is about $6.7 \times 10^5 M_\odot$. This is close to the value of approximately $5.5 \times 10^5 M_\odot$ inferred from multi-mass anisotropic King-Michie models fitted by Meylan (1989). The result just mentioned suggests that the differences between these methods may be rather philosophical than substantive. Indeed, one can think of the model-fitting method of Meylan as simply a different way of constructing a rather arbitrary potential. Only the mass bin containing the giants is directly connected to the observations; the others simply give rise to a potential field sufficient to agree with the kinematic data. The heaviest bins govern the potential at small radii, and successive bins build up the potential well at successively greater radii.

Now let us consider the surface density at 4.6', where deep star counts were obtained by De Marchi & Paresce (1995). By summing their mass function and taking into account the field area, we obtained a surface density in counted main sequence stars of about $305 M_\odot/\text{pc}^2$. This may be compared with the projected density of all matter which we computed from the non-parametric model of Gebhardt & Fischer; this value is approximately $1100 M_\odot/\text{pc}^2$, with a lower limit of about $770 M_\odot/\text{pc}^2$ at 90% confidence.

At face value these data imply a substantial fraction of "missing mass", and we immediately mention some possible explanations.

- *Giants* These were excluded from the counts of De Marchi & Paresce. The mean mass of the stars in their most massive bin was about $0.75 M_\odot$, and we estimate that inclusion of stars between this bin and turnoff ($\sim 0.9 M_\odot$, cf. Hesser et al. 1987) would increase the surface density to about $385 M_\odot/\text{pc}^2$. This implies that the proportion of the projected mass unaccounted for is still at least 50%, and it could be as high as about 70% (Table 3).
- *Low-mass stars* All the remaining mass could be accounted for if, below the least massive stars counted, the mass function varies as $dN(m) \propto m^{-1-x} dm$ with $x \simeq 0.6$. Though De Marchi & Paresce find that the mass function flattens at low masses, this conclusion is controversial (see the paper by Piotto, this volume).
- *M/L relation* This is controversial for stars of low mass, and errors here can lead to large differences in the inferred mass function. Note, however, that the surface density is obtained from the *magnitude* distribution $N(m)$ by the integral $\int M(m) dN(m)$, where $M$ is the mass of a star of magnitude $m$. Several $M/L$ relations are plotted by De Marchi & Paresce, and indicate that the resulting uncertainty in the projected mass density is at most 0.1 dex.
- *Completeness* De Marchi & Paresce claim that, even at the faintest bin, their counts are at least 67% complete. This has been corrected for in the data which we used to compute the projected density.
- *White dwarfs* Previous estimates of the mass fraction in all white dwarfs are given in Table 3.

Though the first five are global estimates, it seems unlikely that at the radius of these observations (about 0.7 half-mass radii) mass segregation could produce a much larger proportion of white dwarfs. Now that white dwarfs can be counted in 47 Tuc (Paresce et al. 1995b) it is worth reexamining the sorts of models listed in Table 3 to check whether the population of

TABLE 3. Mass fraction of white dwarfs in 47 Tuc

| Source | Mass fraction (%) |
|---|---|
| Illingworth & King 1977 | 31 |
| Da Costa & Freeman 1985 | 35 |
| Meylan & Mayor 1986 | 12–35 |
| Meylan 1988 | 27–37 |
| Meylan 1989 | 16–23 |
| This paper | $\lesssim 70$ |

white dwarfs could not perhaps be rather more significant than was previously thought. The result could also illuminate the still rather controversial evidence on the presence of cataclysmic variables in clusters such as 47 Tuc.

## 4. Conclusions

We wish to emphasise that all kinds of dynamical models are useful in this kind of study. All have some advantages, but the list of their limitations is depressingly long. Studying the same cluster by different techniques is an important way of guarding against some of these.

Based on such methods, the studies of the two clusters on which we have concentrated suggest that a large fraction of their mass, around 50%, is still unobserved. It is not implausible, however, that all of this can be accounted for by white dwarfs or low-mass stars, and so we consider that this represents a generous upper limit on the mass fraction of other, more exotic forms of dark matter in these two clusters. In other words, though there are several good reasons for studying the dark matter problem in globular clusters, it is first necessary to improve our knowledge of the abundance of white dwarfs and low-mass stars. At present it is not even clear which of these might dominate. But the time is ripe for renewed study of these low-luminosity components, thanks to the wealth of new observational data.

## 5. Acknowledgements

We thank several participants of the Symposium, too numerous to mention individually, who commented on our paper during the question session and also privately. We are grateful also to Ivan King and Pierre-Yves Longaretti for their comments on an earlier draft of this written version. We have tried to incorporate all these remarks in the present version.

## References

Aguilar L., Hut P., Ostriker J.P., 1988, ApJ, 335, 720
Ashman K., 1992, PASP, 104, 1109
Carr B., 1994, ARAA, 32, 531
Da Costa G.S., 1979, AJ, 84, 505
Da Costa G.S., Freeman K.C., 1985, in Goodman J., Hut P., eds, Dynamics of Star Clusters, IAU Symp. 113. Reidel, Dordrecht, p.69
De Marchi G., Paresce F., 1995, STScI Preprint No.932
Djorgovski S., Piotto G., Phinney E.S., Chernoff D.F., 1991, ApJL, 372, L41
Drukier G.A., Fahlman G.G., Richer H.B., Searle L., Thompson I.B., 1993, AJ, 106, 2335
Drukier G.A., 1994, ApJ, sub.
Fusi Pecci F., Cacciari C., Ferraro F.R., Gratton R., Origlia L., 1994, The Messenger (ESO), 77, 14
Gebhardt K., Fischer P., 1995, AJ, 109, 209
Guhathakurta P., Yanny B., Schneider D.P., Bahcall J.N., 1992, AJ, 104, 1790
Heggie D.C., Griest K., Hut P., 1993, in Djorgovski S.G., Meylan G., eds, Structure and Dynamics of Globular Clusters, ASP Conf. Ser., Vol.50. ASP, San Francisco, p.137
Hesser J.E., Harris W.E., Vandenberg D.A., Allwright J.W.B., Shott P., Stetson P.B., 1987, PASP, 99, 739
Illingworth G., King I.R., 1977, ApJL, 218, L109
King I.R., Sosin C., Cool A.M., 1995, ApJL, 452, L33
Larson R.B., 1984, MNRAS, 210, 763
Leonard P.J.T., Richer H.B., Fahlman G.G., 1992, AJ, 104, 2104
Merritt D., Tremblay B., 1994, AJ, 108, 514
Meylan G., 1988, A&A, 191, 215
Meylan G., 1989, A&A, 214, 106
Meylan G., Mayor M., 1986, A&A, 166, 122
Meylan G., Mayor M., 1991, A&A, 250, 113
Moore B., Silk J., 1995, ApJL, 442, 5L
Oort J.H., 1932, BAIN, 6, 349
Paresce F., De Marchi G., Romaniello M., 1995a, ApJ, 440, 216
Paresce F., De Marchi G., Romaniello M., 1995b, ApJL, 442, L57
Peebles P.J.E., 1984, ApJ, 277, 470
Pryor C., 1992, in Longo G., et al., eds, Morphological and Physical Classification of Galaxies. Kluwer, Dordrecht, p.163
Renzini A., 1987, A&A, 171, 121
Richer H.B., Fahlman G.G., 1989, ApJ, 339, 178
Rosenblatt E.I., Faber S.M., Blumenthal G.R., 1988, ApJ, 330, 191
Silk J., Stebbins A., 1993, ApJ, 411, 439
West M. J., 1993, MNRAS, 265, 755

# PROPERTIES OF GLOBULAR CLUSTER SYSTEM: PRIMORDIAL OR EVOLUTIONAL?

VLADIMIR SURDIN
*Sternberg Astronomical Institute*
*13, Universitetskij Prospect*
*Moscow, 119899, Russia*
*e-mail: surdin@sai.msu.su*

**Abstract.** Some observable relationships between globular cluster parameters appear as a result of long time dynamical evolution of the cluster system. These relationships are inapplicable to the studies of the globular clusters origin.

## 1. Introduction

There are some well known relations between dynamical parameters of Galactic globular clusters and their galactocentric distances ($R_g$). If this relations are primordial, we can obtain an important knowledge on physical conditions at the globular cluster formation epoch. But if the relations are evolutional, we do not have such possibility. Any way, we must understand reasons for each particular relation to solve the problem of globular cluster formation. It is particularly important to find some invariant relations which keep stable along the evolutional path of the cluster system.

The values of ($R_g$) and the diameters containing half of the cluster mass/luminosity in projection ($D_{0.5}$) are mostly invariable characteristics of them. Therefore these are used for the comparison of the predictions of the cluster formation theory with observational data.

However it is easy to show that, in spite of a relative evolutional stability of the particular values of $R_g$ and $D_{0.5}$, the relation between them (Fig. 1)

$$D_{0.5} \propto R_g^{1/2} \qquad (1)$$

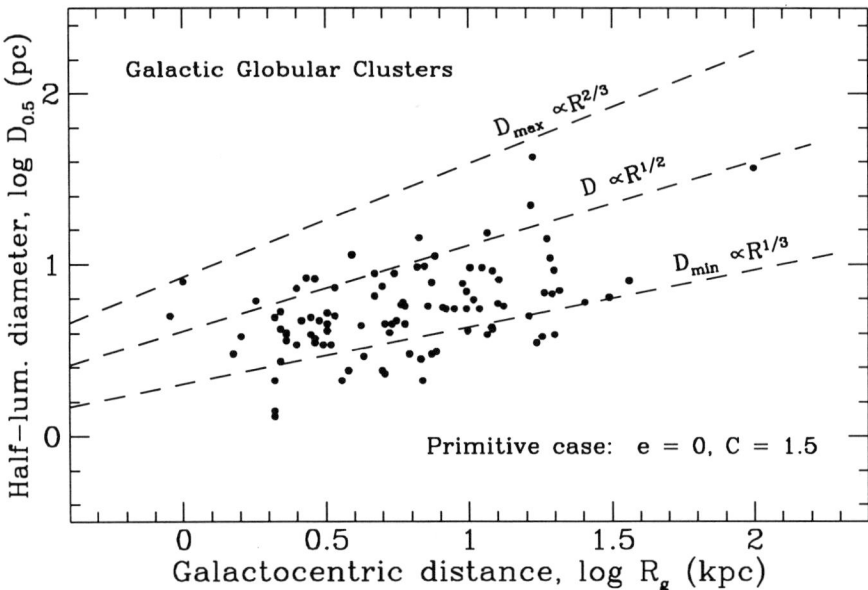

*Figure 1.* Distribution of globular clusters on $R_g$ v. $D_{0.5}$ plot; data by van den Bergh at al., 1991. Theoretical lines of limiting diameters and predicted correlation are shown for the simplest case of circular orbits and equal concentration parameters.

for the globular cluster system could occur as a result of the disruption of some clusters under the action of evaporation provided they are tidally truncated (Surdin, 1994).

## 2. The tidal limitation of the cluster size

Globular clusters are fully relaxed systems whose external radius ($r_t$) is limited by the galactic tidal field (Binney and Tremaine 1987):

$$r_t = R_p \left[ \frac{M}{2g(e) M_G(R_p)} \right]^{1/3}, \qquad (2)$$

where M is the cluster mass, $M_G(R_p)$ is the mass of the Galaxy inside the perigalactic distance of the cluster ($R_p$), and $g(e)$ is a weakly verying function of the cluster orbit eccentricity ($e$). A singular isothermal sphere is an adequate model of the Galaxy in our case:

$$M_G(R_p) = \frac{R_p V_c^2}{G}, \qquad (3)$$

where $V_c$ is the circular velocity. In this case, $g(e) \simeq 2$ (Seitzer, 1985). We can express $D_{0.5}$ in terms of $r_{hP}$, the radius containing half the cluster mass in projection:

$$D_{0.5} = 2r_{hP}, \qquad (4)$$

which simply depends on $r_h$, the half-mass radius of the cluster (Spitzer, 1987):

$$r_{hP} \cong 0.74 r_h, \qquad (5)$$

whose value can be expressed in terms of the core radius ($r_c$) and the tidal cut-off radius for the King model (Fall and Rees, 1977):

$$r_h \cong 0.70\sqrt{r_c r_t}. \qquad (6)$$

These give us the following simple relation:

$$D_{0.5} \cong r_t 10^{-C/2}, \qquad (7)$$

where $C \equiv \lg(r_t/r_c)$ is a concentration index. Finally we obtain:

$$D_{0.5} \cong \left(\frac{GM}{2}\right)^{1/3} \left(\frac{R_p}{V_c}\right)^{2/3} 10^{-C/2}. \qquad (8)$$

Considering the simplest case of circular orbits ($R_g = R_p$) and equal concentrations ($C = 1.5$) of all clusters, we see that the connection between $D_{0.5}$ and $R_g$ depends on the value of $M$, which can not be arbitrary on the $D_{0.5} - R_g$ plane.

2.1. AN UPPER LIMIT ON THE CLUSTER DIAMETER

An upper limit on the cluster mass is determined by the observational limit of $M_{max} = 2.5 \cdot 10^6 M_\odot$, which is caused by the natural exhaustion of the globular cluster luminosity function for large values of $M$. Then the restriction on the cluster diameter follows from the tidal stability condition (8):

$$D_{0.5} \leq D_{max} = 8.5 \left[\left(\frac{R_g}{1kpc}\right)\left(\frac{220km/s}{V_c}\right)\right]^{2/3} pc, \qquad (9)$$

which for $V_c = 220 km/s$ fits well the observable upper boundary of the cluster distribution on the $D_{0.5} - R_g$ plot (Fig. 1).

## 2.2. A LOWER LIMIT ON THE CLUSTER DIAMETER

According to the results of numerical simulations the evaporation time of an isolated cluster is $t_{ev} \simeq 100 t_{rh}$, where $t_{rh}$ is the half-mass relaxation time. The evaporation is more rapid if the size of cluster is tidally limited: $t_{ev} \simeq (20-30) t_{rh}$. Taking into account that the physical diameter of a cluster reaches the tidal one only at the perigalactic point of their orbit, we can adopt $t_{ev} = 70 t_{rh}$ as a reasonable compromise. According to the definition of $t_{rh}$, we obtain

$$t_{ev} = \frac{1}{m}\left(\frac{M r_h^3}{G}\right)^{1/2}, \qquad (10)$$

where $m$ is the mean stellar mass ($m = 0.3 M_\odot$). The evaporation of a cluster is very slow at the beginning, but accelerated to the end of the process. Then we can assume as a probable lower boundary for the cluster mass distribution the value of the mass obtained from equation (10) for a cluster evaporated during time $t_{ev}$:

$$M_{min} = \frac{G M_\odot^2 t_{ev}^2}{10 r_h^3}. \qquad (11)$$

Substituting $M_{min}$ from equation (11) to equation (8), we obtain

$$D_{min} = 0.73 \cdot 10^{-C/4} \left[\frac{G M_\odot t_{ev} R_p}{V_c}\right]^{1/3}. \qquad (12)$$

It is conventional lower limit on $D_{0.5}$ caused by cluster evaporation in the Galactic tidal field. For the simplest case of circular orbits and equal values of $C$ we obtain:

$$D_{min} = 2\left[\left(\frac{R_g}{1 kpc}\right)\left(\frac{t_{ev}}{1.5 \cdot 10^9 yr}\right)\left(\frac{220 km/s}{V_c}\right)\right]^{1/3} pc, \qquad (13)$$

which, for $t_{ev} = 1.5 \cdot 10^9 yrs$ and $V_c = 220 km/s$, is near to observable lower boundary of the cluster distribution (Fig. 1).

## 3. Evolutional? Yes! Primordial? Maybe.

Even with very primitive assumption about circular orbits and equal values of $C$, the effects of the tidal truncation and evaporation of the clusters make it possible to predict the region of their localization on the $D_{0.5} - R_g$ plot. Besides, the upper and lower limits on the cluster diameters are power-law functions of $R_g$ with exponents of 2/3 and 1/3. Then the prediction for the

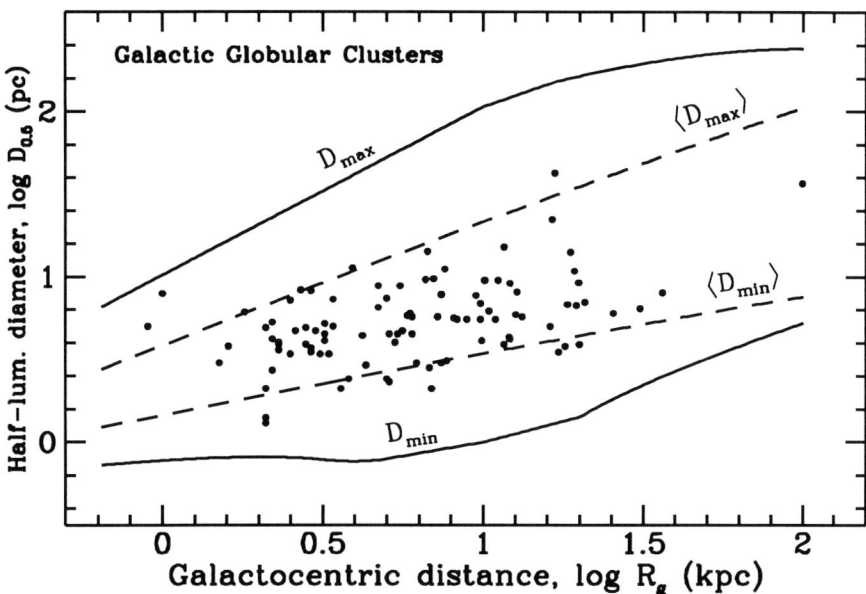

*Figure 2.* The same as in Fig.1, but for more realistic consideration.

relation between these two quantities is $D_{0.5} \propto R_g^{1/2}$. It looks absolutely like the observable correlation by van den Bergh *et al.* (1991).

We considered also these dynamical effects in more detail, taking into account the ellipticity of the cluster orbits and the distribution of their concentration parameters. We made the investigation like this and showed that the agreement of theoretical and observable boundaries becomes more precise, and the form of the correlation between $D_{0.5}$ and $R_g$ survives (Fig. 2). All clusters are located inside the absolute theoretical boundaries ($D_{max}$ and $D_{min}$), and most of them are inside likelihood boundaries ($<D_{max}>$ and $<D_{min}>$).

The similar limitation of globular clusters distribution one can obtain in plot "half-mass density – kinetic temperature" which is used in some scenarios of the clusters formation (Murray and Lin, 1992). In fact, dynamical evolution of any initial cluster population can explain this relationship.

Under development of this consideration, we found an absence of evolutional upper boundary fits observable distribution of clusters on plot "half-differences of origin, and which of them are results of dynamical evolution. Regarding the latter, we understand qualitatively the effects of internal relaxation and external tidal shocks, but our quantitative knowledge of these processes is still inadequate. It is possible to follow a cluster such as NGC 6397 and calculate its detailed evolution, as stars rearrange themselves and some escape, while Galactic tidal forces change from moment to moment—but this has not yet been done, and I think it is very much needed.

2. A great deal can be done, and remains to be done, in fitting equilibrium models of clusters to the observed data. It is clear that good fits can be made, to clusters that are in a pre- or a post-collapse stage, and that the departures from equilibrium found in the Fokker–Planck calculations do not seriously invalidate the models. (A likely exception to this statement, however, is the late stage of collapse, which might be exemplified by the present state of M15.) An interesting corollary of this fitting exercise is the possibility of detecting, through the density perturbations of their gravitational potential, the presence and number of massive remnants at the center of a cluster. Collapsed cores, however, present a frustrating observational problem: they have too few stars in them to allow us to distinguish a pure cusp from one with a small core radius. It may be that a detailed modeling of the entire mass mixture will introduce large enough numbers of stars to make this problem tractable.

3. It is very satisfying to see core oscillations actually demonstrated in an $N$-body calculation. But it remains to examine in detail what happens in this phenomenon, as a function of time and for the individual mass groups. An important related question is, how are we to distinguish between a cluster that has never collapsed and one that has collapsed and re-expanded? At a level of even greater detail, will it be possible to determine at what phase of the collapse-and-re-expand cycle a particular cluster is? —and even more difficult, how many times has it collapsed and re-expanded?

4. Finally, the role of binaries is crucial. We are still debating the workings of the two-body capture mechanism; when will this be settled? Perhaps we can relegate this to a side issue, however; the necessities of dynamical evolution dictate that the collapse must, sooner or later, be turned around by the formation of binaries through *some* process, so that perhaps the how and the when merely affect details of the time scale. A separate important issue, however, is the number of primordial binaries, which is not at all well known and has a very serious effect on the evolutionary path that a cluster will take. A final point—which may reflect only my own lack of understanding—is a lack of clarity about the history of an individual binary as it reacts dynamically with its surroundings, and the statistics of such pairs as the cluster evolves.

## Georges Meylan

### NEW THEORETICAL RESULTS

### GRAVOTHERMAL OSCILLATIONS IN N-BODY CALCULATIONS

For about a decade, Fokker-Planck and conducting-gas-sphere evolutionary models of globular clusters have been computed well into core collapse and beyond, leading to the discovery of possible post-collapse oscillations. Nevertheless, there have been until now some doubts about the presence of such gravothermal oscillations in a pure large N-body calculation, because of its realistically grainy stellar distribution. This IAU Symposium # 174 has seen the first announcement of the long awaited confirmation of this theoretical prediction, a confirmation which is the consequence of hardware and software improvements of N-body codes. J. Makino (these proceedings) does indeed observe gravothermal oscillations in his N-body simulations (32,000 equal-mass bodies). His fundamental and beautiful results are very similar to those predicted by Fokker-Planck calculations.

*Open cluster N-body simulations*

Similar hardware and software improvements of N-body codes have allowed S.J. Aarseth (these proceedings) to simulate open cluster dynamical evolution with increasingly realistic physical details such as, e.g., mass loss by stellar evolution, chaotic tidal interaction, Roche lobe mass transfer, common envelope evolution, magnetic braking, and gravitational radiation. Collision outcomes predict blue stragglers and other exotic objects. This represents the beginning of the micro- and macroscopic understanding of the dynamical evolution of genuine star clusters by using N-body codes, which should soon reach $N = 50,000$.

*Stellar collision simulations*

There is also an increasing number of very detailed studies of stellar flybys, encounters, collisions, and mergers. See the contributions by D.F. Chernoff, M. Davies, R.A. Mardling, S. McMillan, and F. Rasio (these proceedings).

### NEW OBSERVATIONAL RESULTS

The new, very impressive observational results constraining the dynamics of globular clusters, presented at this IAU Symposium # 174, are essentially the consequences of data obtained with two satellite observatories: HST and ROSAT.

*GCs luminosity and mass functions from HST data*

For four of the nearby globular clusters, deep imaging provides luminosity and mass functions for faint low-mass stars, nearly down to the mass which should represent the lower limit of hydrogen burning, i.e. at about 0.08 $M_\odot$. Mass segregation is clearly observed, in a quantitative way, between stars of different masses at different locations in the clusters. For the low-mass stars in the lower part of the main sequence, significant differences exist between the mass functions of NGC 6397 and M15. This could be a direct consequence of the very different galactic orbits followed by these two globular clusters, making NGC 6397 much more susceptible to lose stars via tidal shocking. See the fundamental contributions by I.R. King, A. Cool, G. Piotto, and C. Sosin (these proceedings). See also the beautiful white dwarf sequence observed in M4, the nearest globular cluster (G. Fahlman, these proceedings).

The above results come from the first year of observations with the refurbished HST. This is only the beginning of the use of such data, and there is no doubt that many more studies will follow, with more data and more dynamical interpretations. Star cluster observations are and will be one of the major contributions by HST, leaving far behind the imaging of globular cluster cores from the ground.

*HST and ROSAT observations of compact binaries*

Compact binaries, containing a degenerate companion, such as a white dwarf, or neutron star, have been observed in globular clusters as X-ray sources by ROSAT. See the results concerning individual sources in, e.g., NGC 6397 and 47 Tucanae, and about the general statistics of the population of such sources in galactic globular clusters in the contributions by J.E. Grindlay, F. Verbunt, and M. Shara (these proceedings).

## FROM QUALITATIVE TO QUANTITATIVE ANSWERS

The full interpretation of present and future ground-based and satellite data should allow a systematic improvement in our knowledge of star clusters, by providing quantitative estimates instead of the past qualitative statements (see, e.g., the previous hints about the presence of mass segregation). Great improvements in our understanding of star clusters will come from:

- a precise knowledge of Initial Mass Function and Mass Function for young and old open clusters, respectively;
- a precise knowledge of Mass Function for globular clusters;
- a quantitative estimate of mass segregation;
- taking advantage of the age and mass variety of LMC, SMC and M31 star clusters;

- estimating the precise fraction of binaries, from photometric and radial velocity observations;

- measuring, for large numbers of stars (a few thousands), both radial velocities and proper motions (the latter contain the information of two out of the three components of the stellar spatial velocity);

- interpreting these data with more than one model, i.e., by constraining the observations with King-Michie, Fokker-Planck, non-parametric, and N-body codes; so far, the observations gathered in one cluster have been generally constrained by one kind of model only; the use of more than one model would bring out naturally the pros and cons of each theoretical approach, and would also point out the similarities and differences between globular clusters.

## Frank Verbunt

In my view it is rather surprising that the people, well... most people, who do N-body calculations have always insisted on comparing their results with globular clusters, whereas in fact what they are calculating are open clusters. My advise is to *forget* globular clusters for a while, and to concentrate on open clusters, which have a number of stars that is actually doable.

I would of course not hasard to put forward such an advise if Douglas Heggie had not done so already in his contribution to this meeting; and if I had not known that Sverre Aarseth already is applying his code to the study of open clusters.

The main advantage of concentrating on open clusters is that the observational data of these are much more detailed than for globular clusters, so that a detailed comparison is possible between the theory and the observations. The observations are rapidly improving, and soon it will be possible also to test binary evolution scenarios is detail. I will not believe any calculation on a globular cluster untill the code used has been tested extensively against observations of open clusters. (In the same way, people who use hydrodynamical codes start by doing simple test calculations of, say, flow through a tube.)

The importance of the initial conditions should be tested. What is it that determines whether a cluster becomes open or globular?

As regards globular clusters, our knowledge of the contents of the globular clusters is becoming much better, and comparison with calculations can be made in much more detail. Measurements of the period derivatives of radio pulsars provide very good estimates of the gravitational acceleration, and thereby information on the distribution in the cluster of the total mass, i.e. including white dwarfs and neutron stars. And the HST observations give information on the distributions of the visible stars, including blue stragglers.

**Piet Hut**

Theoretical studies of dense stellar systems have recently seen rapid progress in the development and refinement of many different techniques for modeling star systems. I will briefly mention six classes of techniques, while commenting on the progress to be expected therein during the years to come, until the next IAU Symposium on star cluster dynamics.

1. *N-Body Techniques.* Until the time of this meeting, $N$-body techniques did not go much further than producing toy models. These were certainly interesting and promising, but fell far short of modeling realistic globular clusters. The sheer computational expense was simply too forbidding to come even close to modeling clusters on a star-by-star basis, given that most globular clusters have numbers of stars in the range $10^5 - 10^6$. All this has changed, now that the GRAPE-4 special purpose computer has come online, with its Teraflops speed (see Taiji's contribution). As detailed in Makino's paper, calculations with $3 \times 10^4$ particles have now become routine, and calculations with up to $10^5$ particles are feasible with the new generation of front end computers that will become available in 1996. In addition, the GRAPE designers are now setting their eyes on an even more ambitious goal, namely the development of a machine with a speed a factor $10^3$ higher than that of the current GRAPE-4. Such a Petaflops computer could become operational by the year 2000, if the necessary funding will be found, something that is being actively pursued at present.

2. *Scattering Experiments.* On the other side of the spectrum of possibilities for star-by-star modeling of processes in globular clusters, we find local treatments of the 'microphysics' of close encounters between single stars and binaries. Significant progress has been made in this area, for example through the development of fully automated scattering software. Specification of a few physical parameters is sufficient to start up this software laboratory, allowing the set-up, execution, and on-line analysis of experiments to be carried out without any human intervention. Extensions of this package to include binary-binary scattering is currently underway; see McMillan's contribution to these proceedings for further details. Other approaches to the dynamics of small-$N$ systems include a stability analysis of hierarchical triples (see Kiselova's contribution), the addition of hydrodynamical effects (Davies' paper), and an analysis of the frequency of physical collisions during scattering events (Chernoff and Huang's contribution).

3. *Numerical Approximation Techniques.* Until this year, simulations of globular clusters with realistic particle numbers of $10^5$ and larger could only

be undertaken through a variety of numerical approximation techniques. Between twenty and thirty years ago, various Monte Carlo approaches have been developed for solving the evolution of a star cluster in the Fokker-Planck approximation. Soon afterwards, direct integration techniques have been developed for solving the Fokker-Planck equation. In addition, various conducting gas sphere models have been developed. The combination of all these models, and detailed comparative studies of the results of the different techniques, have proved beneficial for our understanding of the strengths and weaknesses, as well as the limits of applicability, of the individual models. For detailed descriptions, see the talks in these proceedings by Giersz, Lee, Spurzem, and Takahashi, the posters by Drukier, Einsel and Spurzem, and McMillan and Engle, as well as the talk by Heggie, who reports on an interesting technique in which the results from many small-$N$ runs are averaged. Significant further progress can be expected, both in the ongoing development and refinement of individual techniques, as well as in applications in connection with the new GRAPE hardware. Even though realistic $N$-body calculations now are feasible, allowing a star-by-star modeling of globular clusters, these calculations still remain expensive, requiring sometimes weeks and often months for a single run to reach completion. Approximate methods therefore are called upon in order to interpolate between, and perhaps even extrapolate from, the few detailed runs coming out of the GRAPE machinery.

4. *Analytic Techniques.* Although one sometimes gets the impression that current research in star cluster dynamics is almost completely a numerical enterprize, there is plenty of room left for the development of new analytic techniques. Perhaps the most promising area will be that of the development of physically inspired fitting formulas. Rather than tabulating the results of detailed numerical calculations, or applying arbitrary curve fitting to those results, it is often possible to use more physically motivated reasoning in the choice of fitting functions. This has the double advantage of allowing more physical insight in the results obtained as well as providing a measure of confidence in extrapolations beyond the regimes currently tested numerically. An example of this approach is presented in the poster paper by Heggie *et al.*. Others examples of analytical approximation techniques are given in the contribution by Mardling and in the posters by Heggie, and Heggie and Rasio.

5. *Approximate Treatments of Stellar Evolution.* In addition to the four techniques listed above, which can be applied to the purely gravitational $N$-body problem, a more realistic treatment of star cluster evolution requires us to go beyond the point-mass approximation. A first step in this direction is the inclusion of a time dependency of the mass of a single star, for example by taking into account the mass that is lost in later stages of

stellar evolution, towards the formation of a white dwarf, neutron star, or black hole. However, as soon as this step has been taken for single stars, far more complicated issues arise when we want to apply such simple recipes to the case of interacting binary stars, let alone the simultaneous interaction of three or four stars during scattering encounters. How to treat mass overflow, how to discriminate between stable and unstable cases of such overflow, what to do with common envelope phases in their evolution — all these questions require careful consideration, together with the imperative to refrain from too-fancy a type of solution. The challenge to produce a coherent set of recipes that are as simple as possible, but no simpler, across the board is a formidable one. This challenge has recently been taken up by several individuals, as can be seen from the contributions by Aarseth, Eggleton, and Portegies Zwart.

6. *Approximate Treatments of Hydrodynamics.* Another extension that is called for in realistic star cluster modeling, beyond the point-mass limit, is the inclusion of some type of hydrodynamics. Smooth Particle Hydrodynamics forms a natural candidate; see the papers by Davies and Rasio. In the not-too-distant future, it will be feasible to include local SPH calculations as an option in large $N$-body calculations, since the additional computational cost required will become relatively less for larger $N$ values. The challenges in combining stellar dynamics and SPH techniques will mainly take the form of software technicalities, related to the complicated bookkeeping required for the treatment of simultaneous three-body and four-body interactions, and occasional interactions with much larger number of particles. When following $10^5$ stars for a Hubble time, exceptional cases are bound to occur in which, for example, two binaries will be involved in a complex resonance encounter while at the same time encountering yet another binary. The treatment of the inclusion of stars into and escape from such a six-star interaction will be far from straightforward, and has not yet been attempted in all generality.

Other techniques could be mentioned here as well, especially when we include the modeling of dense galactic nuclei. Here the possible presence of black holes, as well as young stars and molecular clouds, create additional complications. Closer to home, the modeling of a proto-planetary nebula again sets different requirements for the physics to be included in a study of the dynamics of this type of dense stellar systems. However, the above six classes of techniques, while not being exhaustive, give at least some taste of the further developments to be expected in the near future.

## Daiichiro Sugimoto

From theoretical sides I would like to discuss five points.

1) *Fundamental Concepts.* For the evolution of globular clusters the targets for the fundamental concepts were as follows. During 1975-84 the concept of Gravothermal Collapse was established which included not only the linear analysis but also the instability of finite amplitude as represented by similarity solution. From the standpoint of the linearized theory the gravothermal instability leading to the contraction can lead also to expansion as well. Thus, it led to the concept of gravothermal oscillation. During 1984-95 the gravothermal oscillation was shown to occur in gaseous and Fokker-Planck models, but it is only at this Symposium when the gravothermal oscillation is clearly shown to occur in $N$-body model. Commonly in these models the gravothermal oscillation takes place as a refrigeration cycle. It implies that the energy input by binary hardening only triggers the oscillation and the non-linear oscillation is maintained, not by the energy input from the binary hardenings but by the gravothermal instability itself. This will explain the post-collapse evolution how the globular clusters are able to survive even after the core collapse. What will be the fundamental concept that we should clarify in the coming 1995-2005? In this Symposium the results of very advanced observations were presented including HST observations and 3-dimensional spectroscopy. They will be used to construct detailed models of evolution of stellar systems including the effect of stellar evolution. At present, however, any central concepts to pursue are not clearly posed yet. The observational results should not simply await for more detailed models. What fundamentals could we extract from them? Or have the fundamental concepts been all clarified and are they only waiting for elaborations?

2) *Gravothermal Oscillation.* In the IAU Symposium No. 113 in 1984 people discussed against Sugimoto and Bettwieser concerning their theory of gravothermal oscillation. There were two points for this discussion. The first was whether such non-linear oscillation takes place or not even in gaseous or continuous model. The second was whether it takes place even in a discrete system such as star clusters. From the theory of self-gravitating system, which had been developed from the theory of stellar structure and evolution, it was self-evident, at least for me, that such oscillation took place as later clarified with different models. However, the second point could not help waiting for a large scale $N$-body calculations, because the computations with several thousand bodies can not extract any clear results because of statistical fluctuations. It became possible only recently when the dedicated computer GRAPE-4 became operational in this year. The result presented in this Symposium by Makino is based on computations with a 32,000 body model as much as 500 Gflops · months on GRAPE-4. The cost for constructing this dedicated machine was only 1.5 Million dollars. If the same amount of computation had been done on a commercially available general-

purpose machine, the machine time for this single computation would have costed ten times the cost for constructing GRAPE-4 machine. This is one of the examples telling us that new means are often needed for a breakthrough.

3) *Seed of Black Hole.* There seem some compelling evidences that there are massive black holes in the active galactic nuclei. In this respect Genzel's presentation in this Symposium was very interesting. Once a high density core containing a black hole is formed, the black hole will grow by eating the stars or by merging with another black hole in the core or in the nucleus of another colliding galaxy. Therefore, the real difficulty lies in the first formation of a seed of the massive black hole so far as we assume the standard model for the gravothermal collapse of single-mass star cluster. The time scale of the gravothermal collapse is simply too long for a large system and, in addition to it, the core formed by the gravothermal collapse is usually too small in mass. We obviously need an idea.

4) *Mass Segregation.* For a mass function of a star cluster, Salpeter's mass function is assumed in many cases which is expressed as $f(m)dm = m^{-2.35}dm$. If we express it as a mass weighted function as $mf(m)dm = m^{-a}d\ln m$, the value of $a$ is as small as 0.35. This implies that possible uncertainties in the value of $a$ are important in the context of evolution of the bulk of the mass contained in the cluster. Mass segregation proceeds so that massive stars contract to form a core consisting mainly of the massive stars. At the same time the less massive stars receive kinetic energies from the massive stars and are prevented from the collapse. When the massive stars explode as supernovae in the core of the cluster, the mass loss takes place from the core, which makes the outer region of the cluster expand. The less massive stars will be tidally stripped off by the external gravitational field exerted in the parent galaxy. This will modify the mass function very much and could be important in the context of producing the seed black hole. Thus, it seems important to simulate the gravothermal collapse with the mass spectrum and stellar evolution taken into account. Such simulations could be compared with young globular clusters in the Magellanic Clouds. The difference in the mass functions according to the different radial distances from the cluster center is now being observed in the Magellanic Clouds. Not only for this point but also for the relation between the flattened and possibly rotating globular clusters it is very important to study more closely the globular clusters in the Magellanic Clouds. They give unique examples with which we can construct a sequence of dynamical evolution for globular clusters.

5) *Stellar Collision.* As discussed in this Symposium the 3-D hydrodynamics of stellar collision is a challenging problem in the views that the direct collision of the stars should take place even in the nucleus of our

Galaxy, and that the matter density in the sub-parsec region of the nucleus of NGC4258 is high enough for stellar collisions if it should consist of the stars instead of a massive black hole. The stellar collision is a rapid process with a large entropy production when seen in the stellar envelope. On the other hand when seen in the stellar core, it is a slow quasi-adiabatic process if the sound speed in the stellar core is higher than the relative velocity of the collision. The result will be dissipation of the stellar envelope and merging of the cores of the stars. The stellar structure just after the merging is important to discuss further evolution of the star itself and of the star cluster as well. It seems rather difficult to calculate accurately the entropy production during the collision processes which determines the fate of the colliding stars. Here, again, the theory of stellar structure will help the situation as it helped in clarifying the concepts of gravothermal collapse and oscillation. For such an approach we need to understand physical processes in general terms rather than simply to believe the results of numerical calculations.

stars. The stellar ages and ZAMS masses of ~600 extinction corrected main-sequence O, B & A stars permit accurate mass-based dynamical calculations.

We constrain the cluster age between 3 Myr and 7 Myr from the presence of WR stars and the lack of red supergiants. The derived WR to O star ratio of 4% could be explained with a cluster age of $3.7 \pm 0.5$ Myr (Parker et al. 1995).

Within our $12.''8 \times 12.''8$ field of view the mass function has the slope $\Gamma = -1.59 \pm 0.10$. The mass function steepens with radial distance from $\Gamma = -1.33 \pm 0.16$ at the cluster center to $\Gamma = -1.63 \pm 0.18$ at 1pc.

We fit the projected mass density distribution to a King (1962) model. Taking all detected stars ($M \gtrsim 4\,M_\odot$) into account, we find $r_c = 0.''97 \pm 0.''07$ (0.24pc). If only the more massive stars ($M \gtrsim 25\,M_\odot$) are considered, the derived core radius decreases down to $0.''2$ (figure 1). Hence the "core" of massive stars is smaller than that of light stars. Calculating the half–mass relaxation time of the O–star population we find $t_{rh(O)} \approx 10^6$ yr, which is several times smaller than the derived cluster age. Dynamical relaxation processes therefore certainly play an important role in producing the observed radial variations of the present day mass function.

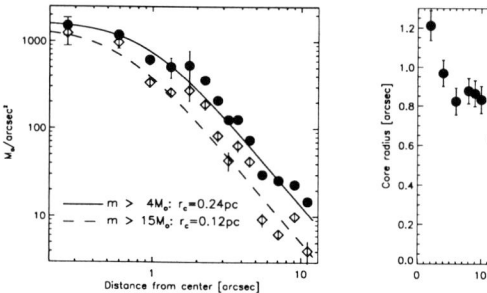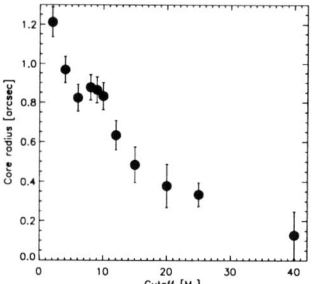

*Figure 1.* Left: Fit to the stellar mass density distribution, excluding the first data point. Right: The core radius as a function of the detected mass range.

## References

Beuzit, J.-L. et al. 1995, A&A, in preparation.
Brandl, B. 1995a, Ph.D.thesis, Ludwig-Maximilians-Universität München.
Brandl, B. 1995b, S&T, August 1995, 14.
Brandl, B., Sams, B.J., Bertoldi, F., Eckart, A., Genzel, R., Drapatz, S., Hofmann, R., Loewe, M., Quirrenbach, A. 1996, submitted to ApJ.
Hunter, D. A., Shaya, E. J., Holtzman, J. A., Light, R. M., O'Neil Jr., E. J., Lynds, R. 1995, ApJ 448, 179.
King, I.R. 1962, AJ 67, 471.
Parker, J. W., Heap, S. R., Malumuth, E. M., 1995, ApJ 448, 705.

# PRELIMINARY STUDY OF THE STELLAR POPULATIONS AND DENSITY PROFILE OF NGC 6624 USING HST

PURAGRA GUHATHAKURTA
*Univ. of California, Lick Obs., Santa Cruz, CA 95064, USA*
BRIAN YANNY
*Fermi National Accelerator Lab., Batavia, IL 60510, USA*
JOHN N. BAHCALL
*Inst. for Advanced Study, Princeton, NJ 08540, USA*
AND
DONALD P. SCHNEIDER
*Dept. of Astronomy & Astrophysics, Pennsylvania State Univ., University Park, PA 16802, USA*

We present preliminary results from an ongoing study of the central 5 arcmin$^2$ of NGC 6624 based on short F336W, F439W, and F555W ($UBV$) Hubble Space Telescope WFPC2 exposures (Yanny et al. 1996). NGC 6624 is a dense, metal rich, post core collapse globular cluster at a distance of 8.1 kpc. Nearly 5000 stars with $V \lesssim 21$ (1.5 mag below the turnoff) are detected within the $34'' \times 34''$ area of the PC1 CCD which imaged the cluster center. Individual stars brighter than $V = 20$ are easily identifiable in the central image section shown in Fig. 1. Image simulations and similar data on the denser cluster M15 indicate that the effects of incompleteness and photometric error are negligible on the $V < 19.5$ sample of stars in NGC 6624. The stellar surface density profile derived from such a sample is well approximated by a power law of index $\alpha = -0.85$ (Fig. 2). The density profile shows no hint of flattening towards smaller radii in the radial range over which it is reliably measured ($r \gtrsim 0''.3$). A $B$ vs. $U-V$ color–magnitude diagram of the $r < 15''$ region (Fig. 3) shows a well defined blue straggler sequence. Our preliminary findings are consistent with an earlier study of this cluster by Sosin and King (1995) using pre-repair FOC images.

### References

Sosin, C. and King, I.R. (1995) *A. J.* **109**, 639.

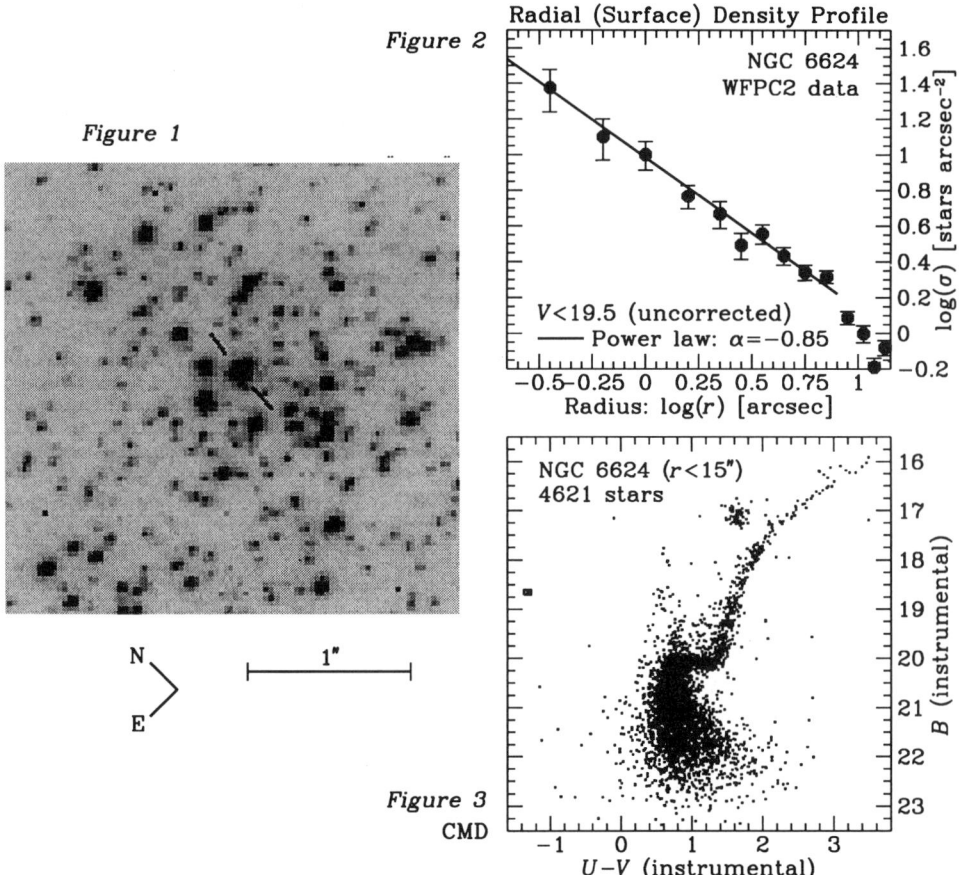

*Figure 1.* A 2″.5×2″.5 section of a $U$ band HST WFPC2 exposure of NGC 6624, centered on the X-ray source 4U 1820–30 (marked by line segments). The cluster centroid lies about 0″.5 south of this object. Individual stars are easily resolved in this 0″.1-resolution image. The number density of stars is high enough to permit reliable determination of the density distribution on subarcsecond scales.

*Figure 2.* The projected density profile of $V < 19.5$ stars in NGC 6624. The solid line shows an $\alpha = -0.85$ power law and it is a good fit to the data. The data do not require a finite core; the 99% upper limit on the core radius is about 2″ (0.08 pc).

*Figure 3.* A $(B, U - V)$ color–magnitude diagram of over 4500 stars within $r < 15''$ of NGC 6624's center. The sample is complete down to $B \sim 21.5$ which is about 1.5 mag below the main sequence turnoff. Note the clear sequence of blue stragglers in the region $18 < B < 19.75$ and $0.4 < U - V < 0.9$. The bold box near the left edge of the plot marks the X-ray bright object 4U 1820–30.

Yanny, B., Guhathakurta, P., Schneider, D.P. and Bahcall, J.N. (1996) *A. J.* in prep.

# CENTRAL DENSITIES OF STAR CLUSTERS IN THE MAGELLANIC CLOUDS

*Comparison with our Galaxy*

M. KONTIZAS
*Section of Astrophysics, Astronomy & Mechanics,*
*Department of Physics, University of Athens*
*Panepistimiopolis GR-157 83, Athens, Greece*

AND

D. GOULIERMIS AND E. KONTIZAS
*Astronomical Institute, National Observatory of Athens,*
*P.O.BOX 20048, 118 10 Athens, Greece*

**Abstract.** The way star cluster systems in galaxies are forming and survive seem to depend on the relation of the central density $\rho$ (at half mass radius) of each cluster with its galactocentric distance $R_{gc}$. It is found that this relation takes the form of:
$$\log \rho = A \times \log R_{gc} + B$$
The cluster systems of our Galaxy and of the two Magellanic Clouds, have been investigated. We have taken the cluster system of the conventional globulars of our Galaxy whereas the young and old systems of clusters in the LMC and SMC were treated separately. The radial distributions of central densities and half mass radii were found for all these systems showing a definite trend which depends on: ($\alpha$) The total mass of the parent galaxy & ($\beta$) The age of the cluster system (young - old). It therefore appears that the total mass and/or the morphology of the parent galaxy plays a major role on the loci where clusters survive and form.

## 1. Introduction

The conventional definition of galactic globular clusters requires them to be dynamically old, populous stellar systems, relaxed under the two body realaxation mechanism. In the MCs very young populous globulars have been found, which didn't have the time to relax under the same mechanism. Therefore the Galactic globulars are first generation stellar systems, whereas the young globulars in the MCs are a second generation cluster population. We represent the central density $\rho$ of a globular cluster as the density of a spherical volume of radius $r_h$ where the half of the total mass of the cluster is contained [4]. This parameter - expressed in $M_\odot/pc^3$ - determines the total lifetime of a stellar system [11] and can provide a criterium for the separation between bound and unbound systems [8].

## 2. Discussion

Using previous catalogues ([5], [6], [7], [10], [1]) we estimated the half mass radii $r_h$ [12] and the central densities $\rho$ for the star clusters of LMC & SMC with known dynamical parameters and we ploted these values versus the galactocentric distances $R_{gc}$ of the clusters. Then we created the same plots for the globular clusters of our Galaxy ([9], [3]). In all three galaxies for all cluster systems considered we found that the densities $\rho$ and the half mass radii $r_h$ are related to galactocentric distance $R_{gc}$ with trends of the form:

$$\rho \propto R_{gc}^{\gamma_\rho} \quad \text{and} \quad r_h \propto R_{gc}^{\gamma_h}$$

where the slopes $\gamma$ are estimated by applying a best fit method.

The slopes $\gamma$ are varying for the various systems. More specifically for our Galaxy the slopes are steeper than those of the MCs with the old ones being steeper for the LMC clusters. There is also a difference in $\gamma$s between the young and old cluster systems in the MCs. The lower density systems are formed and seem to survive at larger galactocentric distances. It is also interesting to note that the maximum observed density values drop with the galaxy's total mass.

Finally we should note that the values of the slopes $\gamma_\rho$ and $\gamma_h$ we found for our Galaxy agree with those from references [2] and [13] respectively.

## References

[1] Chrysovergis M., Kontizas M., Kontizas E.,1989, *A. Ap. Sup. Ser.*, **77**, 357
[2] Djorgovski S. & Meylan G., 1994, *A. J.*, **108**(4), 1292
[3] Harris W.E., New Catalog of Globular Cluster Parameters, anonymous ftp to: physan.physics.mcmaster.ca
[4] King I., 1958, *A. J.*, **63**,110
[5] Kontizas M., 1984, *A. Ap.*, **131**, 58
[6] Kontizas M., Chrysovergis M., Kontizas E., 1987a, *A. Ap. Sup. Ser.*, **68**, 147
[7] Kontizas M., Hadjidimitriou D., Kontizas E., 1987b, *A. Ap. Sup. Ser.*, **68**, 493
[8] Lada, C. & Lada, E. A., 1991, in *Astr. Soc. Pas. Con. Ser.*, Janes K.(ed), **13**, p1.
[9] Mandushev G., Spassova N., Staneva A., 1991, *A. Ap. Sup. Ser.*, **252**, 94
[10] Metaxa M., Kontizas M., Kontizas E., 1988, *A. Ap. Sup. Ser.*, **73**, 373
[11] Spitzer L. Jr., 1958, *Ap. J.*, **127**, 17
[12] Surdin V. G., 1979, *Sov. Astron.*, **23**(6), 648
[13] Van den Bergh S., 1994, *A. J.*, **108**(6), 2145

# A SEARCH FOR VARIABLES IN THE CENTRAL REGIONS OF SOUTHERN GLOBULAR CLUSTERS

J. W. MENZIES
*South African Astronomical Observatory, PO Box 9,
Observatory 7935, South Africa*

**Abstract:** Preliminary results are presented for 3 southern globular clusters that have been searched via CCD photometry for new short-period variables that might be associated with blue stragglers. The clusters considered here are NGC6121, NGC3201 and NGC6809.

## 1. NGC6121

From a number of observational series, one series in each of V (58 frames) and I (65 frames) has been measured. A 3' x 2' region including the cluster centre has been observed. **Three** new contact binaries have been found, with probable periods in the region of 0.3 to 0.4 day. One of them is clearly a blue straggler, the second lies just beyond the turnoff while the third is very red, lying below the subgiant branch.

## 2. NGC3201

A search was made on 2 series of images covering areas of 3' x 2' in the northern (23 V frames) and southerni (19 V frames) halves of the cluster. The cluster centre was on the edge of the frames for each series. Apart from the already known RR Lyrae stars, no variables were found with amplitudes greater than 0.1mag.

## 3. NGC6809

A 3'x 2' field centred on the cluster centre has been surveyed for variables. Apart from the known RR Lyrae stars, **one** new eclipsing variable has been found with a primary eclipse depth of 1.8mag. Altogether, 311 V frames obtained in 5 observing weeks were measured. A possible secondary minimum of depth about 0.1mag was seen Rin one series. No other variables with amplitudes greater than 0.1mag. The eclipsing variable has a period of 1.209day, and lies amongst the blue stragglers in the colour-magnitude diagram.

# CCD PHOTOMETRY OF THE OPEN CLUSTER BE 69

A.K. PANDEY[1], A.K. DURGAPAL[1], B.C. BHATT[2],
VIJAY MOHAN[1], H.S. MAHRA[1]
1. *Uttar Pradesh State Observatory, Manora peak,
Naini Tal, 263129, India*
2. *Indian Institute of Astrophysics, Koramangla,
Bangalore, 560034, India*

**Abstract.** The morphological features of CMDs of Be 69 are better understood in terms of convective overshooting. The comparison of CMDs with the convective overshoot models (Bertelli et al. 1994) produces a good fit for a metallicity Z= 0.008, age = 0.8 - 1.0 billion yr and (m-M)= 14.3, which corresponds to a distance of 2860 pc.

## 1. Observations and Results

With the aim to contribute to the progress in our understanding of intermediate age and old open cluster population we have undertaken an observational program to obtain reliable UBVRI CCD phototmetry of those clusters which are unstudied or poorly studied. In this paper, we present CCD UBVRI photometry for the open cluster Be 69 (C0521+ 326, l = 174°.4, b = -1°.8), for which no previous photometric studies could be found in the literature and compare its colour - magnitude diagram (CMD) with the theoretical ones. The observations were obtained using photometric CCD system at f/13 cassegrain focus of the 104-cm telescope of the Uttar Pradesh State Observatory (UPSO) on four nights during October-December 1990. Clean images have been obtained using the ESO MIDAS software package. The photometric reductions were made using the DAOPHOT profile fitting software (Stetson 1987).

A mean reddening of E(B-V)= 0.65 has been obtained using the colour-colour diagram. We first compare the CMD of Be 69 with the standard evolutionary model of VandenBerg (1985), (figure 1). The best comparison yields an age 0.8 - 1.0 billion yr and (m-M)= 14.0 for Z= 0.006. However, more detailed inspection of the comparison manifests an apparent discrepancy between the shape of the turnoff and the isochrones. This has been noted before by several authors (e.g., Anthony -Twarog et al. 1991, Alfaro et al. 1992) and can be explained as a failure of the standard isochrones to include convective overshoot.

Theoretical isochrones with convective overshoot computed by Bertelli et al. (1994) have also been compared with the CMDs of Be 69. The best fit is obtained for Z= 0.008, (m-M)= 14.3 and age 0.8 -1.0 billion yr. The morphological features of the CMD match very well the theoretical tracks, predicted by this model (figure 2). The availability of two field regions at $\sim 30'$ northward and southward allows us to correct the MS star distribution for contamination due to presence of possible field stars. The integrated luminosity function (ILF) of the cluster has been obtained by subtracting the contribution of field stars. A comparison of the ILFs of Be 69 and Haffner 6 (age 1 billion yr ) taken from Patat and Carraro (1994) manifests that the ILF of Be 69 is consistent with a Salpeeter IMF x = 1.35.

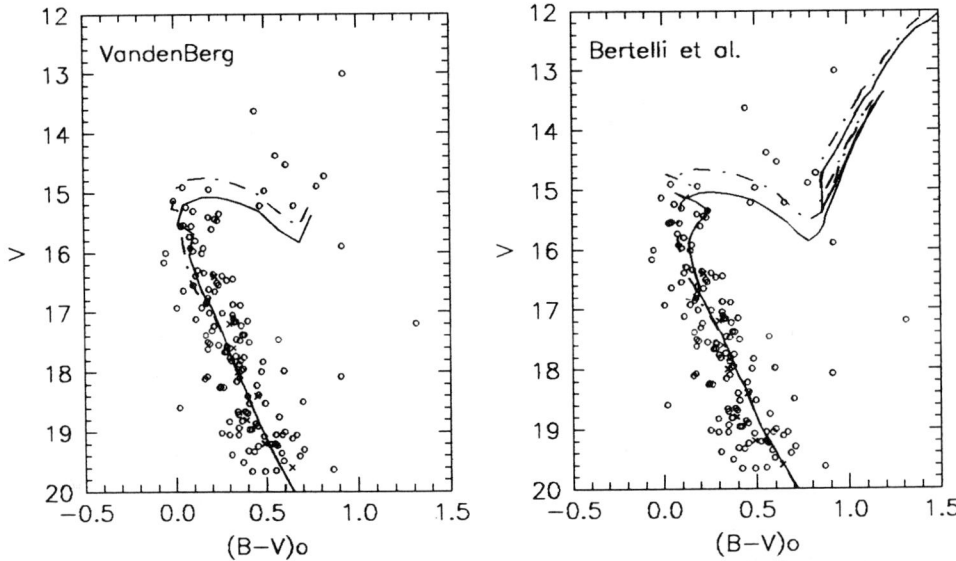

Figure 1. The color-magnitude diagram    Figure 2. The color-magnitude diagram

## 2. Acknowledgements

This work is partly supported by the Department of Science and Technology (India) under the grant SP/S2/O -07/93. AKP is thankful to the Indian Natioanl Science Academy and LOC, IAU 174 for providing partial grants to attend the symposium.

## 3. References

Alfaro, E.J., Aparicio, A. (1992),*Astron. J.* **103**, 204.
Anthony-Twarog, B.J., Heim, E.A., Twarog, B.A., Caldwell, N.(1991),*Astron. J.* **102**, 1056.
Bertelli, G., Bressan, A., Chiosi, C., Fagotto, F., Nasi, E. (1994), *Astron. Astrophys. Supp. Ser.* **106**, 275.
Patat, F., Carraro, G.(1994), *M.N.R.A.S.* (in press)
Stetson, P.B. (1987), *Pub. Astron. Soc. Pac.* **99**, 191.
VandenBergh, D.A. (1985), *Astrophys. J. Supp. Ser.* **58**, 711.

# PAL 1: ANOTHER YOUNG GLOBULAR CLUSTER?

A. ROSENBERG
*Osservatorio Astronomico di Padova, Italy.*

I. SAVIANE, G. PIOTTO AND S. ZAGGIA
*Università di Padova, Dip. di Astronomia, Italy.*

AND

A. APARICIO
*Instituto de Astrofisica de Canarias, Spain.*

**Abstract.** We present a color magnitude diagram (CMD) in the V and I bands reaching $\sim 4$ magnitudes below the turn off (TO) for the galactic globular (?) cluster Pal 1. A comparison with other well-observed clusters and theoretical models suggests that Pal 1 has an age of $8\pm2$ Gyrs, which would make it the youngest globular cluster of our Galaxy.

**The age.** V and I band frames centered on the cluster core, for a total exposure time of 7260 s and 3630 s respectively, have been collected at the 2.5m INT (Roque de los Muchachos Observatory, La Palma) for a total field coverage of $11.2 \times 10.3$ arcmin$^2$. A further field at 10 arcmin from the cluster center has been secured in order to evaluate the background-foreground contamination.

Fig.1 *(left)* presents the CMD of the internal 1.35 arcmin, where the main features of the diagram are clearly defined. The same figure displays the fiducial sequence for NGC 1851 *([Fe/H]=-1.29, Age=17 Gyrs, Saviane et al. (1995))*, Rup 106 *([Fe/H]=-1.69, Age=12-13 Gyrs, Buonanno et al. (1993)*, and M67 *([Fe/H]=-0.09, Age=5 Gyrs, Montgomery et al. (1993))*. The red giant branch (RGB) falls between those of Rup 106 and M 67. If it's still uncertain metallicity ([Fe/H]=-1.01, as quoted by Webbink (1985)) is close to that of NGC 1851, Pal 1 should be one of the youngest GC's in our Galaxy. In the attempt to put some constraints on the age, we have used Bertelli et al. (1994) models. Due to the impossibility to have an independent estimate of the distance modulus, we have compared the theoretical isochrones with our fiducial points, using the color difference between the

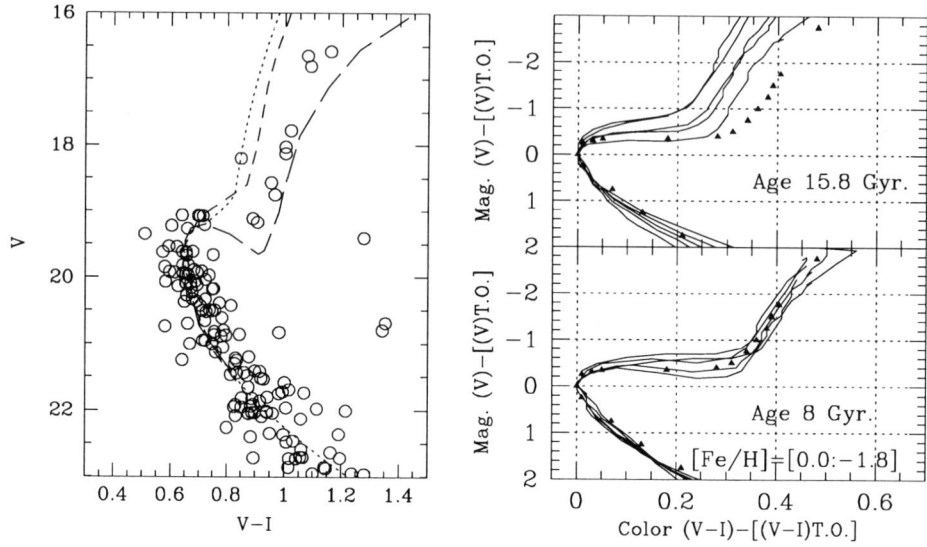

Figure 1.

TO and the SGB ($\delta(V-I)_{TO}^{SGB}$). In particular, in this case the difference has been measured fixing the TO position in color and magnitude (($V-I)_{TO}$ and $V_{TO}$), and taking a reference point on the SGB $\delta$ V magnitudes above the TO, (($V-I)_{SGB}$ and $V_{SGB}$), so that $V_{SGB} = V_{TO} - \delta$ V, and $\delta(V-I)_{TO}^{SGB}$ = $(V-I)_{SGB} - (V-I)_{TO}$. With different values of $\delta$ V (0.8<$\delta$ V<1.6) we find that Pal 1 is consistently very young, precisely, we obtain ages between [6.5-9.5 Gyrs] for [Fe/H]=[0.0:-1.8]. It is also shown in Fig.1 *(right)* that a standard 16 Gyr age isochrone cannot match Pal 1 fiducial points for any metallicity in the interval (0.0:-1.8) *(upper panel)*, while a good match with an isochrone of 8 Gyr can be done *(lower panel)*.

**The morphological parameters.** The density profile of Pal 1 has been obtained from star counts, corrected for completeness and background subtracted. For the first time we have been able to accuratelly correct the counts for the background contamination. Dynamical parameters were obtained by a single-mass isotropic King model fit, giving $c = 1.6$ and $r_c = 14''$ making Pal 1 like a normal globular cluster. The central surface brightness is $\mu_V = 20.85$.

## References

Bertelli G., Bressan A., Chiosi C., Fagotto F. & Nasi E., 1994, A&AS, 106, 275.
Buonnano R., Corsi C.E., Pecci F.F., Richer H.B. & Fahlman G.G., 1993, AJ, 105, 184
Montgomery K.A., Marshall L.A. & Janes K.A., 1993, AJ, 106, 181
Saviane I., Piotto G., Fagotto F. & Capacciolli M. (This meeting.)
Webbink R.F., 1985, IAU SYMP 113, 541.

# DEEP HST/FOC OBSERVATIONS OF THE CENTER OF M15[1]

CRAIG SOSIN AND IVAN R. KING
*Department of Astronomy, University of California,
Berkeley, CA 94720-3411, USA*

**Abstract.** We present preliminary results of the analysis of a set of *Hubble Space Telescope*/Faint Object Camera images of the center of M15, the prototypical post-core-collapse globular cluster. We rule out, at the 95% confidence level, the $2\farcs 2$ core claimed to be detected in pre-repair *HST* imaging. We also measure a mass function in a field $20''$ from the center.

The advent of the repaired *Hubble Space Telescope (HST)* has made it possible to study the dense cores of globular clusters in unprecedented detail. The telescope's high spatial resolution allows us to observe the distribution of *faint* stars in these crowded regions—a capability that is crucial is we wish to understand these clusters' dynamical state.

We observed three $7 \times 7''$ fields in M15 on 27 September 1994; two near the cluster center, and one at $r \simeq 20''$. (A portion of the former images is reproduced in King's paper in this volume.) All were observed in the FOC equivalents of $B$ (F430W) and $V$ (F480LP), for $\sim$2000 seconds in each color. The resulting images show stars down to $V \simeq 22$ in the inner fields, and down to $V \simeq 24$ ($\sim 0.5 M_\odot$) in the outer field.

We used the standard DAOPHOT software to produce lists of stellar positions and magnitudes, with a few additions—such as an algorithm to reject false detections in diffraction rings—that will be described more fully in an upcoming paper (Sosin & King, in preparation).

The completeness-corrected surface-density profile of stars with $V$ magnitudes between 18.5 (just above the main-sequence turnoff) and 20.0 is shown in Figure 1. All of these 839 objects have nearly the same mass. The sample is $> 90\%$ complete over most of its radial range. Improvements to

---

[1]Based on observations with the NASA/ESA *Hubble Space Telescope*, obtained at the Space Telescope Science Institute, which is operated by AURA, Inc., under NASA contract NAS5-26555. This work was supported by NASA grant NAG5-1607.

# 344 CRAIG SOSIN AND IVAN R. KING

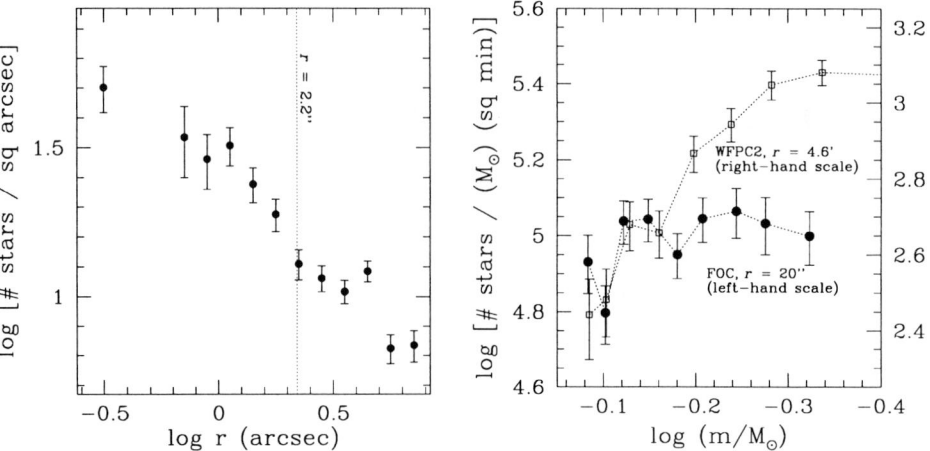

**Figure 1.** (*left*) The surface-density profile of stars with $18.5 \leq V \leq 20.0$.
**Figure 2.** (*right*) The mass function at $r = 20''$, and at $r = 4'.6$.

the bad-object rejection algorithm are currently being developed, and it should soon be possible to extend the analysis to fainter magnitudes.

The surface-density profile clearly continues to climb steadily within $2''$. The $2''.2$ core found in pre-repair *HST* work by Lauer *et al.* (1991) (and subsequently questioned by Yanny *et al.* [1994]) is *not* seen in these data; using a maximum-likelihood method, we rule out a $2''$ core at the 95% confidence level. We cannot distinguish at present between a pure power-law profile and a very small core.

The lesser degree of crowding in the $20''$ field allows us to measure a mass function (MF), shown in Figure 2. The counts have been corrected for incompleteness, and the magnitudes converted to masses via the mass-luminosity relation of D'Antona & Mazzitelli (1995). The best-fitting power-law slope of the MF at $r = 20''$ is $-0.99 \pm 0.5$, where the Salpeter value is 1.35.

We also show the mass function at $r = 4'.6$, as measured in a WFPC2 image (Piotto, Cool, & King, in preparation). The zero points of the two plots (*i.e.*, the left- and right-hand scales) are chosen so that the two MFs appear to overlap at the high-mass end, to emphasize their substantial difference at lower masses—the result of dynamical mass segregation.

## References

D'Antona, F. & Mazzitelli, I. 1995, preprint, "Stellar models and luminosity functions for the Population II main sequence down to its low end"
Lauer, T. *et al.* 1991, ApJ, 369, L45
Yanny, B., Guhathakurta, P., Bahcall, J., & Schneider, D. 1994, AJ, 107, 1745

# $M_V-\sigma$ RELATION: A UNIVERSAL LAW?

S.R. ZAGGIA
*Dip. di Astronomia, Univ. di Padova, Italy*

Globular clusters (GC) of different galaxies are thought to be similar in their intrinsic nature, despite the very different environments in which they reside. This statement is mainly based on the luminosity function of GC that appear to be similar for galaxies structurally very different (Secker and Harris 1993). To support the above conclusion only few other GC parameters have been used: integrated colors and metallicity index (Harris 1991). Now, there is a growing body of GC velocity dispersions of other local group galaxies which can be used to perform a useful comparison. The tool for such a comparison is the correlation between GC total luminosity, $M_V$, and velocity dispersion, $\sigma$, found by Djorgovski (1991). This is one of the best non trivial correlation of GC structural parameters that can give important answers to the problems of the formation and evolution of GCs.

To perform the comparison between local group GCs, we started from the velocity dispersions database for 56 Milky Way GCs (MW) of Pryor and Meylan (1993). It was updated to a total of 60 GCs, adding measures of $\sigma$ coming from our survey of velocity dispersion cusps in GCs (Zaggia et al. 1995), and some recently published measures (Peterson et al. 1994, 1995). Then, we added the data of Dubath et al. (1993a) for 10 old LMC clusters and for NGC 121, the only known old GC in the SMC (Stryker et al. 1985). Magellanic Cloud cluster total luminosities are from van den Bergh (1981). Finally, $\sigma$ and $M_V$ for 3 Fornax GCs were measured by Dubath et al. (1993b). For M31 we used the only existing published set of 12 GCs velocity dispersions (Peterson 1989). We selected only the clusters with reliable $\sigma$ according to Peterson (1989), using for two of them (Bo 193, 225) an updated velocity dispersion given by Peterson (1993).

The total of 86 GCs in local group galaxies are plotted in Fig. 1: it is impressive to see that all the clusters follow the same correlation. The continuous line is drawn from the fit to the MW GCs sample, while the dashed line is the fit to the not-MW clusters: the two fits are almost iden-

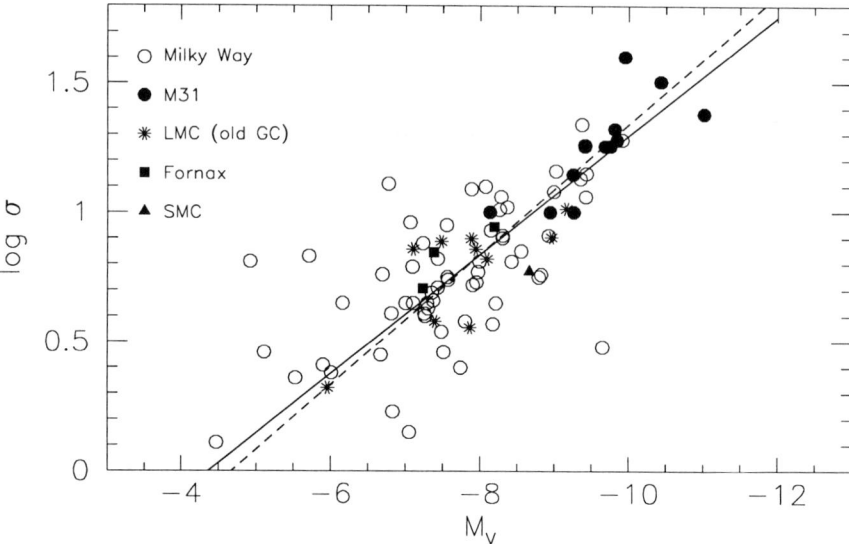

*Figure 1.* GC from different Local Group galaxies. See text for reference on data sources.

tical. The best fit to all GCs gives an $\alpha = 1.68 \pm 0.11$, where $\alpha$ is defined by the relation: $L \propto \sigma^\alpha$ (Djorgovski 1991). The only clear bias present in the correlation is the lack of low luminosity M31 clusters.

In summary, the correlation between $M_V$ and $\sigma$ of GCs shows that the structure of a GC is independent from galaxy environments: a spiral, a dwarf elliptical, or a magellanic galaxy. If confirmed to be an intrinsic characteristic of GCs by larger sample of clusters (from M31, for example), the $M_V$–$\sigma$ correlation can be one of the basis for answering the problem of the formation of globular clusters in galaxies.

## References

Djorgovski S., 1991, *ASP Conf. Ser.*, 13, 112
Dubath P., Mayor G., and Meylan G., 1993a, *ASP Conf. Ser.*, 48, 557
Dubath P., Meylan G., and Mayor G., 1993b, *ApJ*, 400, 510
Harris W.E., 1991 ARAA, 29, 543
Peterson R.C., 1989, in *Dynamics of Dense Stellar Systems*, ed. D. Merrit, Cambridge University Press, p.161
Peterson R.C., 1993, *ASP Conf. Ser.*, 48, 463
Peterson R.C., and Cudworth K.M., 1994, *ApJ*, 420, 612
Peterson R.C., Rees R.F., and Cudworth K.M., 1995, *ApJ*, 443, 124
Pryor C., and Meylan G., 1993, *ASP Conf. Ser.*, 50, 357
Secker J., and Harris W.E., 1993, *ApJ*, 105, 1358
Stryker *et al.*, 1985, *ApJ*, 367, 528
van den Bergh S., 1981, *A&A Sup. Ser.*, 46, 79
Zaggia S.R., Capaccioli M., Piotto G., 1995, *A&A*, *in preparation*

# ON THE RETENTION OF GLOBULAR CLUSTER NEUTRON STARS

G.A. DRUKIER
*Institute of Astronomy, University of Cambridge*
*Madingley Rd., Cambridge CB3 0HA, UK*

## 1. Introduction

Although globular clusters are known to contain a population of neutron stars, the recent finding by Lyne and Lorimer (1994) that field neutron stars are formed with a mean "kick" velocity of 450 km s$^{-1}$ implies that globular clusters would retain very few neutron stars. The number of neutron stars is important when discussing cluster mortality and the formation rate of millisecond pulsars.

Here, I estimate the number of neutron stars retained by globular clusters assuming the that Lyne and Lorimer kick-velocity distribution is correct. Cluster evolution is followed using a direct numerical integration of the Fokker-Planck equation. I calculate the retained fraction of neutron stars as a function of kick velocity $f_{ret}(v_k)$ by a Monte Carlo technique and the retained number by integrating this over the kick velocity distribution. This work is described more fully in a paper submitted to *MNRAS*.

## 2. Results

*Scenario 1, all neutron star progenitors are single:* For the most part, less than 1% are retained, only the models with mass $5 \times 10^6 M_\odot$ retain more than 3%. The fraction retained increases with the depth of the cluster potential.

*Scenario 2, all neutron star progenitors are massive binaries:* The results under this scenario are based on those of Brandt and Podsiadlowski (1995) (especially §3.2) who estimated the effects of the Lyne and Lorimer kick velocities on the evolution of massive binaries. For each of the three possible

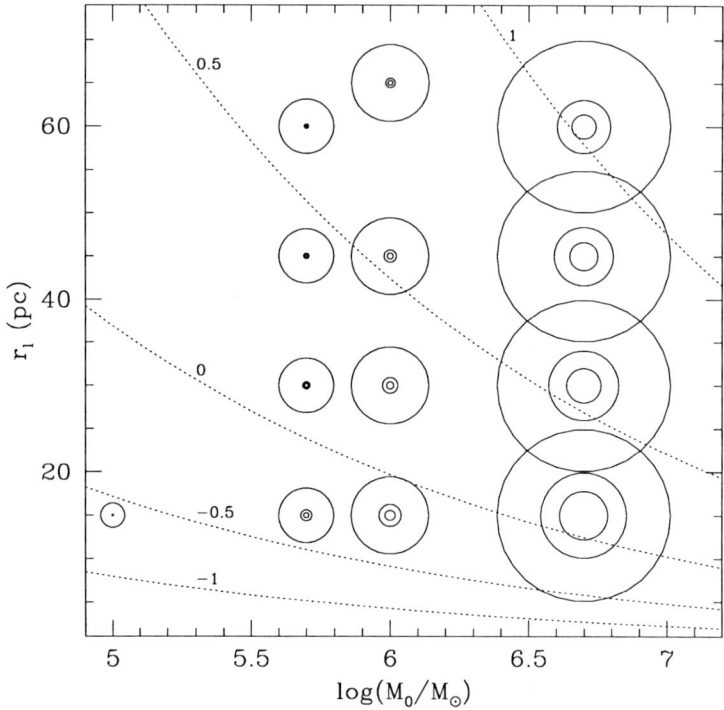

*Figure 1.* For $W_0 = 7$ and $x = 1$, the figure shows the number of neutron star progenitors and retained neutron stars under the two scenarios as a function of $M$ and $r_l$. The areas of the circles are proportional to the number of stars. For each model, the largest circle gives the number of massive stars initially in the cluster. The smallest circle gives the number of neutron stars retained under the single-star scenario and the medium circle gives the number under the binary scenario. The dotted curves are curves of constant half-mass relaxation time as labelled. The units are log Gyr.

outcomes after the first supernova (2 unbound stars, binary with new orbit, or merged system) the velocity distributions of the resulting stars or systems were combined with $f_{ret}(v_k)$ to give the retained numbers. Most of the retained neutron stars come from the $\sim \frac{1}{4}$ of systems which remain bound. It can be argued that most of these systems will pass through a high-mass x-ray binary phase leaving a single neutron star in the cluster. The number retained under this scenario is generally 2 to 5 times larger than the number retained in the first scenario.

## References

Brandt N., Podsiadlowski P. 1995, MNRAS, 274, 461
Lyne A. G., Lorimer D. R. 1994, Nature, 369, 127

# HST/FOS DISCOVERY OF PROBABLE CVS IN NGC 6397

JONATHAN E. GRINDLAY
*Harvard-Smithsonian Center for Astrophysics*
*60 Garden Street*
*Cambridge, MA 02138, USA*

AND

ADRIENNE M. COOL
*Department of Astronomy*
*University of California*
*Berkeley, CA 94720, USA*

Our recent HST/FOS spectra (Grindlay et al 1995; GC95) have spectroscopically identified the 3 central H$\alpha$ candidates (Cool et al 1995; CG95) as the probable cataclysmic variable (CV) counterparts of the the three central dim x-ray sources (Cool et al 1993; CG93) in the nearby core collapsed globular NGC 6397. Here we present the x-ray vs. optical properties of these three objects, derived from the FOS spectra presented in GC95 and the x-ray data of CG93, which strengthen the case that they are CVs and not quiescent LMXBs. Additional data and discussion are presented in the accompanying paper by Grindlay (these proceedings).

CG95 plotted (their Fig. 6) the apparent correlation between the x-ray/optical flux ratio and the equivalent width of the H$\beta$ emission line. As discussed in Patterson and Raymond (1985; PR), such a correlation is found for CVs locally in the Galaxy. The "standard" correlation between x-ray/optical flux ratio, $f_x/f_V$, and emission line strength uses the equivalent width of H$\beta$. Our WFPC1 images provided only H$\alpha$ and R band magnitudes so that $f_V$ was estimated from assuming (V-R) = 0.5, while EW(H$\beta$) was assumed equal to EW(H$\alpha$). With our FOS spectra now actually covering both H$\beta$ and the V band continuum (cf. GC95), the ratio can be measured (cf. Fig. 1). Compared with the WFPC1-derived points of CG95 (plotted here as open symbols), the FOS points lie significantly closer to the correlation line (solid) of PR and in fact within the factor of 3 variation region (dashed).

The x-ray/optical flux ratio vs. EW(H$\beta$) plot may help to further discriminate quiescent LMXBs from CVs for the cluster dim sources (cf. GC95). For CVs, as the accretion rate increases onto the WD, the innermost accretion disk or boundary layer must become increasingly optically thick so that the continuum radiation is both increased and the Balmer lines eventually go into absorption. Conversely, at low accretion rates, the disk surface or corona contributes more of the visible flux and the relatively optically thin higher temperature gas ensures the lines are in emission. Thus in quiescent LMXBs, with their very much lower accretion rate (for dim sources), the lines are expected to be relatively strong and the optical continuum weak. Indeed, taking the ROSAT detection of the quiescent LMXB Cen X-4 (Verbunt et al 1994) as corresponding to the optical continuum and H$\beta$ flux values of McClintock and Remillard (1990), the point is well above the CV correlation line in Figure 1.

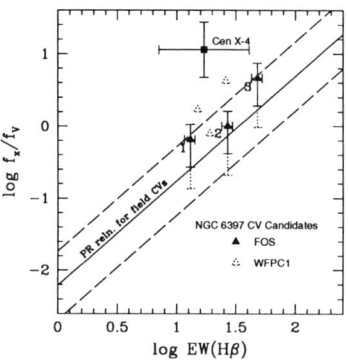

*Figure 1.* H$\beta$ emission strength (equivalent width) vs. x-ray/optical flux for 3 CV candidates in NGC 6397 vs. field CVs and the quiescent LMXB Cen X-4.

The FOS spectra therefore support the identification of the cluster dim sources with CVs. The presence of moderately strong He I and He II lines, with EW(He II) $\sim$ 0.3–0.5 $\times$ EW(H$\beta$) are very similar to the values found by Silber (1992) for magnetic (DQ Her type) CVs as discussed in more detail in the accompanying paper by Grindlay.

This work was supported by grants HST GO-5497 and NASA NAGW-3280.

## References

Cool, A. et al 1993, ApJ, 410, L103 (CG93).
Cool, A. et al 1995, ApJ, 439, 695 (CG95).
Grindlay, J., Cool, A. et al 1995, ApJ, 455, L47 (GC95).
McClintock, J. and Remillard, R. 1990, ApJ, 350, 386.
Patterson, J. and Raymond, J. 1985, ApJ, 292, 535 (PR).
Silber, A. 1992, *Ph.D. Thesis*, MIT.
Verbunt, F. et al 1994, A&A, 285, 903.

# BINARY-SINGLE ENCOUNTERS OF 10 $M_\odot$ BLACK HOLES INCLUDING THE EFFECTS OF GRAVITATIONAL RADIATION

K. N. KONG & H. M. LEE
*Pusan University*
*Department of Earth Sciences, Pusan 609-735, Korea*

**Abstract.** We compute the outcomes of close encounters between a binary and a single black holes including the effects of gravitational radiation reaction. All masses of individual black holes are assumed to be 1 $M_\odot$. We found that merger of two black holes takes place during the encounters in some cases. Thus the gravitational radiation can act as a mechanism for the dissipation of energy of a cluster mainly composed of 10 $M_\odot$ black holes which are produced by the evolution of high mass stars. The merger probability depends on many parameters in a complex way. Our preliminary calculations show that about 10 % of the strong encounters (i.e., $r_p \sim a$) between a binary of hardness 100 and a single lead to mergers of two black holes in the stellar system of one-dimensional velocity $\sigma = 100$ km/s.

## 1. Introduction

Evolution of high mass stars leads to the formation of black holes of around 10 $M_\odot$ (Brown & Bethe 1994). If such black holes are present in a dense cluster of low mass stars ($m \sim 0.7\ M_\odot$), many interesting dynamical phenomena occur (Lee 1995). Black holes form a a subcluster within the central parts of the stellar system as a result of dynamical friction. The subsystem then undergoes core collapse and the density of black hole cluster grows rapidly. Binaries between black holes can be formed efficiently via three-body processes. These binaries release energy to the cluster through the interactions with single black holes.

During the course of interactions between a hard binary and a single, any two objects can become very close. A merger occurs if the amount of gravitational radiation is large. Here we report some preliminary results of numerical simulations of three-body encounters including the effects of gravitational radiation.

## 2. Basic Scheme and Numerical Method

The numerical method we adopt here is is based on the three-body regularization scheme developed by Aarseth & Zare (1974). The general relativistic effects are included as perturbations in this regularization scheme. The dynamical effect of gravitational radiation back reaction was computed by adopting the 2.5 post-Newtonian correction formula given by Damour (1987). We also incorporate the first post-Newtonian term, but the general results are not sensitive to this term. Our numerical program was tested against the analytical result by Peters (1964) for the orbital decay of an isolated binary.

## 3. Results

For our preliminary calculations, we have fixed the eccentricity at $e = 0.67$. We also fixed one-dimensional velocity dispersion of the background cluster at 100 km/sec. We have sampled 112 random directions for initial relative velocity between the single and the center of mass of the binary for a given set of $v_{rel}$, $e$, $x$, $r_p$ (minimum distance between binary and single), and $\sigma$. The orbital phase of the binary at the beginning of the calculation was fixed in all 112 samplings. About 5 to 20 % of strong encounters ($r_p \sim a$ where $a$ is the semi-major axis of the binary orbit) between a hard binary of $x = 100$ and a single are found to experience mergers when $v_{rel} \sim \sigma$. The merger fraction becomes about 1% or less if $x \sim 10$, and $r_p \sim a$. More extensive calculations with larger number of samplings and wider parameter range are being carried out and the results will be published elsewhere.

## Acknowledgements

This research was supported in part by the Basic Science Research Institute Program, Ministry of Education Project No. BSRI-95-2413 and in part by KOSEF under grant No. 941-0200-001-2.

## References

Aarseth, S. J., & Zare, K., 1974, Celestial Mechanics, 10, 185.
Brown, G. E & Bethe, H. A., 1994, ApJ, 423, 659.
Damour, T., 1987, in Three Hundred Years of Gravitation, ed. S. Hawking & W. Israel (Cambridge: Cambridge University Press), 128.
Lee, H. M., 1995, MNRAS, 272, 605.
Peters, P. C., 1964, Phys. Rev., 136, B1224.

# ASCA OBSERVATION OF ω CENTAURI

*Energy Spectra of Einstein IPC Sources*

H. NEGORO[1] AND K. KAWASHIMA[2]
1. Institute of Space and Astronautical Science
3-1-1 Yoshinodai, Sagamihara, Kanagawa 229, Japan
2. Osaka University
1-1 Machikaneyama, Toyonaga, Osaka 560, Japan

## 1. Introduction

The number of compact stars such as white dwarfs, neutron stars, and black holes in globular clusters gives us information on binary formation rates and dynamical history of the clusters.

In *Einstein* IPC observations of ω Centauri, 5 dim ($L_X \sim 10^{32-33}$ erg/s) X-ray sources were discovered toward the center of the cluster; IPC sources $A$–$E$ (Hertz & Grindlay 1983). Recently, sources $A$ and $D$ were optically identified as foreground dMe stars (Cool et al. 1995). *ROSAT* PSPC observations revealed that there were three point sources near the region of extended source $B$ (Hartwick et al. 1982), and that source $C$ within the cluster core was two sources separated by 44" (Johnston et al. 1994).

Despite these advanced studies, the nature of sources $B$ and $C$ are still unknown mainly because of the lack of direct information from these sources. *ASCA* (Tanaka et al. 1994) observation of ω Centauri provides wide energy range 0.5 – 10 keV images and energy spectra of these sources for the first time. We present their first results obtained with the GIS.

## 2. Observation and Results

ω Centauri was observed on 1994 August 1–2 by *ASCA*. IPC sources $A$–$D$ were detected with the GIS (gas scintillation imaging proportional counters) in a soft X-ray band (Fig.1 *left*). Furthermore, several X-ray sources were discovered, for instance, on the NW side of $C$, and on the west side of $B$. In a hard X-ray band, $C$ is the brightest source in the FOV (Fig.1 *right*). The NW source of $C$ is not separated from $C$ because of the *ASCA*'s spatial resolution of $\sim 1'$, but its effect is also clearly seen.

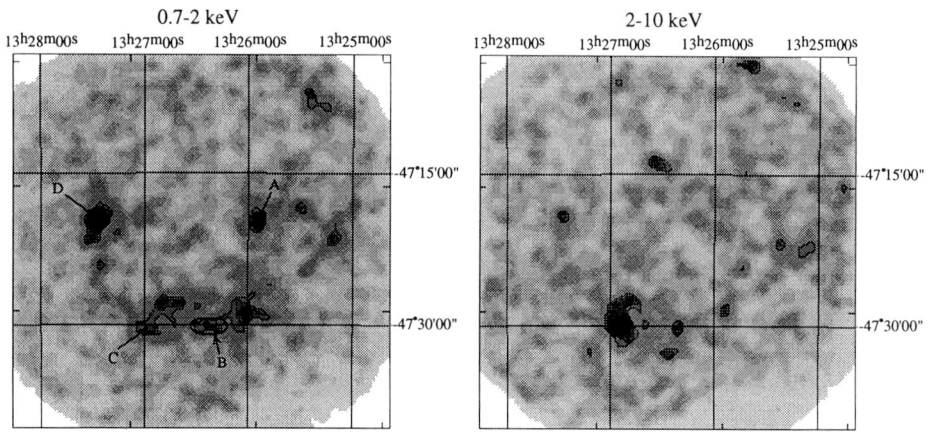

Figure 1. GIS2+3 images of ω Centauri in the energy range of 0.7–2 keV (*left*) and 2–10 keV (*right*). Both the images has been smoothed with a Gaussian of $\sigma = 0.5$'.

These images indicate that $C$ has a very hard spectrum. In fact, its energy spectrum taken with the GIS can be represented by a thermal bremsstrahlung model with $kT > 10$ keV. While, fits to GIS energy spectra of $A$ and $D$ with a Raymond Smith model spectrum with *no* interstellar absorption give $kT \lesssim 1$ keV, which are marginally consistent with (but higher than) typical coronal temperature of dMe stars (Agrawal et al. 1986). In the region of $B$, we found a source with a hard spectrum (probably No. 7 source of Johnston et al.). More detailed information of these and the other sources will be reported elsewhere.

## 3. Discussion

The spectrum of $C$ is compatible with those of Galactic CVs (Ishida 1992; Mukai & Shiokawa 1993), and much harder than those of the other X-ray sources with $L_X \sim 10^{31-33}$ erg/s possible to exist in globular clusters (Verbunt et al. 1994). This strongly suggests that source(s) $C$ is CVs. Their relatively high luminosity of $\sim 10^{32}$ erg/s, compared with those of CVs, also suggests that only high luminous CVs in the cluster were detected.

### references

Agrawal, P. et al., 1986, MNRAS, **219**, 225
Cool, A. et al., 1995, ApJ, **438**, 719
Hertz, P. et al. 1983, ApJ, **275**, 105
Hartwick, F. et al. 1982, ApJ, **254**, L11
Ishida, M. 1992, Ph.D.thesis, Univ. Tokyo
Johnston, H. et al. 1994, A&A, **289**, 763
Mukai, K. et al. 1993, ApJ, **418**, 863
Tanaka, Y., et al. 1994, PASJ, **46**, L37
Verbunt, F. et al. 1994, ASP Conference Series, A. Shafter (ed), **56**, 224

# THE DEPLETION OF GIANTS IN GLOBULAR CLUSTERS

SIMON F. PORTEGIES ZWART
*Astronomical Institute Utrecht, Postbus 80000,
3508 TA Utrecht, The Netherlands*

**Abstract.** A significant depletion of red giants is observed in the central regions of post-collapsed globulars (Djorgovski et al., 1993) like M 15 (Stetson, 1991). A simple model shows that the depletion of red giants in the high-density cores of globular clusters can be understood in terms of mutual stellar collisions. Slightly outside the core stellar collisions are not frequent enough to explain the reduction in the observed number of red giants.

## 1. The model

The stellar system is represented by a computational box (with a density of $10^6 \star /\mathrm{pc}^3$ and size $0.1\,\mathrm{pc}$) which is static, homogeneous and in thermal equilibrium. Each of the 40000 star in the simulation is given a mass from the initial mass-function and a velocity according to equipartition. Since only the core of a collapsed cluster is modeled i use a fit to the mass function derived from observations from the core of 47 Tuc. (see Meylan, 1989). In the model the radius of a star is a function of its mass and age using parameterized evolutionary-tracks (Eggleton et al., 1989 and Portegies Zwart and Verbunt, 1996). The size, mass and velocity of the stars are used to determine the collisions rate in the modeled cluster-core (for details see Portegies Zwart et al., 1996). A Monte-Carlo technic is used to decide when and which two stars merge conservatively as a result of a collision. Since it is unrealistic to let the stars be born in a high-density cluster, i switch on the dynamics at $T = 8$ Gyr and stop the simulation at $T = 16$ Gyr.

## 2. Results

Due to their large geometric- and gravitational-focusing cross-section the encounter probability for a giant is considerably larger than for a main-

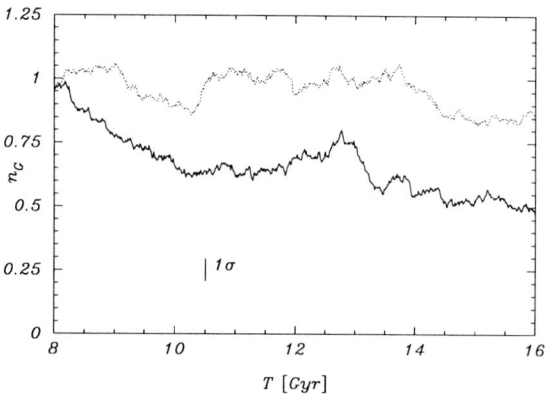

*Figure 1.* The total number of giants from two model computations are presented as fraction of the total number of giants expected from a computation where no collisions took place. The lower line is from the high-density model, the upper line is from a model with a core radius of 0.5 pc and a density of $2 \cdot 10^5$ $\star$/pc$^3$. As a consequence of the small number of giants in the simulations, the error is large (see one-$\sigma$ Poissonian error-bar).

sequence star. However, their small number and short lifetime puts strong limits on the total number of giants that experience an encounter. A collision between a giant and another star generally reduces the lifetime of the collision product and results is a depletion of giants in the high-density cluster-core (see Fig. 2 lower line) where a lower density cluster does not show a significant depletion of giants (upper line). Only in a stellar system with an extremely high stellar-density the depletion of giants can have a collisional origin. Outside the core of a high-density globular, collisions between stars are not frequent enough to explain the depletion of giants.

**Acknowledgments.** The visit to this conference was supported by Shell under a grant for foreign exchange for PhD. students and by the Netherlands Organization for Scientific Research (NWO) under grant PGS 78-277.

## References

Djorgovski, S., Piotto, G., Phinney, E. S., & Chernoff, D. F. (1993) Modification of stellar populations in PCC globular clusters *Astrophys. J. Lett.* **Vol. no. 372**, pp. 41–44

Eggleton, P.P., Fitchet, M.J., Tout, C.A. (1989) The distribution of visual binaries with two bright components *Astrophys. J.* **Vol. no. 347**, pp. 998–1011

Meylan, G. (1989) Studies of dynamical properties of globular clusters. *Astron. & Astroph. J.* **Vol. no. 214**, pp. 106–112

Portegies Zwart, S.F., and Verbunt, F. (1996) Population synthesis of high-mass binaries *Astron. & Astroph. J.* in press

Portegies Zwart, S.F., Hut, P., and Verbunt, F. (1996) *in preparation*

Stetson, P. (1991) in Precision Photometry: *Astrophysics of the Galaxy*, eds. A.G.D. Philip, A.R. Upgren, & K. Janes, (Schenectady: L. Davis Press), pp. 69

# ENVIRONMENTAL INFLUENCE ON STELLAR EVOLUTION: THE HORIZONTAL BRANCH OF NGC 1851

I.SAVIANE, G.PIOTTO AND M. CAPACCIOLI
*Dipartimento di Astronomia, Università di Padova – Italy*

AND

F. FAGOTTO
*Instituto de Astrofisica de Canarias – Spain*

The bimodal nature of the horizontal branch (HB) of NGC 1851 is known since Stetson (1981). In order to better understand the properties of its HB, we collected a set of data at the ESO-NTT telescope, which provides a full coverage of the cluster area. Additional archive images from the HST-WFPC camera have been used in order to study the central region. The resulting c–m diagram (CMD) for 20500 stars is presented in Fig. 1 (*left*). Despite its metallicity ([Fe/H]=−1.3), NGC 1851 presents a well defined blue HB tail, besides the expected red clump. The observed CMD has been compared with the synthetic ones. The bimodal HB can be reproduced assuming that there are two stellar populations in the cluster, with an age difference of ∼ 4 Gyr, hypothesis not supported by other properties of the CMD. On the other side, if we assume that the stars in NGC 1851 are 15 Gyr old (as suggested by the difference between the HB and the TO luminosities), only a bimodal mass loss can reproduce the HB morphology: only stars with higher than standard mass loss rate are able to populate the blue-HB (BHB) tail (Fig. 1,*left*). There are no observational evidences for a bimodal distribution of other parameters (He, CNO, *etc.*).

There is another observational evidence which might shed some light on the nature of the peculiar morphology of the NGC 1851 HB: the radial distribution of the HB stars. Fig. 1 (*right*) displays the ratio blue- over red-HB stars at four different radial positions. The ratio is clearly increasing towards the center of the cluster as consistently shown by three independent data sets. The blue HB stars are more centrally concentrated than the red ones, as furtherly confirmed by the fact that the ratio $N_{BHB}/N_{SGB}$ increases towards the cluster center, while the ratio $N_{RHB}/N_{SGB}$ is constant. NGC 1851 is a high concentration cluster. Color and population gradients

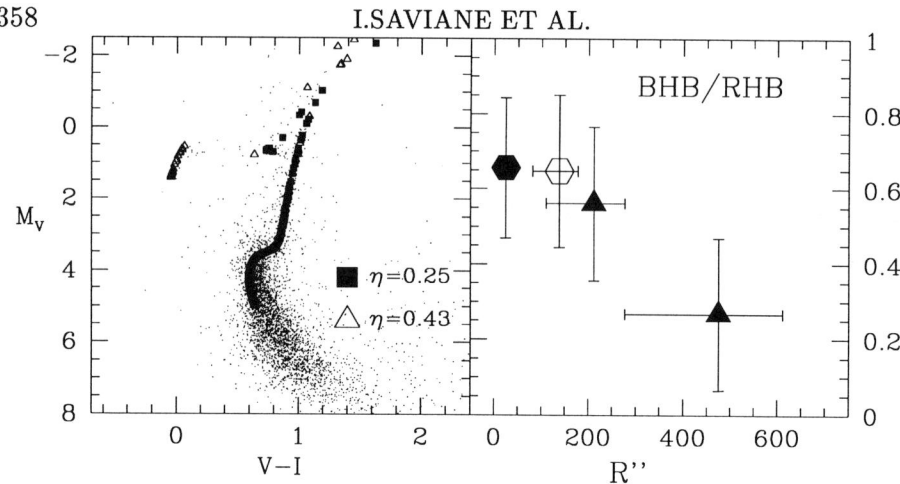

*Figure 1.* **Left panel.** Synthetic CMDs are compared with the observed CMD of NGC 1851. The simulations are obtained adopting the isochrones by Bertelli et al. (1994). Assuming an age of 15 Gyr, the RHB can be reproduced by a mass loss efficiency of $\eta = 0.25$ while the BHB by $\eta = 0.43$. **Right panel.** The ratio of BHB over RHB stars from the NTT (triangles), Walker (1992) (open circle) and HST/WFPC (filled circle) data. The ratio BHB/RHB increases towards the center of the cluster.

(in the sense of bluer centers) in most of the post–core–collapse have already been found (Djorgovski and Piotto 1993, and references therein). Stars in the core of high concentration clusters are likely to live in an environment where stellar encounters are common; encounters can lead to stripping, particularly of the outer expanded envelope of a red giant (Castellani 1994), which could lead to a population of very blue HB stars. An interesting working hypothesis is that the mass loss in the evolving stars in NGC 1851 could be influenced by its high central concentration. The idea that the HB morphology might be related to the cluster concentration has been developed also in Fusi Pecci *et al.* (1993), where observational evidences of bluer and extended BHB tails in high concentration clusters are presented. Here, for the first time, we show that the environment could be responsible also of the bimodality in the HB of a few clusters. The statistical significance of the gradients is not high, as usual for the GC population gradients, due to the small number of evolved stars. For this reason, our hypothesis needs to be tested on the other bimodal HB GCs.

## References

Castellani V., 1994, Mem.SAIt, 65, 649
Djorgovski S. & Piotto G., 1993, ASPCS, 50, 203
Fusi Pecci, F. *et al.*, 1993, AJ, 105, 1145
Stetson P.B., 1981, AJ, 86, 687
Walker A.R., 1992, PASP, 104, 1063

# ON THE ROLE OF DK AND DM STARS IN PILING UP THE TOTAL MASS OF STAR CLUSTERS

G. SZÉCSÉNYI-NAGY
*Department of Astronomy, L. Eötvös University*
*H-1083 Budapest VIII. Ludovika tér 2., Hungary*

## 1. Red Dwarfs in the Solar Vicinity

Although many projects were aimed at identifying nearby red dwarfs, the catalogues may be complete up to a limiting distance of 5 pc only and may be reliable to 10 pc . From recent observations it is evident that their absolute visual brightness ($M_V$) ranges from about 6 to at least 18.5 what means that an average dK0 star is at least 100,000 times brighter than red dwarfs of the latest spectral subclasses. The photon detecting efficiency of our instruments is to be increased that these can measure late dM stars from the same distance from which dK stars can be analysed now. **Another solution is offered by the flare-active red dwarfs which may be 1000 times brighter during their flare ups than usually.**

It is known (cf.Rodonó 1986) that the mass range of the dwarfs extends from about $0.07 M_\odot$ to $0.7 M_\odot$ and that K-type main sequence stars may reach $0.8 M_\odot$. These seemingly low values may suggest that red dwarfs are negligible contributors to the total mass of the Milky Way. In spite of that the incredibly high percentage of red dwarfs among the stars of the Galaxy (about 87% in the solar vicinity, cf.Benest 1983) shows, that these tiny objects may be the most generous donors to its total mass.

## 2. Red Dwarfs in the Open Clusters

There are only a few open clusters with reliable membership data including red dwarf cluster members too. The farthest of all is most probably the *Pleiades*, which has long been scrutinized also for its variables (Szécsényi-Nagy 1990a), especially for flare stars. More than 500 objects of this type have been discovered in a rather limited field of the cluster. These red dwarf stars are concentrated in the central part of M45, a very important fact, which clearly demonstrates that *these flare stars are physically associated with the cluster itself*. Albeit membership probabilities published for some of these stars are inconsistent, it is likely that the flare stars were also born in company with the other (more luminous) members of the system.

Based on results of more than 3000 hours of photometric observations it was possible to calculate that almost 70% of the flare stars of the core of the cluster had been catalogued and that *the whole volume of the Pleiades system contained 1000 or even more flare-active red dwarfs* (Szécsényi-Nagy 1990b). Adopting $0.4 M_\odot$ for the mean mass of dM stars and $0.75 M_\odot$ for that of K dwarfs, **the combined mass of the flare star subsystem of the Pleiades must fall in the 450 $M_\odot$ - 500 $M_\odot$ mass range, and the total mass of M45 is about twice as large as the value derived from previous star counts.**

## 3. dK/dM Stars in Globular Clusters

It is known that the shape of the faint end of the stellar luminosity function is the same everywhere in the Galaxy. Furthermore, since massive stars evolve very quickly, younger clusters are more rich in bright and high-mass blue/white stars, and in older systems the red dwarf population may have definitely higher relative weight in the combined mass of the cluster. It may happen that *the relative mass density of dK/dM class stars in globular clusters (GCs) exceeds the value derived for the much younger open clusters and that of the solar vicinity*. Although this conclusion is probably correct, it is lacking in observational evidences. Until now it has been practically impossible to find any red dwarf members of GCs because of the huge distance moduli of these objects. Late dM stars of GCs remained unobservable and unidentifiable but we have to try to catch late K or early M dwarfs of the nearest clusters, while **the discovery of flare stars in GCs would be a true sensation allowing the extension of the level of activity vs. age relation for these kind of objects.**

The author gratefully acknowledges the support of this research by the NSRF of Hungary (grant no.: OTKA-T 7595) and travel support from the same Foundation (grant no.: OTKA-U 18843) and from the International Astronomical Union.

## References

Benest, D. 1983, *IAU Coll. 76*, eds. A. G. D. Philip & A. R. Upgren (Schenectady: L. Davis Press Inc.) p25
Rodonó, M. 1986, *NASA SP-492*, eds. H. R. Johnson & F. R. Querci, p409
Szécsényi-Nagy, G. 1990a, Publ. Astron. Dep. of the Eötvös Univ. No. 9., Budapest, p308
Szécsényi-Nagy, G. 1990b, *IAU Sym. 137*, eds. L. V. Mirzoyan, B. R. Pettersen & M. K. Tsvetkov (Dordrecht:KAP), p71

# THE FATE OF BINARY SYSTEMS AFTER THE EXPLOSION OF SNE 1993J AND 1994I

HITOSHI YAMAOKA
*Department of Physics, Faculty of Science, Kyushu University,
4-2-1 Ropponmatsu, Chuo-ku, Fukuoka 810, Japan*

**Abstract.** The kinematics of the binary systems to which supernovae(SNe) 1993J and 1994I belonged are studied. The kick velocity caused by supernova(SN) asymmetry will highly influence the fate of the system. The survival rates with assumed kick velocity distribution are discussed.

## 1. Introduction

The progenitors of type Ib/Ic/IIb SNe are considered to be massive stars in close binary systems, of which the outer envelope are torn off. Because of large mass ejection, SN in the binary will destroy the system in most case. But if the SN explosion is asymmetric and/or if mass ejection is small, the binary system can survive. They can be the origin of the Be/X-ray binary or the binary pulsers, and so on.

Type IIb SN 1993J and type Ic SN 1994I are heavyly investigated and the progenitor models are gotten in hand. With these models and appropriate assumptions, we can construct the models of the binary systems.

## 2. Models

The models for the progenitors are taken from Shigeyama *et al.* (1993) and Nomoto *et al.* (1994), and their properties are summarized in Table 1. The mass of the produced neutron star are assumed to be 1.36 $M_\odot$ for both case. For SN 1993J system, the mass transfer should be nearly conservative, so we take 20 $M_\odot$ star as the companion. Its radius is taken as 6 $R_\odot$. For SN 1994I system, we consider three cases: the companion is (A) 1.4 $M_\odot$ neutron star, (B) 1 $M_\odot$ white dwarf, (C) 1 $M_\odot$ main sequence star.

Assumimg that the orbit just before the explosion is circular and the Roche lobe is filled up by either the progenitor or the companion, the orbital radius $a_0$ is determined. The SN ejecta which hits the companion will give the kinetic energy to the companion, and the efficiency $\gamma$ is assumed to be unity (see Yamaoka et al., 1992). The kick velocity $v_k$ is expected to be isotropic, and $v_k^2$ are assumed to follow $\chi^2$ distribution.

## 3. Results and Discussion

For SN 1993J system, small $v_k$ will break the binary for any kick direction, because the orbital velocity before explosion is rather small. The maximum $v_k$ for the system to survive is $v_{k,max} \sim 140$ km s$^{-1}$. On the other hand, if the explosion is symmetric (i.e., $v_k = 0$), the system will not be destroyed.

For SN 1994I, case (A) and (B) are similar to the case of SN 1993J. Symmetric explosion will not destroy the system, also. But, because of the large orbital velocity before explosion, $v_{k,max}$ are very large. Case (C) is quite different from others: the reaction caused by ejecta is quite large, the system cannot be bound with symmetric explosion. The system can survive only if 1100 km s$^{-1} < v_k < 2150$ km s$^{-1}$, but the possibility is very small.

Integrating with the $v_k$ distribution, we can derive the "probability of survival" of the binary as the function of mean value of $v_k$ ($v_{ave}$), which are taken as 150 km s$^{-1}$ (Lane et al., 1982) and 450 km s$^{-1}$ (Lane and Lorimer, 1994). Results are summarized in Table 1 as $P(v_{ave})$. If $v_{ave}$ is large, SN 1993J system scarcely survive. Thus, Be/X-ray binary are not likely formed by this scenario. Also, the low-mass X-ray binary can scarcely be formed such the scenario as 1994I(C).

|         | $M_p$ | $R_p$ | $a_0$ | $v_{k,max}$ | $P(150)$ | $P(450)$ |
|---------|-------|-------|-------|-------------|----------|----------|
| SN93J   | 4.89  | 300   | 1250  | 140         | 0.118    | 0.006    |
| SN94I(A)| 2.28  | 0.23  | 0.55  | 2520        | 0.995    | 0.752    |
| SN94I(B)| 2.28  | 0.23  | 0.51  | 2410        | 0.986    | 0.700    |
| SN94I(C)| 2.28  | 0.23  | 3.24  | *2150       | 1.8e−32  | 1.4e−5   |
|         | ($M_\odot$) | ($R_\odot$) | ($R_\odot$) | (km s$^{-1}$) |   | * See text. |

Table 1. Summary of inputs and results.

## References

Lyne, A.G., Anderson, B. and Salter, M.J. (1982) *MNRAS* **201**, 503–520.
Lyne, A.G. and Lorimer, D.R. (1994) *Nat.* **369**, 127–129.
Nomoto, K. et al. (1994) *Nat.* **371**, 227–229.
Shigeyama, T. et al. (1994) *Ap. J.* **420**, 341–347.
Yamaoka, H., Shigeyama T. and Nomoto K. (1992) *A. Ap.* **267**, 433–438.

# PRE-COLLAPSE EVOLUTION OF GALACTIC GLOBULAR CLUSTERS

TOSHIYUKI FUKUSHIGE
*College of Arts and Science, University of Tokyo*
AND
DOUGLAS C. HEGGIE
*University of Edinburgh*

**Abstract.**
We investigated collisionless aspects of the early evolution of model star clusters. The effects of mass loss through stellar evolution and of a steady tidal field are modelled using $N$-body simulations. Our results (which depend on the assumed initial structure and the mass spectrum) agree qualitatively with those of Chernoff & Weinberg (1990), who used a Fokker-Planck model with a spherically symmetric tidal cutoff. For those systems which are disrupted, the lifetime to disruption generally exceeds that found by Chernoff & Weinberg, sometimes by as much as an order of magnitude.

## 1. Basic Concept

Recently, the focus of our interest has moved to the study of the evolution of more realistic models of globular clusters. For example, stars in a real globular cluster have different masses, and they change mass in response to their internal evolution. Moreover, most globular clusters exist within galaxies, and are influenced by the effects of the galactic tide.

These processes are linked. Each star (above the turnoff mass corresponding to the age of the system) loses a certain fraction of its mass near the end of its own evolution, and this process decreases the total mass of the cluster. Therefore the potential of the cluster weakens and the cluster expands. As a result some stars flow over the tidal boundary, and so further mass is lost by the cluster. The time scale of stellar evolution is $\sim 10^7$ years for a $10M_\odot$ star, which is roughly comparable to the dynamical time scale (or crossing time) of the cluster. If the cluster contains a large fraction of massive stars initially, the dynamics of the cluster is greatly affected by the mass loss due to stellar evolution.

Chernoff and Weinberg (1990; hereafter CW) investigated these and other aspects of the evolution of globular clusters using Fokker-Planck models. Their study included the following realistic effects: the spectrum of stellar masses, mass loss due to stellar evolution, and a tidal cutoff to model the effect of the galactic tidal field. They performed an extensive survey of models differing with regard to the initial mass function, the central potential of the cluster, and the galactocentric distance. For example, they obtained the result that the mass loss during $5 \times 10^9$ yr is sufficiently strong to disrupt weakly bound clusters with a Salpeter initial mass function.

The main purpose of the present paper is to check the results of CW using a model which should be an improvement in several respects. We use direct $N$-body calculations, which allow us to include processes taking place on the dynamical timescale. These are neglected by CW, who used an orbit-averaged method, and because the time scale for mass loss and the dynamical time are not well separated, it is not clear *a priori* that their approximation is justified. Like CW, we also include the following "realistic" effects: a spectrum of stellar masses, mass loss due to stellar evolution, and the tidal field of the galaxy. This last feature differs from CW's tidal cutoff, because it is not spherically symmetric, and also because it affects stars even while they remain inside the tidal radius. Like CW, we performed a survey of King models which differed in respect of the dimensionless central potential, $W_0$, the slope ($\alpha$) of the initial stellar mass function (assumed to be a power law), and the galactocentric distance, $R_g$, of the cluster. In order to calculate the gravitational forces we used a special purpose computer for the gravitational $N$-body problem: GRAPE-3A (Okumura et al. 1993).

In general we obtained qualitatively similar results to those of CW: the less concentrated clusters ($W_0 \lesssim 4$) and/or ones that contained a greater proportion of massive stars initially ($\alpha \gtrsim -2.5$) are disrupted before reaching the stage at which core collapse can begin. The main quantitative difference is that the lifetime to disruption (for those clusters which do not survive until core collapse) may be much longer than that found by CW. For more details, see Fukushige and Heggie (1995).

**References**

Chernoff D. F., and Weinberg M. D., 1990, ApJ, 351, 121.
Fukushige, T., and Heggie, D. C., 1995, MNRAS, 276, 206.
Okumura, T., Makino, J., Ebisuzaki, T., Fukushige, T., Ito, T., Sugimoto D., Hashimoto, E., Tomida, K., and Miyakawa, N., 1993, PASJ, 45, 329.

# TIME-SYMMETRIZED KUSTAANHEIMO-STIEFEL REGULARIZATION

YOKO FUNATO AND JUNICHIRO MAKINO
*University of Tokyo*

PIET HUT
*Institute for Advanced Study*

AND

STEVE MCMILLAN
*Drexel University*

**Abstract.**
In this paper, we show a new algorithm to integrate the orbits of binaries. Our new algorithm has the good properties of both symmetrized timesteps and KS regularization: (1) no secular error in either energy or angular momentum; (2) a constant number of timesteps per orbit for a binary with arbitrary eccentricity (Funato et al., 1995).

## 1. Introduction

The long-term numerical integration of the binary orbit is important in the theoretical study of the evolution of dens stellar clusters. Binary evolution is also important from the observational point of view, since many interesting objects, such as X-ray sources, millisecond pulsars, high-velocity stars, and blue stragglers, are believed to be the result of binary interactions. In order to study the evolution of clusters, self-consistent $N$-body simulation is a most useful tool. However, its computational requirements would be prohibitive; in addition, both truncation and round-off error would be unacceptably large.

Here we present a new time-integration algorithm which has no secular error in either the binding energy or the eccentricity, while allowing variable stepsize. By contrast, the stabilization technique, which has been widely used in the field of stellar dynamics, conserves energy very well but does not conserve angular momentum.

## 2. Symmetrized KS Hermite Scheme with Variable Stepsize

We integrate the equations of motion of the KS binary and its specific energy (Aarseth, 1985) using the time-symmetrized Hermite scheme (Hut, Makino and McMillan, 1995). The procedure for a single step of the time integration is as follows.

[1] Predict the positions and velocities of all particles.
[2] Evaluate all accelerations and jerks using predicted positions and velocities.
[3] Calculate the fourth and fifth derivatives of the relative position of the binary components in KS coordinates at both the beginning and the end of the step.
[4] Evaluate the new stepsize ($\Delta\tau_{new}$) using the stepsize calculated at the end of the step ($\Delta\tau_e$) and the beginning of the step ($\Delta\tau_b$).
[5] Correct the integrated values using the new stepsize $\Delta\tau_{new}$.
[6] Repeat procedures [2]–[5] until both the stepsize and the integrated variables converge.

## 3. Numerical Tests

We have carried out integration of orbits of a binary with initial eccentricity $e = 0.9$. We compared the results of three schemes; symmetrized, stabilized and "plain". All calculations were done in double precision (16-digit accuracy). The result shows that the eccentricity of the binary is conserved by the symmetrized integrator, but not by either the plain or the stabilized schemes. The plain scheme conserves angular momentum but not energy, while the stabilized scheme conserves energy but not angular momentum. Thus neither scheme preserves the eccentricity. Only the symmetrized scheme conserves both energy and eccentricity up to round-off error.

We also experimented the case of a hierarchical triple in which the eccentricity of inner binary is 0.9 and that of outer is 0.0. Our result for the hierarchical triple case shows that our scheme can follow the evolution of weakly perturbed binary for a long time.

## References

Aarseth, S. J., 1985, "Multiple Time Scales" Academic Press.
Funato, Y., Hut, P., McMillan, S., and Makino, J., 1995, *AJ*, submitted.
Hut, P., Makino, J., and McMillan, S. : 1995, *ApJLetter*, **443**, L93–96.

# A NUMERICAL APPROXIMATION FOR HIERARCHICAL TRIPLES

DOUGLAS C. HEGGIE
*Department of Mathematics and Statistics, University of Edinburgh, King's Buildings, Edinburgh EH9 3JZ, U.K.*

**Abstract.** This paper describes a numerical method for following the evolution of the orbit of a perturbed binary (e.g. the inner binary of a hierarchical triple) by means of averaging.

## 1. Introduction

Interactions between primordial binaries in star clusters frequently give rise to long-lived hierarchical triple systems. These are a troublesome feature of $N$-body simulations. In many cases the relative motion of the inner components cannot be treated as unperturbed: perturbations by the outer body can radically alter the probability of a physical collision (Marchal 1990; this paper, Fig.1). In this paper we analytically average over the fast motion of the binary. Then it is necessary only to integrate numerically the equations for the secular evolution.

## 2. Outline and Illustration of the Method

If the method of averaging is applied to the motion of the inner binary, its semi-major axis, $a$, is constant (Marchal 1990). Therefore the orientation and shape of its orbit are determined by its angular momentum vector $\mathbf{h}$ and the Laplace vector $\mathbf{e}$, whose magnitude is the eccentricity, $e$.

Let $m_3$ be the mass of the third body, and $\mathbf{R}$ its position vector relative to the barycentre of the binary. Then in the quadrupole (tidal) approximation, the average rate of change of $\mathbf{h}$ is given by $\langle \dot{\mathbf{h}} \rangle = \langle r_2^2 \rangle f_{23} \mathbf{u}_1 - \langle r_1^2 \rangle f_{13} \mathbf{u}_2 + (\langle r_1^2 \rangle - \langle r_2^2 \rangle) f_{12} \mathbf{u}_3$, where $\langle r_1^2 \rangle = a^2(1/2 + 2e^2)$, $\langle r_2^2 \rangle = a^2(1 - $

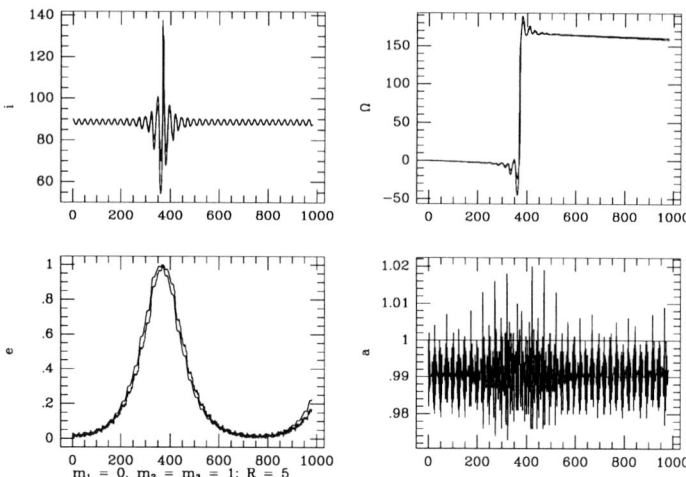

*Figure 1.* Illustration of the method. The masses are as indicated. Initially $e = 0$, $a = 1$ and the orbital planes are orthogonal. The third body is *forced* to move on a circular orbit of radius 5. Each graph shows the variation of one element of the inner binary with time, calculated in two ways: (i) the averaged equations described here, and (ii) an "exact" integration of the equation of motion of the binary, with the exact perturbation by $m_3$. Where the two graphs can be distinguished, the latter is the one with high-frequency oscillations. *Upper left*: the inclination between the orbital planes. Large oscillations occur when $e \simeq 1$, but the two integrations are generally in satisfactory agreement. *Upper right*: the longitude of the line of intersection of the orbital planes. At the time when $e \simeq 1$ the motion of the inner binary changes sense. *Lower right*: the semi-major axis. The systematic offset could be corrected by taking into account the periodic oscillations in $a$ in setting up the initial conditions. *Lower left*: the eccentricity. When $e \simeq 1$ a collision between the components of the inner binary is possible.

$e^2)/2$, the unit vectors $\mathbf{u}_i$ are parallel to $\mathbf{e}$, $\mathbf{h}$ and $\mathbf{h} \times \mathbf{e}$ (respectively), and $f_{ij} = 3Gm_3 R_i R_j/R^5$ if $i \neq j$.

Derivation of the simplest form of the method is completed by carrying out a similar treatment of the Laplace vector $\mathbf{e}$. In fact, however, two further developments are necessary before a satisfactory method is obtained: inclusion of the octupole perturbation, and allowance for periodic perturbations when setting up the initial conditions for $\mathbf{e}$ and $\mathbf{h}$.

## Acknowledgements

I am grateful to Piet Hut for encouraging me to develop this method.

## References

Marchal C., 1990, The Three-Body Problem. Elsevier, Amsterdam.

# EXCHANGE CROSS SECTIONS FOR HARD BINARIES

DOUGLAS C. HEGGIE

*University of Edinburgh, King's Buildings,*
*Edinburgh EH9 3JZ, U.K.*

PIET HUT

*Institute for Advanced Study, Princeton, NJ 08540, U.S.A.*

AND

STEPHEN L.W. MCMILLAN

*Drexel University, Philadelphia, PA 19104, U.S.A.*

**Abstract.** We present results on the exchange cross section for the interaction between a hard binary and a field of single stars, for arbitrary masses. The results are based partly on extensive numerical scattering experiments, and partly on analytic estimates of the mass-dependence of the cross section. They can be used to estimate the rate of exchange in an arbitrary mixture of masses, provided that the binary is hard.

## 1. Introduction

In globular clusters, exchange interactions between binary stars and single stars are a plausible mechanism for the formation of low-mass X-ray binaries. The rate of formation depends on the mass function of the stars and binaries present. Relatively little is known about the relevant cross sections for unequal masses. If they are computed by three-body scattering experiments, a fresh set of experiments must be performed for each mass function. Here we summarise a forthcoming paper (Heggie et al. 1995) in which we present approximate cross sections which are valid for arbitrary masses if the binary is hard. Then the rate of formation is given by a simple integration over the mass function.

## 2. Analytical Estimates

For comparable masses the hard binary exchange cross section is easily estimated by correcting the geometrical area of the binary for gravitational focusing. The result is $\Sigma \sim GM_{123}a/V^2$, where $M_{123}$ is the total mass of the three participating stars, $a$ is the initial semi-major axis of the binary, and $V$ is the speed of the intruder relative to the barycentre of the binary, while they are still far apart.

The next step is to investigate how this formula is modified in various extreme mass regimes. For example, suppose the mass of the intruder, $m_3$, much exceeds that total mass of the binary, $M_{12}$. Then the binary can be unbound by a tidal interaction, and the corresponding cross section is quite readily estimated. Surprisingly, all these estimates (in the various extreme mass regimes) can be summarised in a single formula, which is given by the coefficient of the exponential in the equation below.

## 3. Inclusion of Numerical Data

We have used a scattering package within the Starlab environment (McMillan & Hut 1995) to compute accurate cross sections for exchange. Initial conditions were chosen appropriate to hard binaries with a Jeans eccentricity distribution, for a wide variety of masses.

The numerical results have been combined with our analytic estimate in the following semi-numerical fitting formula. Let $x = m_1/M_{12}$, $y = m_3/M_{123}$ and $M_{ij} = m_i + m_j$, where $m_1$ is the ejected component. Then

$$\Sigma \simeq \frac{GM_{123}a}{V^2} \frac{m_3^{7/2} M_{23}^{1/6}}{M_{12}^{1/3} M_{13}^{5/2} M_{123}^{5/6}} \exp(3.70 + 7.49x - 1.89y - 15.49x^2 -$$
$$- 2.93xy - 2.92y^2 + 3.07x^3 + 13.15x^2y - 5.23xy^2 + 3.12y^3).$$

This formula fits 75% of the measurements to better than 20%; larger discrepancies are restricted to mass ratios where the cross sections are probably too small to be of importance in applications. We have also compared our formula with results by other authors, and again the agreement is satisfactory in general. In addition, these comparisons suggest that the above formula is applicable also to circular binaries, except in some parameter ranges where the cross section is very small anyway.

## References

Heggie D.C., Hut P., McMillan S.L.W., 1995, preprint
McMillan S.L.W., Hut P., 1995, preprint

# A NUMERICAL SCHEME FOR COLLISIONS OF STARS

*Godunov-type Particle Hydrodynamics*

SHU-ICHIRO INUTSUKA
*Division of Theoretical Astrophysics*
*National Astronomical Observatory*
*Mitaka, Tokyo 181, Japan*

## 1. INTRODUCTION

Since Lucy [7] and Gingold & Monaghan [3] invented smoothed particle hydrodynamics (SPH), it has been used to study a variety of astrophysical problems. A broad discussion of the method can be found in a review by Monaghan [8]. However, the most serious disadvantage of SPH remains to be overcome: it cannot handle strong shocks accurately. Especially in three dimensional calculation, particles often penetrate each other, and there are typically large post-shock oscillations.

In this paper, a new method to handle shocks in particle hydrodynamics is presented [4, 5]. In this method a smoothing kernel is used to interpolate various quantities in space. This aspect is similar to SPH [8]. However, the integration of the equations of motion is completely different. Here we make use of a Riemann solver which is the main ingredient of Godunov's method [2]. The method may thus be regarded as a multi-dimensional Lagrangian sequel to Godunov's method. Hereafter we call it Godunov-type particle hydrodynamics (GPH).

## 2. METHOD & EXAMPLES

A brief explanation of the method can be found in Inutsuka 1994 [4, 5]. Similar procedures as MUSCL (van Leer 1979) or PPM (Colella & Woodward 1984) are used to make the accuracy of the method higher order in space. In this case, we require monotonicity algorithms similar to the type of van Leer's[9].

GPH is tested on a variety of 1D, 2D and 3D problems. Here we describe only one of them. It is called "Fifteen Degree Wedge Channel Flow" problem. This is considered by Levy, Powell and van Leer [6], and this is a useful

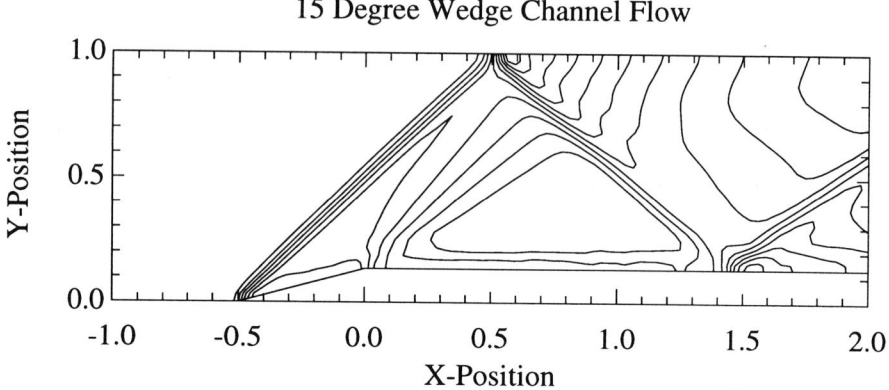

*Figure 1.* Results of GPH on "15 degree wedge channel flow" problem. Initially $32 \times 96$ particles are flowing inside the channel. Mach-number contours from 1.0 to 2.0 are plotted. A variable smoothing length is used.

test to calibrate our method. The geometry is a two-dimensional channel with a 15 degree wedge on the lower wall. A 15 degree expansion corner is also included. The inflow Mach number is two. As apparently noticed, however, this kind of problem is very badly suited for particle method, because (1) we have to realize fixed boundary condition at the wall of the tunnel, and (2) we have to flow particles through the left side boundary continuously. Contrary to the common belief that particle methods such as SPH are not good at this kind of problem, results are excellent.

The present version of GPH has many advantages over standard SPH, especially with strong shocks. In some *astrophysical* problems such as collisions of stars where void fraction is large and free boundary condition is useful, GPH may be still better suited than the "state of the art" finite difference method. Therefore GPH will be a strong tool for *astrophysical* fluid dynamics.

## References

1. P. Colella and P. Woodward, *J. Comput. Phys.* **54**, 174 (1984).
2. S. K. Godunov, *Mat. Sb.* **47**, 271(1959).
3. R. A. Gingold and J. J. Monaghan, *Mon. Not. R. Astr. Soc.* **181**, 375 (1977).
4. Inutsuka, S., *Mem. Soc. Astron. It.* **65**, 1027 (1994).
5. Inutsuka, S., 1994, submitted to *J. Comput. Phys.*
6. D. W. Levy, K. G. Powell, and B. van Leer, *J. Comput. Phys.* **106**, 201 (1993).
7. L. Lucy, *Astron. J.* **82**, 1013 (1977).
8. J. J. Monaghan, *Ann. Rev. Astron. Astrophys.* **30**, 543 (1992).
9. B. van Leer, *J. Comput. Phys.* **23**, 276 (1977).
10. B. van Leer, *J. Comput. Phys.* **32**, 101 (1979).

# DYNAMICAL EVOLUTION OF ROTATING GLOBULAR CLUSTERS

P.-Y. LONGARETTI
*Observatoire de Grenoble*
*BP 53X 38041 Grenoble Cedex 9 - FRANCE*

AND

C. LAGOUTE
*Observatoire Midi-Pyrénées*
*14 avenue Edouard Belin 31400 Toulouse - FRANCE*

## 1. Objectives and method

We have computed simplified globular cluster evolutionary tracks which take into account the effects of internal relaxation, of the cluster rotation, of the galactic tidal field, and, in a cruder way, of stellar evolution and of gravitational shocking. The objectives are first to quantify the influence of rotation in the dynamical evolution of globular clusters; and second, to investigate the evolution of globular cluster angular momentum and flattening (Lagoute and Longaretti 1995a, Longaretti and Lagoute 1995b,c).

In the evaporation phase (when rotation is expected to affect most the dynamics), globular clusters' evolution is well represented by a sequence of equilibrium distribution functions (King, 1966, Wiyanto et al. 1985). We have therefore solved the evolutionary problem in two steps. First, following Michie (1963), we have shown that "rotating" King models are an approximate solution of the Fokker-Planck equation for rotating clusters in the evaporation phase (prior to core collapse). Second, we have computed the evolution through the time dependence of the four parameters on which the distribution function depends from the use of the Fokker-Planck losses of mass, energy and angular momentum and of the tidal constraint [see also King (1966), Chernoff et al. 1986 and Chernoff and Shapiro (1987)]. About 150 000 evolutionary tracks have been computed.

## 2. Conclusions:

Rotation *decreases* the critical concentration at which the gravothermal instability sets in, but this decrease is rather modest (15% for the most rapidly rotating clusters).

Rotation *increases* the mass loss rate by up to a factor of 4, but the total mass loss (integrated over the whole evolution) decreases at most by a factor of 2, due to angular momentum losses during the evolution.

The evolution of the concentration is less affected, because it uncouples from the evolution of the mass as soon as the rotational energy of the cluster exceeds $\sim 5\%$ of its kinetic energy, i.e., *even for slowly rotating clusters.*

Rotation, like gravitational shocking, reduces the domain of survival of globular clusters to higher concentrations.

Globular clusters loose on average about half their initial angular momentum during their evolution, although their flattening often decreases much more dramatically.

Internal relaxation results in a *decrease* of the cluster flattening during the evolution, whereas gravitational shocking induces an increase of the flattening. Combined with the dependence of the relaxation time-scale on the cluster mass, these effects provide a simple explanation for the correlations between flattening and relaxation time (Davoust and Prugniel 1990) and flattening and luminosity (Van den Bergh and Morbey 1984) observed in the Galactic globular clusters, and the flattening/age relation found by Frenk and Fall (1982) in the Magellanic Clouds.

## 3. References

Davoust, E., Prugniel, P., 1990, *A&A*, **230**, 67
Chernoff D.F., Kochanek C.S., Shapiro S.L., 1986, *ApJ*, **309**, 183
Chernoff, D.F., Shapiro, S.L., 1987, *ApJ*, **322**, 113
Frenk, C.S., Fall, S.M., 1982, *MNRAS*, **199**, 565
King, I., 1966, *AJ*, **71**, 64
Lagoute, C., Longaretti, P.-Y., 1995a. Accepted in *A& A*.
Longaretti, P.-Y., Lagoute, C., 1995b. Accepted in *A& A*.
Longaretti, P.-Y., Lagoute, C., 1995c. To be submitted.
Michie, R.W., 1963 *MNRAS*, **125**, 127
Van den Bergh, S., Morbey, C.L., 1984, *ApJ*, **283**, 598
Wiyanto, P., Sato, S., Inagaki, S., 1985, *PASJ*, **37**, 715

# THE DARK SIDE OF GLOBULAR CLUSTERS

P.-Y. LONGARETTI
*Observatoire de Grenoble*
*BP 53X 38041 Grenoble Cedex 9 - FRANCE*

AND

R. TAILLET, P. SALATI
*LAPP*
*B.P. 110, 74941 Annecy-le-Vieux - FRANCE*

## 1. Objectives and method

Searches of low-mass stars have become possible in globular clusters, and the first results suggest that the mass function turns up below $\sim 0.4$ $M_\odot$ (Fahlman et al. 1989; Richer et al., 1990; Richer et al. 1991; G. Piotto, these proceedings). This conclusion is independently supported by the non-parametric dynamical analysis of four clusters by Gebhardt and Fisher 1995.

We have tried to explore how much mass can be hidden in globular clusters in the form of low mass stars and brown dwarves without violating the known observational constraints, and to devise observational strategies to uncover their presence. To this effect, we have constructed series of multi-mass King models of globular clusters with fixed mass function between 0.2 and 1.3 $M_\odot$ (the observationally constrained mass range), and varying power-law mass function below 0.2 $M_\odot$, down to some arbitrary cutoff. The ratio of light to massive star masses varies between 0 and 10 in these models. For definiteness, all our models have the same central surface brightness in the V band (10 000 $L_\odot/$ pc$^2$), the same total luminosity ($3\times 10^5$ $L_\odot$), and dispersion parameter for the red giants (7 km/s). The mass to luminosity relations are taken from D'antona 1987 and Burrows et al. 1993, except for the red giant bin where it is assumed to be equal to 10 $L_\odot/M_\odot$ (Lupton et al. 1987). Finally, the squares of the velocity dispersion parameters in each mass bin are assumed to be inversely proportional to the mass. This assumption is commonly made in dynamical analyses of globular cluster data, and is partially supported by the rapidity of the thermalization process in cluster cores. However, thermalization is at best incomplete in the outskirsts of globular clusters, so that this approximation *maximizes* the quantity of low mass stars which can be hidden in globular clusters for a given set of observational constraints.

## 2. Conclusions:

Complete results obtained from this modelling can be found in Taillet *et al.* 1995a& b. The most interesting conclusions are the following:

Evaporation is too inefficient to eject low-mass stars from globular clusters. However, whether low-mass stars can survive gravitational shocking remains to be seen.

Luminosity profiles are basically unaffected by the presence of low-mass stars over about three decades in luminosity. On the other hand, velocity dispersion profiles flatten out when low-mass stars are added; unfortunately, these profiles are at present too poorly determined to be discriminant.

Thermalisation tends to confine heavy stars in the inner parts of the cluster and low-mass stars in the outskirts. These low-mass stars could be detected there in the infrared by ISO or SIRTF if they constitute more than half the cluster mass and if the thermalisation process is efficient enough.

Low-mass stars can also induce several microlensing events per year on background stars. Some clusters are present in the observation windows of the OGLE collaboration, but the statistics are too scarce to draw any conclusion at the moment.

## 3. References

Burrows, A. Hubbard, W. B., Saumon, D., and Lunine, J. I., 1993, *ApJ*, **406**, 158.
D'Antona, F. 1987, *ApJ*, **320**, 653.
Fahlman, G. G., Richer, H. B., Searle, L. and Thompson, I. B. 1989, *ApJ*, **343**, L49.
Gebhardt, K., and Fischer, P. 1995, *AJ*, **109**, 209.
Lupton, R. H., Gunn, J. E., and Griffin, R. F., 1987, *AJ*, **93**, 1114.
Richer, H. B., Fahlman, G. G., Buonanno, R., and Fusi Pecci, 1990, *ApJ*, **359**, L11.
Richer, H. B., Fahlman, G. G., Buonanno, R., Fusi Pecci, Searle, L., and Thompson, I. B., 1991, *ApJ*, **381**, 147.
Taillet, R., Longaretti, P.-Y., and Salati, P. 1995a, *To be published in* Astropart. Phys.
Taillet, R., Salati, P. and Longaretti, P.-Y. 1995a, *To be published in* ApJ.

# ARE GRAVOTHERMAL OSCILLATIONS GRAVOTHERMAL?

STEPHEN L. W. MCMILLAN AND KIMBERLY A. ENGLE
*Department of Physics, Drexel University,
Philadelphia, PA 19104, U.S.A.*

**Abstract.** We examine critically the properties of the large-amplitude oscillations seen in Fokker–Planck simulations of globular clusters, with both continuous and stochastic binary heating, and compare them to the defining characteristics of gravothermal oscillations.

## 1. Introduction

Gravothermal oscillations were first reported in gas-sphere models of star clusters by Sugimoto & Bettwieser (1983). They have since been observed in both analytical (Goodman 1987) and Fokker–Planck (e.g. Murphy et al. 1990) studies of post-collapse cluster evolution. Very recently, Makino (1996) has described similar core oscillations in $N$-body simulations. The essential features of these oscillations are: (1) they are driven by a temperature inversion in the inner halo that drives energy into the core, causing it to expand, (2) their long-term behavior is largely unrelated to the heat source triggering them, and (3) once established, they require no central heat source to sustain them. We examine these features in Fokker–Planck cluster models incorporating a variety of binary heating mechanisms.

## 2. Continuous and Stochastic Binary Heating

With continuous binary heating (as described by Murphy et al.), we find that the early stages of every expansion phase occur *without* any temperature inversion, suggesting that binary heating is the initial driving mechanism. Further, when the heating is terminated early enough in this phase, the expansion does *not* continue—the system recollapses, even if a small temperature inversion has already appeared. Only if there is a substantial inversion when the heating stops does further expansion occur.

Figure 1 shows the time variation of the central density for a model in which binaries form stochastically, then heat the cluster at a constant rate until they are ejected from the system after depositing $100\,kT$ of energy. The periods during which binaries are heating the system are marked in grey. In almost every expansion cycle, recollapse begins as soon as binary heating stops, and binary heating is more than sufficient to account for the gross energetics of the core expansion. Only when binaries drive a particularly large expansion does a true gravothermal "coasting" phase ensue.

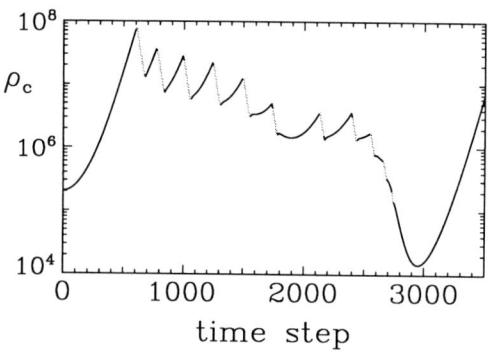

**Fig. 1:** Time variation of the central density $\rho_c$ for stochastic binary heating. The abscissa shows time step, rather than time, to bring out the structure of the peaks. In reality, the expansion episode at right occupies almost the entire interval shown.

## 3. Conclusions

We find that the large-scale core oscillations observed in Fokker–Planck models are not entirely gravothermal in nature. Parts of the expansion are not associated with any temperature inversion at all, and the expansion is closely tied to the continuation of heating in the core. Indeed, the core expansion is initially driven entirely by binary heat input. There appears to be a truly gravothermal phase in some, but not all, cycles, but the temperature inversion driving it develops quite slowly, and may not have time to take effect before heating stops. The most important indicator of gravothermal expansion is not the existence of a temperature inversion, but rather continued core expansion once the heat source is removed.

We thank H. Cohn for allowing the use of his Fokker–Planck code. This work was supported in part by NASA grant NAGW-2559 and NSF grant AST-9308005. It is described in more detail by McMillan & Engle (1996).

## References

Goodman, J. G. 1987, ApJ, 313, 576
Makino, J. 1996, these proceedings
McMillan, S. L. W., & Engle, K. A. 1996, in preparation
Murphy, B. W., Cohn, H. N. & Hut, P. 1990, MNRAS, 245, 335
Sugimoto, D. & Bettwieser, E. 1983, MNRAS, 204, 19p

# GALACTIC DISK SHOCKS ON GLOBULAR CLUSTERS

N. RAMAMANI
*Department of Mathematics and Statistics,*
*University of Edinburgh, UK,*
*Present address: 5 The Hythe, Littleport, Ely, UK*

D.C. HEGGIE
*Department of Mathematics and Statistics,*
*University of Edinburgh, UK*

AND

S.J. AARSETH
*Institute of Astronomy, University of Cambridge, UK*

**Abstract.** We study tidal shocking on globular clusters by $N$-body simulation. The results, which cover a range of cluster and disk parameters, are compared with the impulsive approximation.

## 1. Introduction

Disk shocking is thought to be a significant mechanisms affecting the dynamical evolution of globular clusters. Usually it is treated with an impulsive approximation, but here we use $N$-body simulations. About 150 crossings (with various parameters) have been studied using the special-purpose computers GRAPE-3A and HARP-2.

## 2. Model Parameters and the Impulsive Approximation

We adopted the model of Kuijken and Gilmore (1989) for the acceleration at a distance $z$ from the plane of the disk, i.e. $a = -2\pi Gkz/(z^2 + D^2)^{1/2}$. Guided by data in Mihalas and Binney (1981) we have used the values $k = 50, 100$ and $200$ $M_\odot$ for the disk surface density, $D = 90, 180$ and $360$ pc for the disk scale height and $V_z = 50, 100$ and $200$ km/s for the perpendicular component of the space velocity of the cluster. Based on the ranges for real clusters in Webbink (1985), we adopted the values $M = 10^4, 10^5$ and

$10^6$ $M_\odot$ for the cluster mass, $W_0 = 4, 6$ and 8 for the scaled central potential of the cluster (taken to be a King model) and $4 \leq R_{vir} \leq 22$ pc for the cluster virial radius.

For each simulation we measured $\Delta E_n/E$, the fractional change in the internal energy of the cluster, and the fractional mass loss by escape. The number of stars in each simulation, $N$, was chosen large enough to reduce statistical errors to an acceptable level ($N = 1024$ equal-mass particles on GRAPE-3A and $N = 4096$ on HARP-2). In order to suppress two-body relaxation on time scales up to $\simeq 200$ $t_{cr}$, and to obtain satisfactory results on GRAPE-3A, we employed a standard softening with radius of $R_{vir}/64$. The clusters started at $z = 2D$ and the integration was ended after four disk crossing times, $t_{sh} = D/V_z$.

The change in energy may also be calculated theoretically using the impulsive approximation (Binney and Tremaine 1987). This gives $\Delta E_{ia} = 8(\pi G k/V_z)^2 \sum m_i z_i^2$, where the summation is over all the stars in the cluster. As a measure of impulsiveness we define $T = t_{sh}/t_{cr}$, the ratio of the disk crossing time and the crossing time of the stars in the cluster ($t_{cr}$).

## 3. Results and conclusions

For each simulation the value of $\gamma = \Delta E_n/\Delta E_{ia}$ was calculated. For $W_0 = 4$ and 6 and for constant $V_z^2/KD$, $\gamma$ approaches zero asymptotically. For $W_0 = 8$, $\gamma$ decreases linearly with $T$ up to $T \simeq 140$.

The relative mass loss $\Delta M/M$ was obtained from stars with positive energy with respect to the cluster centre after the encounter. We find that $\Delta M/M = (0.185 \pm 0.004)\Delta E_n/E$ and $(1.126 \pm 0.004)\Delta E_n/E$ for $W_0 = 4$ and 6 respectively. For $W_0 = 8$, we get $\Delta M/M = (0.3635 \pm 0.02)\Delta E_n/E + (0.216 \pm 0.05)(\Delta E_n/E)^2$.

The agreement with the impulsive approximation is good and poor, respectively, for fast and slow encounters, as expected. The mass loss depends only on the energy changes, whose functional form depends on $W_0$.

## Acknowledgements

This work was initially supported by SERC/PPARC grant GR/H93941. N.R. thanks the Institute of Astronomy for providing computer facilities.

## References

Binney J., Tremaine S., 1987, Galactic Dynamics. Princeton UP, Princeton.
Kuijken K., Gilmore G., 1989, MNRAS, 239, 571.
Mihalas D., Binney J., 1981, Galactic Astronomy. Freeman, San Francisco.
Webbink R.F., 1985, in Goodman J., Hut P., eds, Dynamics of Star Clusters, IAU Symp 113. Reidel, Dordrecht, p. 541.

# THE ORBITAL ECCENTRICITIES OF BINARY MILLISECOND PULSARS IN GLOBULAR CLUSTERS

FREDERIC A. RASIO
*Dept of Physics, MIT*

AND

DOUGLAS C. HEGGIE
*Dept of Mathematics and Statistics, Univ. of Edinburgh*

**Abstract.** Low-mass binary millisecond pulsars are born with very small orbital eccentricities, typically of order $e_i \sim 10^{-6}$–$10^{-3}$. In globular clusters, however, higher eccentricities $e_f \gg e_i$ can be induced by dynamical interactions with passing stars. Using both analytical perturbation calculations and numerical integrations, we have shown (Heggie & Rasio 1996) that the cross section for this process is much larger than previously estimated. This is because, even for initially circular binaries, the induced eccentricity $e_f$ for an encounter with pericentre separation $r_p$ beyond a few times the binary semi-major axis $a$ declines only as a power-law, $e_f \propto (r_p/a)^{-5/2}$, and *not* as an exponential. We find that all currently known low-mass binary millisecond pulsars in globular clusters must have been affected by interactions, with their current eccentricities being at least an order of magnitude larger than at birth (Rasio & Heggie 1995).

## References

Heggie D.C., Rasio F.A., 1996, MNRAS, submitted
Rasio F.A., Heggie D.C., 1995, ApJ, 445, L133

# ITINERANCY OF QUASIEQUILIBRIA IN ONE-DIMENSIONAL GRAVITATING SYSTEMS

T. TSUCHIYA
*National Astronomical Observatory, Mitaka, 181, Japan*

N. GOUDA
*Department of Earth and Space Science, Osaka University, Toyonaka, 560, Japan*

AND

T. KONISHI
*Department of Physics, Nagoya University, Nagoya, 464-01, Japan*

One-dimensional self-gravitating many-body systems consist of $N$ identical parallel sheets which have uniform mass density $m$ and infinite in extent in the $(y, z)$ plane. We call the sheets *particles* in this paper. The particles are free to move along $x$ axis and accelerate as a result of their mutual gravitational attraction. The Hamiltonian of this system has a form of

$$H = \frac{m}{2}\sum_{i=1}^{N} v_i^2 + (2\pi G m^2)\sum_{i<j} |x_j - x_i|, \tag{1}$$

where $m$, $v_i$, and $x_i$ are the mass (surface density), velocity, and position of $i$th particle respectively.

This system has been supposed to have three stages of evolution. (1) A system in a non-virial equilibrium experiences the violent relaxation and settles into one of the virial equilibria. (2) Small fluctuations of mean potential change energies of individual particles and the system tends toward *equipartition*, but the global shape does not change in this phase. (3) Collisions among the particles take place and the global shape changes toward the *thermal equilibrium*. Some authors suggested the existence of the three stages(Luwel and Severne, 1985; Sevene and Luwel, 1986), but we were the first to find the above evolutions numerically(Tsuchiya et al., 1994). We call the relaxation processes which take place in the second and the third stage, the *microscopic* and *macroscopic* relaxations, respectively. In the second paper(Tsuchiya et al., 1995) we found the time scales of the two relaxation: $T \sim Nt_c$ for the microscopic relaxation and $T \sim 4 \times 10^4 \, Nt_c$ for the macroscopic relaxation, where $N$ is the number of particles and $t_c$ is the crossing time.

In usual gas dynamics, there is a unique relaxation, thermalization, and its mechanism is the diffusion of the phase point in the phase space. We found all the microscopic processes which happen in the microscopic relaxation shows the evidences of thermalization except for the fact that the global distribution does not change. Thus we continue our investigation to clarify what happens in the microscopic and macroscopic relaxations.

We studied the state of the thermal equilibrium in detail, and found the thermal equilibrium is defined as a time average of different quasi-equilibria, over hundreds of the macroscopic relaxation time, $T_{\text{macro}}$. In a time scale of $T_{\text{macro}}$, the system changes the state from one quasi-equilibrium to another. From the fact we can surmise the phase space dynamics as follows.

1. The microscopic relaxation.   It is a diffusion driven by mean field fluctuations. It takes place on the time scale $T_{\text{micro}} \sim N\,t_c$. This kind of diffusion is also studied by Miller (1991).
2. Quasi-equilibria.   The diffusion of the microscopic relaxation is restricted in a region for a long time, then the orbit in the phase space travels all over the region. Hence the system shows the "mock" ergodicity. Its lifetime is $4 \times 10^4 Nt_c$ for the water-bag distribution, for example.
3. The macroscopic relaxation.   After long time, the orbit escapes from the quasi-equilibrium region. This happens when the orbit find a small window to the outside of the wall which restricts the orbit in the region. Then the orbit visits all the regions in the phase space, the system attains the thermal equilibrium.

The detailed analyses will be published elsewhere.

## References

Luwel, M. and Severne, G. (1985) Collisionless mixing in 1-dimensional gravitational systems initially in a stationary waterbag configuration, *Astron. Astrophys.*, **Vol. 152**, pp. 305.

Miller, B.N. (1991) Gravity in One Dimension: Diffusion in Acceleration, *J. Stat. Phys.*, **Vol. 63**, pp. 291.

Severne, G. and Luwel, M. (1986) Violent relaxation and mixing in non-uniform one-dimensional gravitational systems, *Astrophys. and Space Sci.*, **Vol. 122**, pp. 299.

Tsuchiya, T., Tetsuro, K. and Gouda, N. (1994) Quasi-equilibria in one-dimensional self-gravitating many body systems, *Phys. Rev. E*, **Vol. 50**, pp. 2706.

Tsuchiya, T., Gouda, N. and Tetsuro, K. (1995) Relaxation processes in one-dimensional self-gravitating many-body systems, *Phys. Rev. E*, submitted.

# THE ENERGY EXCHANGE BETWEEN DIFFERENT MASSES IN AN EXPANDING GRAVITATING SYSTEM

YUAN ZHOU[1,2], SHOKEN M. MIYAMA[1]
1. Division of Theoretical Astrophysics,
National Astronomical Observatory,
Mitaka 181, Tokyo, Japan
2. Yunnan Observatory, Academia Sinicai,
Kunming, P. R. China

## Abstract

We investigate whether or not the energy exchange occurs between two species groups of particles in an expanding two-component gravitating system, and we derive the relaxation time scale for energy exchange in such a phase. This is accomplished by solving the dynamic equation coupled with Poisson's equation. We derive a characteristic time determined by the various mass and velocity ratios. When the expansion time does not exceed the characteristic time, energy exchange between the two components is possible and depends on the mass ratio. Once the characteristic time is exceeded, there is virtually no relaxation at all in system. When $m_2 \gg m_1$, the transfer of energy becomes inefficient. Therefore, energy exchange between two species of particle in an expanding two-component gravitating system depends not only on the mass ratio but also on the expansion time.

## 1. Equipartition time in an expanding System

In this section, we will derive the collision integral for gravitating particles in an expanding system, and solve the evolution equation for the distribution function. In this case, the density of particle in the system decreases as a function of time $t$. This is a main difference in comparison to all preceding discussions. In accordance with the usual convention(Spitzer 1969; Binney and Tremaine 1987), we suppose $m_2 > m_1$.

In discussing a non-rotating and spherically symmetric gravitating system composed of two species of particles, the distribution function cannot be described by the multiplication of two single particle distributions or by the simple algebraic addition used by Merriti (1981). It is necessary, at least, to consider the two-body correlation function $g(1,2)$ (Sugimoto 1985),

$$f(1,2) = f_1 f_2 + g(1,2), \tag{1}$$

here $f_1$ and $f_2$ are the single particle distribution functions for two species of particles with masses $m_1$ and $m_2$ respectively. We also suppose that the particles with the mass $m_1$ and particles with the mass $m_2$ both have Maxwellian velocity distributions, but with different kinetic temperatures $\Theta_1$ and $\Theta_2$. We obtain a analytical solution:

$$\tau_{12} = \frac{9\pi^{1/2}(<v_1^2> + <v_2^2>)^{3/2}}{2^{1/2}G(m_1 + m_2)\Lambda_{12}} t^2, \tag{2}$$

$<v^2>$ is the mean square velocity dispersion of the particles. This formula expresses the equipartition time between two species of particles in an expanding gravitating system. It also illustrates the evolution of the time scale of energy exchange between the different masses with the decreasing density of the particles. $\tau_{12}$ is a function of the expansion timescale $t$.

## 2. The Energy Exchange

In the present section, we discuss the condition for energy exchange in an expanding, two-component gravitating system. During the process of collision, the change in the momentum $\Delta \mathbf{p}$ in terms of the angle $\theta$ is

$$\Delta \mathbf{p} = -m_{12}(1 - cos\theta)(\mathbf{v}_1 - \mathbf{v}_2), \tag{3}$$

here $m_{12}$ is the reduced mass. If $m_1 \ll m_2$, the momentum may reverse its sign, $|\Delta \mathbf{p}/\mathbf{p}|_{max} = 2$. Then we have the relation $\mathbf{v}_1' = -\mathbf{v}_1 + 2\mathbf{v}_2$. In this case the energy exchange becomes

$$\Delta E = \frac{m_1}{m_2}(\mathbf{v}_1 - \mathbf{v}_2) \cdot \Delta \mathbf{p}, \tag{4}$$

and we can find that the transfer of energy becomes very inefficient when $m_1 \ll m_2$. When the masses are comparable, that is if $m_1 \approx m_2$, then the momentum may be lost completely. The energy exchange becomes $\Delta E \simeq (\mathbf{v}_1 - \mathbf{v}_2) \cdot \Delta \mathbf{p}$, this means that the transfer of energy is efficient.

## References

Binney, J., and Tremaine,S.(1987) *Galactic Dynamics,* Published by Princeton University Press.
Merritt,D.(1981)*Astron.J.*.**Vol. 86**,p.318
Spitzer,L.(1969)*Ap.J.Letter***Vol.158**,p.L139
Sugimoto,D.(1985)*IAU Symopsium 113,Dynamics of Star Clusters,* ed.J.Goodman and P.Hut(Dordrecht: Reidel), p.207

# ON THE DYNAMICS OF YOUNG OPEN STAR CLUSTERS IN THE JOINT FIELD OF THE GALAXY AND OF A STAR FORMATION REGION

V.M.DANILOV

*Astronomical Observatory of the Ural State University*
*Lenin ave. 51, 620083 Ekaterinburg, Russia*
*danilov@astro.urgu.e-burg.su*

The young open star clusters (OCl) with ages of $10^7 - 10^8$ years in the overwhelming majority of cases are located in the regions of a star formation which are the massive and extensive gas - star complexes (GSC) in our Galaxy. The typical sizes of GSCs are near 600 pc in projection at the Galaxy plane and the masses of such complexes are in range $\sim (10^5 - 10^7) M_\odot$ (Efremov (1989)). It is well known that the old open clusters avoid the regions of an active star formation in our Galaxy. As a result the structural and dynamical parameters of OCls depend on their ages and their locations relatively the nearest GSCs centres.

Two following dependences between the OCls parameters have been recently found by means of the new statistical method of the estimates of open star cluster parameters with nearly 10 % errors (Danilov, Seleznev (1994)): 1-the maximal OCls sizes in the fractions of clusters radii in the Galaxy field increase with the ages of OCls; 2-the maximal young OCls sizes in the fractions of the cubic root of the masses of these clusters decrease inside the complexes with the distance from the GSC centre and such OCls sizes increase outside the complexes.

By the theoretical and numerically - experimental estimates of OCls tidal radii in the joint field of the Galaxy and of a gas - star complex for different GSC models it was shown that both of these dependences may be fully conditioned by the action of the tidal force field of a complex on the young OCls formed in this GSC (Danilov (1991, 1994); Danilov, Beshenov (1992)). Both of these dependences are the arguments in favor of the fact that the complexes are the higher mass density regions in our Galaxy.

The numerical experiments for OCl's models with the number of stars $N = 500$ for clusters moving in the field of forces of a stationary GSC and of the Galaxy were performed in the paper of Danilov, Beshenov (1992). The tidal radii estimates for OCl models for the set of time moments $t$ in that paper were performed by means of the method of Allen, Richstone (1988). Such estimates of clusters tidal radii in an external field, performed in numerical experiments, lead to the $R_t$ values agreed with the corresponding theoretical estimates of the clusters tidal radii. There are two regions of the higher instability of the solutions for a coarse grained phase density function with respect to small variations of the initial phase coordinates of stars in the OCl models. These instability regions are the central part of OCl and the wide region near the tidal boundary of a cluster. Such instability in the central part of a cluster is conditioned by the stellar encounters action. Such instability near the tidal boundary of a cluster is conditioned by the joint action of stellar encounters and of tidal instability of star orbits in the regular field of the cluster and of the Galaxy.

Main features of OCls dynamical evolution in stationary and nonstationary GSC models were dicussed in papers: Danilov, Beshenov (1992), Danilov (1994). The conditions of the stability of OCls to their disintegration in the tidal field of the Galaxy and of the both stationary and nonstationary GSC were used for the estimations of the total masses of some GSCs from the solar vicinity on the basis of the data of young OCls radii and of the parameters of the Galaxy field for these regions of the star formation (Danilov, Seleznev (1995)). Such "dynamical" estimates lead to the GSCs masses in range of $\sim (10^5 - 10^7) M_\odot$ the same as on the basis of the data of the optical and radio observations of GSCs.

## References

Allen A.J., Richstone D.O. (1988) Tidal limitation of stellar systems on both circular and elongated orbits *Astropys. J.* **325**, pp. 583–595.

Danilov V.M. (1991) The tidal sizes of open clusters in the combined field of the Galaxy and a gas - star complex *Astron. Zh.* **68**, pp. 487–500.

Danilov V.M. (1994) Estimation of tidal sizes of open star clusters in star - formation regions in the Galaxy *Astron. Zh.* **71**, pp. 220–227.

Danilov V.M., Beshenov G.V. (1992) The numerical experiments for estimation of the tidal sizes of open star clusters in the joint field of the Galaxy and of a gas - star complex *Astron. Zh.* **69**, pp. 238-249.

Danilov V.M., Seleznev A.F. (1994) The catalogue of structural and dynamical characteristics of 103 open star clusters and the first results of its investigations *Astron. Astrophys. Transactions* **6**, pp. 85–155.

Danilov V.M., Seleznev A.F. (1995) Dynamical estimates of gas - star complex total masses *Astron. Astrophys. Transactions* **7**, pp. 113–116.

Efremov Yu.N. (1989) Sites of star formation in galaxies: star complexes and spiral arms, Nauka, Moscow.

# SPATIAL STRUCTURE OF THE GLOBULAR CLUSTER SYSTEM AROUND NGC 4472

EUNHYEUK KIM, MYUNG GYOON LEE
*Department of Astronomy, Seoul National University*
*Seoul, 151-742, Korea; ekim@astro.snu.ac.kr*

AND

DOUG GEISLER
*KPNO/NOAO, USA*

NGC 4472 (M49) is a giant elliptical galaxy in the Virgo cluster, and has a rich GCS. Radial surface density profiles of the GCS in this galaxy are well-known(Cohen 1988, Harris 1986). However, little information is available for the ellipticity ($e$) and position angles of the GCS in NGC 4472. In this study we have investigated in detail the spatial structures of the GCS in NGC 4472 (the surface density, the ellipticity, the color, and the position angle) in comparison with those of the halo, using a large sample of the globular clusters located in a wide field ($16' \times 16'$) of NGC 4472.

In the color-magnitude diagram of the point sources in the NGC 4472 image (see in Lee, Kim & Geisler, 1995), the objects with the colors of $1.0 < (C-T_1) < 2.3$ are considered to be globular clusters in NGC 4472. To avoid the possible contamination in the faint end due to the faint background galaxies we have selected only the GCs brighter than $T_1 = 22.5$ mag for the following analyses. The total number of GCs in our sample is $\sim 1,300$. The sample of the GCs has been divided into two groups ( the metal-poor GC ([Fe/H] $< -0.5$) and the metal-rich GC([Fe/H] $> -0.5$) ) to investigate the differences between the two groups.

Radial distribution of the surface density of the GC shows that the metal-rich GC is spatially more concentrated than the metal-poor GC ( the core radii of the metal-rich GC and the metal-poor GC are measured to be $\sim 140''$ and $\sim 240''$, respectively). The ellipticity of the metal-rich clusters decreases from $e \sim 0.8$ to $e \sim 0$ within $R = 150''$ as the galactocentric

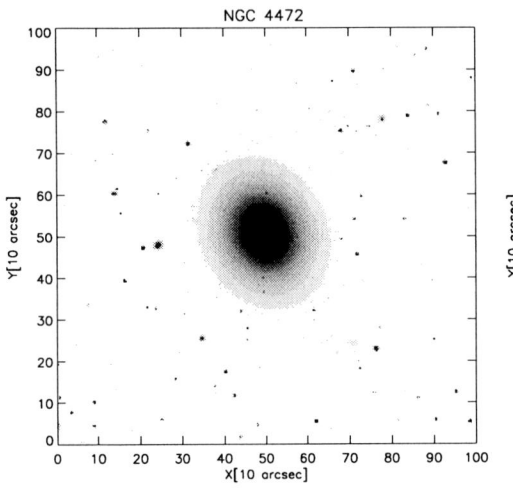
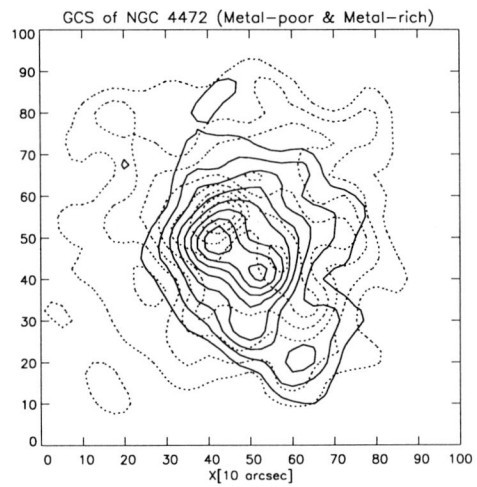

*Figure 1.* A grayscale map of $T_1$ ($T_{exp} = 60$ sec) image of NGC 4472. North is at top, and east is to the right.

*Figure 2.* Displayed is the surface density contour map of the metal-poor GCs (dotted line) and the metal-rich GCs(solid line). Note that the distribution of the metal-rich GCs is spatially more concentrated than that of the metal-poor GCs.

radius increases, and increases to $e \sim 0.4$ outward beyond $R = 150''$, while the ellipticity of the metal-poor clusters stays approximately constant at $e \sim 0.4$. The position angles (PA's) of the two groups show a significant difference in the inner region ($R < 300''$), while the PA's of the two groups agree approximately (PA $\sim 140°$) in the outer region ($R > 300''$). These differences suggest that they have different formation histories.

The galaxy halo light shows quite different features from the GCs. The surface brightness of the galaxy halo light decreases more steeply than that of the GC. The median colors of the GCs get bluer outward. On the contrary the colors of the halo get bluer by a small amount outward in the inner region ($R < 180''$) and become much redder outward beyond $R = 180''$. The mean color of the halo is much redder than that of the GC. The ellipticity of the halo remains almost constant with a value of $e \sim 0.2$, and the PA of the halo decreases slightly outward (PA $\sim 160°$). These results imply that the halo and the GCs in NGC 4472 formed and evolved via different processes.

## References

Cohen, J. G., 1988, AJ, 95, 682
Harris, W. E., 1986, AJ, 91, 822
Lee, M. G., Kim, E., & Geisler, D., 1995, this proceeding, in press

# METALLICITY AND LUMINOSITY FUNCTIONS OF THE GLOBULAR CLUSTERS IN NGC 4472

MYUNG GYOON LEE, EUNHYEUK KIM
*Department of Astronomy, Seoul National University*
*Seoul, 151-742, Korea; mglee@astrog.snu.ac.kr*

AND

DOUG GEISLER
*KPNO/NOAO, USA; dgeisler@noao.edu*

NGC 4472, the brightest elliptical galaxy in the Virgo cluster, has a rich globular cluster system. We present a study of the metallicity and luminosity functions of a large number of globular clusters in NGC 4472. Deep Washington $CT_1$ photometry of a wide ($16' \times 16'$) field of NGC 4472 was obtained using Tek 2048 × 2048 CCD at the KPNO 4m telescope.

The color-magnitude diagram of ∼9,500 measured point sources (Fig. 1) shows two strong vertical structures in the color range of $1.0 < (C-T_1) < 2.3$ which consist mostly of globular clusters, and a dominant horizontal structure fainter than $T_1 \approx 23$ mag most of which are unresolved faint background galaxies.

We have estimated the metallicity of ∼1,300 globular clusters brighter than $T_1 = 22.5$ mag from the $(C-T_1)$ colors. The metallicity distribution of the bright globular clusters shows two strong peaks at [Fe/H] = −1.3 dex and −0.1 dex (Fig. 2). The metal-rich globular clusters are spatially more concentrated than the metal-poor globular clusters (see also Kim *et al.* 1995). The mean metallicity of the globular clusters is decreasing as the galactocentric radius is increasing. These results are consistent with the merger model for the formation of giant elliptical galaxies.

The luminosity function of the globular clusters shows clearly a peak at $T_1 = 23.3 \pm 0.1$ mag (Fig. 3). Comparing this with the value for the galactic globular clusters($M_R = -7.9$ mag), we derive a distance modulus of $(m-M)_0 = 31.2 \pm 0.2$ mag ($d = 17.4 \pm 1.6$ Mpc). This value is very similar to the distances to M87 ($d = 16.8$ Mpc) and M100 ($d = 17.1$ Mpc) in the same cluster (Whitmore *et al.* 1995; Freedman *et al.* 1994). Then we

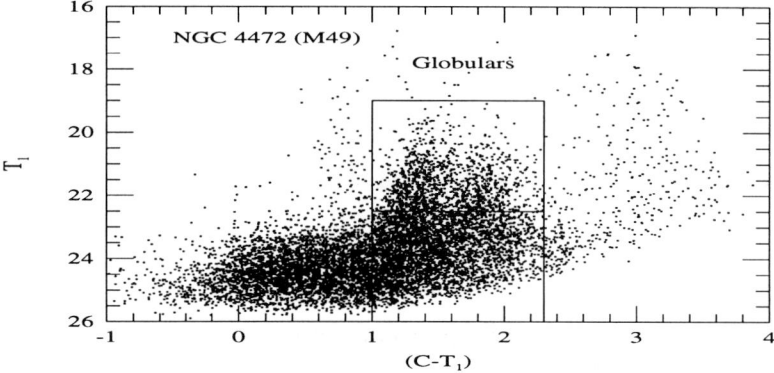

*Figure 1.* $T_1-(C-T_1)$ color-magnitude diagram of ~9,500 measured point sources in NGC 4472 image.

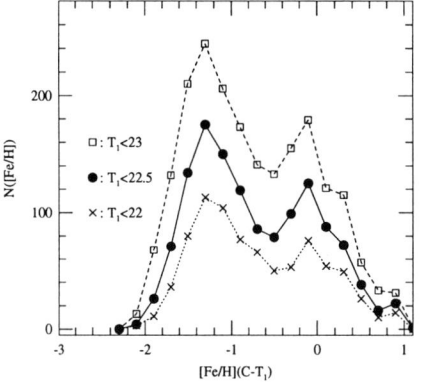

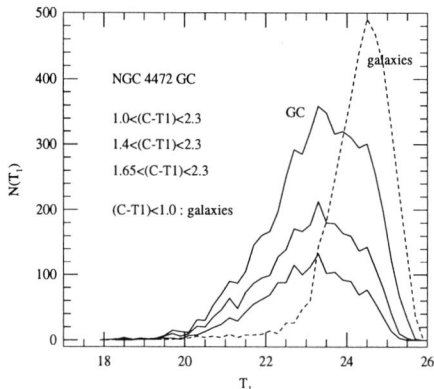

*Figure 2.* Metallicity distribution of the bright globular clusters in NGC 4472.

*Figure 3.* $T_1$ luminosity functions of the globular clusters in NGC 4472 (the solid line) and background galaxies (the dashed line).

estimate the Hubble constant to be $H_0 = 68 \pm 14$ km/s/Mpc and $79 \pm 17$ km/s/Mpc for the cosmic expansion velocity of the Virgo $v = 1179$ km/s (Jerjen & Tammann 1993) and 1380 kms/s (Mould et al. 1995), respectively.

## References

Freedman, W.L. et al. 1994, Nature, 371, 757
Jerjen, H., & Tammann, G. 1993, A&A, 276, 1
Kim, E., Lee, M.G., & Geisler, D. 1995, this proceedings, in press
Lee, M.G., & Geisler, D. 1993, AJ, 106, 493
Mould, J. R., et al., 1995, ApJ, 449, 413
Whitmore, M. et al. 1995, preprint

# GRAPE-SPH SIMULATIONS OF THE CHEMODYNAMICAL EVOLUTION OF DWARF GALAXIES

MASAO MORI[1], YUZURU YOSHII[2,3], TAKUJI TSUJIMOTO[4] AND KEN'ICHI NOMOTO[1,3]

[1] *Department of Astronomy, University of Tokyo, Japan*
[2] *Institute of Astronomy, University of Tokyo, Japan*
[3] *Research Center for the Early Universe,*
*The University of Tokyo*
[4] *National Astronomical Observatory, Japan*

## 1. Introduction

The central concentration of the luminosity distribution in dwarf elliptical galaxies ( dEs ) is weaker than that of giant elliptical galaxies ( Es ). In other words, the luminosity profiles in Es follow the de Vaucouleurs' law whereas dEs have exponential luminosity profiles. Athanassoula(1994) describes the one dimensional simulations of the formation of dEs that include the feedback effects from supernovae. The model with no dark matter halo is shown to be much better agreements with the observations than with dark matter halo. However no attempt has yet been made to reproduce much lower heavy element abundances in dEs than in Es. We calculate the chemodynamical evolution of a less massive gas cloud with an SPH+N-body three dimensional simulation code to explore the luminosity profile and chemical abundances in dEs.

## 2. Models

The code is based on the Smoothed Particle Hydrodynamics. This code assumes that stars are formed in regions both converging and Jeans-unstable. Supernovae are treated as sources of heat, mass, and heavy elements. The gravitational forces are calculated with GRAPE-3AF. We assume that a protogalaxy is initially an isolated isothermal gas sphere, and carry out simulations including gas and stars, but no dark matter halo.

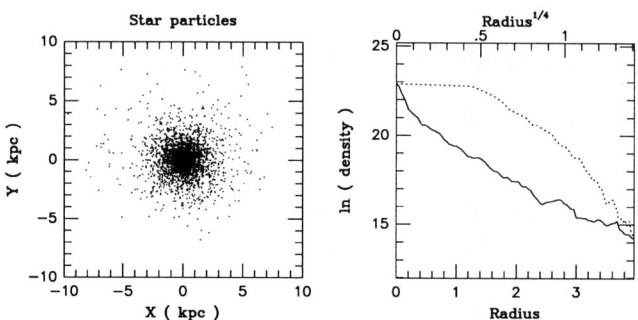

*Figure 1.* Left : Snapshot of stellar distribution. Right : Projected stellar density profile. The solid and the dotted lines are density profiles against the radius and the quartic root of radius, respectively.

## 3. Results and Conclusions

Our calculation indicates that the protogalaxy cloud begin to collapse due to its self-gravity. The collapse is halted by supernova driven-wind. Then the system loses a significant fraction of mass because of its small escape velocity. Figure 1, which is a snapshot after the system reaches a quasi stationary state, shows that the projected stellar surface density profile is approximately exponential as observed in dEs. We also find that heavy element abundance of the system is as low as [Fe/H] $\sim -1.5$. If the gas cloud has a smaller mass and a higher star formation rate, the stellar system would become unbound to disrupt. For a larger mass and a lower star formation rate, enough stars form before a significant mass loss from the system occurs, and a bound stellar system results eventually. Since about a half of the initial gas is blown away, the resulting stellar distribution has a weak central concentration. The chemical evolution as well as the effects of dark matter will be published elsewhere.

### Acknowledgements

This work has been supported in part by the Grant-in-Aid of the Ministry of Education, Science, and Culture ( 05242102, 06233101, 07CE2002 ).

### References

Deckel, A. and Silk, J. (1986) *ApJ* **303**, 39.
Athanassoula, E. (1994) in *ESO/OHP Workshop on Dwarf Galaxies.*, Meylan, G. & Prugniel, P., eds., ESO, Garching, p. 525.

# GRAPE-SPH SIMULATIONS OF GLOBULAR CLUSTER FORMATION

N. NAKASATO[1], M. MORI[1], T. TUJIMOTO[2], G. MATHEWS[3], AND N. NOMOTO[1]

[1] *Univ. Tokyo*, [2] *National Astronomical Observatory, Japan* and [3] *Univ. Notre Dame*

**Abstract.** We examine two globular cluster formation scenarios with SPH simulations, using GRAPE-3AF. First model is the spontaneous collapse, which is found to induce a disruption of the system. However, the system can survive when sufficiently high external pressure exists. Second model is the collision of the two proto-cluster clouds, which can form a bound stellar system due to efficient cooling and a burst of star formation.

## 1. Scenario

At the formation of globular clusters, the star formation history is important in two respects (e.g., Murray et al. 1993). First, the metallicity of globular clusters is quite homogeneous. This property suggests that many of the member stars of the cluster formed in a very short period. Second, massive stars, which formed in proto-cluster clouds, evolve quickly and emit UV radiation. Such radiation heats the surrounding gas to cause its expansion. If the star formation rates (SFR) is low, most of the remaining gas will be brown off due to this heating before the gravitationally bound stellar system is formed. If there is external pressure, however, the expanding gas is decelerated and thus additional star formation may occur.

Proto-cluster clouds can be divided into two types depending on their masses. The clouds whose mass exceed the Jeans mass are gravitationally unstable, thus spontaneously collapsing to form stars. The clouds which are less massive than the Jeans mass are stable until some instabilities are introduced. A cloud-cloud collision or cloud-disk collision can trigger such an instability. When such collisions occur, clouds would become thermally unstable by efficient cooling. This cooling would lead to a burst of star formation because of the formation of a very dense region.

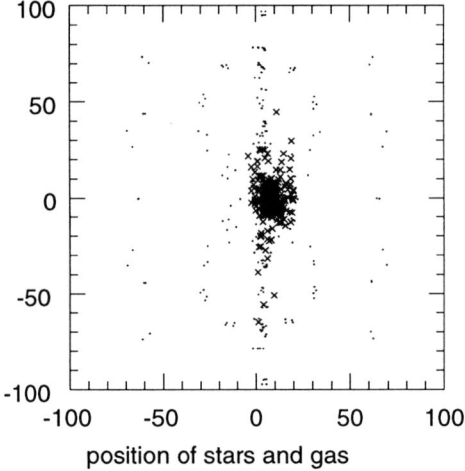 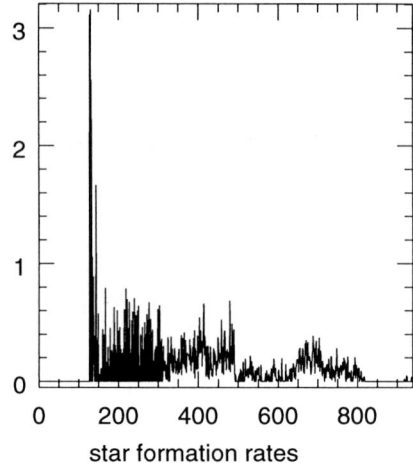

*Figure 1.* Left: positions of stars(cross) and gases(dot),(x, y both in pc). Right:star formation rates(x:$10^4$yr, y:$M_\odot\text{yr}^{-1}$).

## 2. Result

We calculate three simple cases using GRAPE-3AF (Sugimoto et al. 1990). The initial conditions are as follows.

- case I : spontaneous collapse for $M = 10^6 M_\odot$ and R = 50 pc
- case II : same as case I but with external pressure
- case III : collision of the two clouds in case I with $V_\text{collsion} = 100$ km/s

In case I, most gases are driven out of the system due to heating and the resulting stellar mass fraction is only $\sim 10\%$. In case II, the SFR is higher than in case I, so that the stellar mass fraction amounts to $\sim 50\%$. For case III, the position of the formed star particles at $\sim 10$ Myr and star formation history are shown in Figure 1. A violent star formation occurs in a short period and the bound stellar system forms. This short period is consistent with the observational properties. In case II and III the stellar mass fraction exceeds 50% so that the system will survive(Lada et al. 1984).

This work has been supported by grants # 05242102, 06233101, 07CE200.

## References

Murray, S.D., Lin, D.N.C., 1993, ASP Conf. Ser. Vol. 48, 738
Sugimoto, D., Chikada, Y., Makino, J., at al., 1990, Nature, 345, 33
Lada, C., Margulis, M., Dearborn, D., 1984, ApJ, 285, 141

# DYNAMICS OF COLLAPSING SHELLS

CHRISTIAN THEIS
*Inst. of Astronomy and Astrophysics,*
*Olshausenstr. 40, 24098 Kiel, Germany*

## Introduction

In the last decade several scenarios for the formation of globular clusters have been suggested. One of these models starts with an OB-association exploding near the center of a molecular cloud (Brown et al., ApJ **376**, 115 (1991)): The expanding shell sweeps up the cloud material and later the expansion will be decelerated or stopped by the external pressure of the ambient hot gas. The shell itself can break into fragments and form stars. If the total energy of these stars is negative, they will recollapse and eventually form a bound system. According to this idea the dynamics of a thin stellar shell has been studied.

## Initial conditions and numerical scheme

Initially a unit mass was homogeneously distributed within the radial range [0.9,1.0]. The velocity dispersion of the $N = 10\,000\,(100\,000)$ equal mass particles was chosen to give a virial coefficient $\eta_{\rm vir} = 0.05$. The softening parameter $\epsilon$ was set to 0.05. A smaller softening $\epsilon = 0.01$ did not change the results, whereas a larger value of $\epsilon = 0.1$ leads to a more spherical configuration. The equations of motion were integrated with a leap-frog scheme using a fixed timestep $\Delta t = 0.005$. This gives an energy conservation of typically 0.1-0.2% over the whole integration time. All simulations were performed with the direct summation on a GRAPE3 board.

## Results

The dynamics of the shell shows three stages. During the first stage ($t < \tau_{\rm ff}(\rho_{\rm sh})$) the shell is slowly contracting and small inhomogeneities start to grow (Fig. 1 upper right). Already in this early stage the particles in the shell are strongly mixed because of the radially *decreasing* (global) free-fall time in the shell. In the second phase ($\tau_{\rm ff}(\rho_{\rm sh}) < t < 1 - 2\tau_{\rm ff}({\rm shell})$) these clumps become bound subsystems (Fig. 1 lower left), which merge after the shell's free-fall time $\tau_{\rm ff}({\rm shell}) = 1.59$ that is 40% larger compared to the collapse of a sphere with the same mass and outer initial radius.

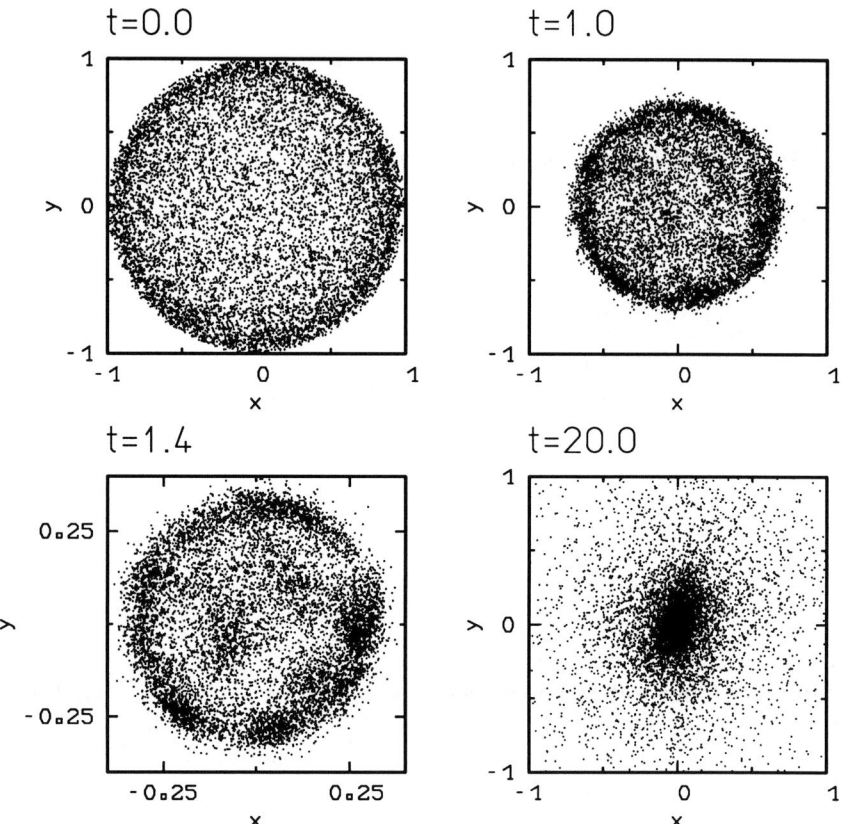

*Figure 1.* Projections of the collapsing shell on the x-y plane at different times: a) during the collapse (t=0.0, 1.0, 1.4) and b) after reaching an equilibrium state (t=20.0).

Finally, a radially anisotropic triaxial system has been formed (Fig. 1 lower right). The mass-loss of 8% is by a factor of 3.5 smaller than for spheres whereas the anisotropy $1 - \sigma_\theta^2/\sigma_r^2$ is reduced by 0.15. Additionally, the half-mass radius is increased by 50% and the 90%-radius is a factor of 6 smaller compared to collapsing spheres. Therefore, *violent relaxation seems to be less efficient for collapsing shells.*

From the simulations one can conclude that the resulting ellipticity is significant larger for collapsing shells than for spheres. Hence, the shape of young clusters might be used to check the scenario of collapsing shells.

**Acknowledgements.** I want to thank the members of the GRAPE-team as well for their hospitality during my stay in spring 95 in Tokyo as for their efforts related to our GRAPE installation in Kiel that was made possible by the DFG-grant Sp345/5-1. The participation at this conference was kindly supported by the *IAU* and the *Friedrich-Flick-Förderungsstiftung.*

# EVOLUTION OF THE GALACTIC GLOBULAR CLUSTER SYSTEM

E.VESPERINI
*Scuola Normale Superiore*
*Piazza dei Cavalieri 7, 56126 Pisa-Italy*

## 1. Introduction

Recent surveys of the observational properties of galactic globular clusters have shown the existence of interesting correlations and trends between structural parameters and between structural parameters and location inside the Galaxy (Chernoff & Djorgovski 1989, Djorgovski & Meylan 1994). The origin of most of these correlations is not clear yet and it is not clear to what extent they reflect the primordial conditions or the result of evolution. We have carried out a set of simulations following the evolution of the properties of a globular cluster system (mass function, spatial distribution, correlations between structural parameters) starting from given initial conditions. The evolution of each individual cluster has been followed by the same method applied by Chernoff *et al.* (1986) and Chernoff & Shapiro (1987). The effects of internal relaxation, disk shocking and dynamical friction have been considered. *The main goal of the analysis is that of establishing the role of initial conditions and evolutionary processes in determining the present observational properties.*

## 2. Initial conditions

Three different initial mass functions for the globular cluster system have been chosen in the survey carried out in this work.

1. A truncated power-law mass function $f(M) \propto M^{-2}$ with lower cut-off $M_{low} = 10^{4.5} M_\odot$ and upper cut-off $M_{up} = 10^6 M_\odot$ (POW).
2. A Gaussian distribution in the logarithm of the mass with mean value $< \log M > = 4.5$ and variance $\sigma^2 = 1.0$ (GAU1).
3. A Gaussian distribution in the logarithm of the mass with mean value $< \log M > = 5.0$ and variance $\sigma^2 = 0.25$ (GAU2).

The number density profile of clusters in the Galaxy has been chosen to be $n(R_g) \propto R_g^{-3.5}$. As for the initial distribution of concentrations of clusters, $c = \log r_t/r_c$, both initial conditions with $c$ correlated to the mass of clusters and with $c$ and mass uncorrelated have been investigated in order to determine whether the present correlation between these two quantities observed for galactic globular clusters is primordial or the result of evolution.

## 3. Results

1. The observed correlation between concentration and mass of clusters is primordial and not induced by evolutionary processes (see also Bellazzini *et al.* 1995).
2. In the POW run evolutionary processes change the power-law initial mass function into a bell shaped mass function resembling a Gaussian in $\log M$ or into a two-component power-law distribution in $M$ flatter for low values of mass.
3. In the GAU1 run the gaussian shape of the initial mass function is preserved but shifted to a higher mean value. The low-mass side is almost entirely destroyed by the evolutionary processes. The final mass function is very similar to the observed one.
4. In the GAU2 run the initial mass function is equal to the observed one. Although a significant disruption of clusters occurs also in this run, the shape and the parameters of the mass function are essentially preserved during the entire evolution showing that a significant effect of evolutionary processes does not necessarily imply a strong difference between initial and final mass function.
5. In all the runs evolutionary processes deplete significantly the number of clusters in the central regions of the Galaxy giving rise to a flattening of $n(R_g)$ in the inner regions.
6. Evolution gives rise to a trend for more concentrated clusters to dominate near the galactic centre.

A more detailed description of these results is in Vesperini (1994, 1995a,b).

## References

Bellazzini M., Vesperini E., Ferraro F.R., Fusi Pecci F. (1995) *MNRAS* in press
Chernoff D.F., Kochanek C.S., Shapiro S.L. (1986) *Ap.J.* **309**, 183
Chernoff D.F., Shapiro S.L. (1987) *Ap.J.* **322**, 113
Chernoff D.F., Djorgovski S.G. (1989) *Ap.J.* **339**, 904
Djorgovski S.G., Meylan G. (1994) *A.J.* **108**, 1292
Vesperini E. (1994) Ph.D. Thesis, Scuola Normale Superiore
Vesperini E. (1995a) submitted to *MNRAS*
Vesperini E. (1995b) submitted to *MNRAS*

# TRUNCATION OF THE BINARY DISTRIBUTION FUNCTION IN GLOBULAR CLUSTER FORMATION

E.VESPERINI
*Scuola Normale Superiore*
*Piazza dei Cavalieri 7, 56126 Pisa-Italy*

AND

D.F. CHERNOFF
*Department of Astronomy*
*Space Science Building, Cornell University 14853 Ithaca, (New York) USA*

## 1. Introduction

Recent observational searches suggest that the frequency of primordial binaries in globular clusters may reach $\sim 10\%$ (see Hut *et al.* 1992 for a review). Several different treatments conclude that primordial binaries are effective in halting core collapse, supporting the core and driving the post-core collapse expansion phase (Goodman & Hut 1989, McMillan *et al.* 1990, 1991, Gao *et al.* 1991, Heggie & Aarseth 1992, McMillan & Hut 1994). The abundance and binding energy distribution have a direct impact on observable characteristics of globular clusters such as the size of the core radius (Vesperini & Chernoff 1994). In this analysis we have investigated how the initial binary distribution function may be altered in the formation of a cluster; the key question we have addressed is whether it is possible for binaries, assumed to be primordial, to survive the birth of the cluster. A detailed description of our results is in Vesperini & Chernoff (1995).

## 2. Results

We focus on an assumed phase of violent relaxation immediately after a large, tidally truncated gas cloud has formed stars. If the initial ratio of kinetic to potential energy of the system is smaller than 1/2 then the gross properties (radius, velocity dispersion) of the system undergo damped oscillations until a virialized state is reached. When binaries are present and suffer collisional interactions within the cluster, their internal binding energy is an additional sink or source for changes in stellar translational energy. For the available binding energy to play a dynamical role in the cluster evolution, binaries must interact collisionally. Collisional interactions occur most rapidly at the point of maximum contraction when the background stellar density and velocity dispersion are largest. As the velocity dispersion reaches very high values at the point of maximum contraction, even binaries hard in the virialized system are soft and can be disrupted more easily.

Using $N$-body simulations and analytic calculations we have examined the binary destruction process during violent relaxation. The main conclusions are:

1. There is no significant change in the gross cluster properties at the end of violent relaxation due to the interaction between internal (binary) degrees of freedom and translational degrees of freedom.
2. It is possible to identify a characteristic binding energy below which the initial distribution is truncated. This cut-off binding energy, $\epsilon_{cut}$, scales with the number of particles in the system as $\epsilon_{cut} \sim N^{-1.19}$ and the hardness parameter for the cut-off energy as calculated in the virialized system falls in all cases in the soft regime ($x < 1$).
3. Analytical calculations are in good agreement with results from $N$-body simulations and show that ionization is the main destructive process.
4. Some hardening of binaries with $\epsilon > \epsilon_{cut}$ is observed in the $N$-body simulations. This is not described by the thermally averaged rate coefficients. Cold initial conditions used for the simulations mean some binaries are bound to the closest single particle. An interaction between the single particle and the binary is responsible for at least part of the observed hardening.

## References

Gao B., Goodman J., Cohn H., Murphy B. (1991) *Ap.J.* **370**, 567
Goodman J., Hut P. (1989) *Nature* **339**, 40
Heggie D.C., Aarseth S.J. (1992) *MNRAS* **257**, 513
Hut P., McMillan S.L., Goodman J. et al. (1992) *Publ. Astron. Soc. Pacific* **104**, 981
McMillan S., Hut P., Makino J. (1990) *Ap.J.* **362**, 522
McMillan S., Hut P., Makino J. (1991) *Ap.J.* **372**, 111
McMillan S., Hut P. (1994) *Ap.J.* **427**, 793
Vesperini E., Chernoff D.F (1994) *Ap.J.* **431**, 231
Vesperini E., Chernoff D.F. (1995) *Ap.J.* in press

# AUTHOR INDEX

Aarseth, S. J., 161, 381.
Ajhar, E., 53.
Anderson, S. B., 181.
Aparicio, A., 341.
Bahcall, J. N., 19, 333.
Bell, R. A., 39, 193.
Bertoldi, F., 331.
Bhatt, B. C., 339.
Bolte, M., 39, 193.
Bond, H. E., 39, 193.
Brandl, B., 331.
Byun, Y.-I., 53.
Capaccioli, M., 357.
Chernoff, D. F., 263, 403.
Cool, A. M., 71, 349.
Danilov, V. M., 389.
Davies, M. B., 243.
Djorgovski, S. G., 9.
Drapatz, S., 331.
Dressler, A., 53.
Drukier, G. A., 347.
Durgapal, A. K., 339.
Ebisuzaki, T., 141.
Eckart, A., 331.
Eggleton, P. P., 213.
Einsel, C., 363.
Engle, K. A., 379.
Faber, S., 53.
Fagotto, F., 357.
Fahlman, G. G., 39, 193.
Fletcher, J. M., 193.
Fukushige, T., 141, 365.
Funato, Y., 367.
Gebhardt, K., 53.
Geisler, D., 393.
Genzel, R., 81, 331.
Giersz, M., 101.
Gouda, N., 385.
Gouliermis, D., 335.
Grillmair, C., 53.
Grindlay, J. E., 171, 349.
Guhathakurta, P., 19, 333.
Harris, W. E., 39, 193.

Heggie, D. C., 131, 303, 365, 369, 371, 381, 383.
Hesser, J. E., 39, 193.
Hofmann, R., 331.
Huang, X., 263.
Hut, P., 121, 303, 319, 367, 371.
Ibata, R. A., 193.
Ibata, R. I., 39.
Inutsuka, S., 373.
Ivanans, N. C., 39, 193.
Kawashima, K., 353.
Kim, E., 391, 393.
King, I. R., 29, 71, 319, 343.
Kiseleva, L. G., 233.
Kong, K. N., 351.
Konishi, T., 385.
Kontizas, E., 335.
Kontizas, M., 335.
Kormendy, J., 53.
Kulkarni, S. R., 181.
Löwe, M., 331.
Lagoute, C., 375.
Lauer, T., 53.
Lee, H. M., 293, 351.
Lee, M. G., 391, 393.
Lin, D. N. C., 283.
Longaretti, P.-Y., 375, 377.
Mahra, H. S., 339.
Makino, J., 141, 151, 367.
Mandushev, G., 193.
Mardling, R. A., 273.
Mathews, G., 397.
McMillan, S. L. W., 223, 379.
McMillan, S., 367.
Mcclure, R. D., 193.
Mcmillan, S. L. W., 371.
Menzies, J. W., 337.
Meylan, G., 61, 319.
Miyama, S. M., 387.
Mohan, V., 339.
Mori, M., 395, 397.
Murray, S. D., 283.
Nakasato, N., 397.
Negoro, H., 353.

Nomoto, K., 395, 397.
Pandey, A. K., 339.
Piotto, G., 71, 341, 357.
Pryor, C., 39, 193.
Quirrenbach, A., 331.
Ramamani, N., 381.
Rasio, F. A., 253, 383.
Richer, H. B., 39, 193.
Richstone, D., 53.
Rosenberg, A., 341.
Salati, P., 377.
Sams, B. J., 331.
Saviane, I., 341, 357.
Schneider, D. P., 19, 333.
Shara, M. M., 49.
Sosin, C., 343.
Spurzem, R., 111, 363.
Stetson, P. B., 39, 193.
Sugimoto, D., 1, 141, 319.
Surdin, V., 313.
Szécsényi-Nagy, G., 359.

Taiji, M., 141.
Taillet, R., 377.
Takahashi, K., 91.
Theis, C., 399.
Tremaine, S., 53.
Tsuchiya, T., 385.
Tsujimoto, T., 395.
Tujimoto, T., 397.
Vandenberg, D. A., 39, 193.
Verbunt, F., 183, 319.
Vesperini, E., 401, 403.
Wijers, R. A. M. J., 203.
Wood, M. A., 39, 193.
Yamaoka, H., 361.
Yanny, B., 19, 333.
Yoshii, Y., 395.
Zaggia, S. R., 345.
Zaggia, S., 341.
Zhou, Y., 387.
Zwart, S. F. P., 355.